# Professional Image Editing Made Easy with Affinity Photo

Apply Affinity Photo fundamentals to your workflows to edit, enhance, and create great images

**Jeremy Hazel**

<packt>

BIRMINGHAM—MUMBAI

# Professional Image Editing Made Easy with Affinity Photo

**Group Product Manager**: Rohit Rajkumar

**Publishing Product Manager**: Vaideeshwari Muralikrishnan

**Senior Content Development Editor**: Rakhi Patel

**Technical Editor**: Saurabh Kadave

**Copy Editor**: Safis Editing

**Project Coordinator**: Sonam Pandey

**Proofreader**: Safis Editing

**Indexer**: Tejal Daruwale Soni

**Production Designer**: Arunkumar Govinda Bhat

**Marketing Coordinators**: Anamika Singh, Nivedita Pandey, and Namita Veglekar

First published: June 2023

Production reference: 1040523

Published by Packt Publishing Ltd.
Livery Place
35 Livery Street
Birmingham
B3 2PB, UK.

ISBN 978-1-80056-078-9

www.packtpub.com

*I would like to dedicate this book to my amazing wife, whose patience, support, and encouragement have made my entire career possible. For all the nights, weekends, and holidays spent with me locked away behind a keyboard and camera, it would not have been possible without you.*

*– Jeremy Hazel*

# Contributors

## About the author

**Jeremy Hazel** started 7th Season Society in 2017 and has over 150,000 students in various programs, such as Designer, Photo, and Procreate. He teaches on many popular platforms, including Udemy, Skillshare, and his own platform, where he is in the top-ranked courses with an average rating of 4.6 out of 5 stars.

In addition to teaching, Jeremy is a T-shirt designer and printer and also works as a brand consultant for creative-based businesses. Jeremy has partnered with the absolute best of the best in the field to bring his courses and education style to the masses, including Video School Online and Lindsay Marsh Design, and recently, his YouTube channel has been featured in Affinity Spotlight.

*I want to take this opportunity to thank Phil Ebiner; without the chance you gave me, none of this would have happened. Thanks for taking a chance on a newbie.*

## About the reviewer

**Michael Burton** is an experienced illustrator and graphic designer who evolved into multimedia. Since 2000, he's worked in branding and decorated apparel; he uses Photoshop and alternative software for digital imaging, screen-printing, drawing, and painting. He has designed for hundreds of schools and local businesses, including Chicago public schools, and the Illinois High School Association. Colleges and pro league clients include the WNBA player Candace Parker, and Brian Urlacher of the Chicago Bears. In addition, he works in vector art and loves creative writing, music, video editing, and the spoken word for graphic storytelling. He has been featured with a 4x Grammy artist and is the author of the creative memoir, *Let Me Paint a Picture*. He has earned two associate's degrees and a BA in graphic design and media arts at SNHU.

# Table of Contents

## Part 1: Foundational and Navigation Basics for Affinity Photo

### 1

### Fundamentals of Vector versus Raster Art and Basics of the Interface

### 2

### Opening Your First Document                                                                             19

# 3

## Layer Fundamentals – The Heart of Affinity Photo     35

# 4

## The Basics of Masking in Affinity Photo     59

# 5

## Selection and How to Achieve It in Affinity Photo    85

# Part 2: Fundamental Concepts Used to Create a Simple Edit

# 6

## Cropping and Composition    113

# 7

## Basics of Workflows and Balancing Dark and Light    137

# 8

## Blend Mode Fundamentals    153

# 9

## Basics of Stock Brushes in Affinity Photo    177

# 10

## Working with Color in Affinity Photo    199

# Part 3 : The Practical Applications of Affinity Photo

## 11

## 12

# 13

## Advanced Color Concepts and Grading    277

# 14

## Destructive Filters and Tools in Affinity Photo    299

# Part 4: Finishing Your Edit and Building Your Own Artistic Palette

# 17

## Editing in Other Personas in Affinity Photo　　　　　373

# 18

## Exporting and Artist Efficiency Tips　　　　　399

## Index　　　　　423

## Other Books You May Enjoy　　　　　436

# Preface

This book is a bottom-up practice-based exploration of the Affinity Photo program, in which you will learn about the tools while making popular edits to your photos. This book teaches the techniques and fundamentals in coordination with the workflow concept. There is no long, drawn-out explanation of the tools, but we will learn about the tools and techniques through their application and also explore *why* it works, not just *how* it works.

So, this is not a technical manual, nor is it a workbook; it is a hybrid of project-based education that leads to the application of the tools and a deeper understanding of how to wield the elements to achieve a result

By the end of this book, you will have a good body of work and be able to evaluate edits you would like to make and feel confident in putting the pieces together to achieve whatever you want to do.

## Who this book is for

This book is for beginners to any form of photo editing and those who were utilizing Photoshop and making a switch from Adobe. Also, small business owners who may utilize stock photos and want to do some brand-specific editing but don't have a ton of experience doing photo-based edits. It is perfect for beginner editors, photographers, and people looking to get into product photography and advertising.

## What this book covers

*Chapter 1, Fundamentals of Vector versus Raster Art and Basics of the Interface*

In this chapter, we will explore the differences between vector and raster art and the structure of the Affinity interface.

*Chapter 2, Opening Your First Document*

This chapter teaches you the fundamentals of documents and how to get started as soon as you open the program.

*Chapter 3, Layer Fundamentals – The Heart of Affinity Photo*

This chapter is an exploration of the layer structure available in Affinity Photo.

*Chapter 4, The Basics of Masking in Affinity Photo*

In this chapter, you will learn the fundamentals of masking, including what a mask layer is, and how we apply it in Affinity Photo.

*Chapter 5, Selection and How to Achieve It in Affinity Photo*

In this chapter, you will learn the fundamentals of selection and the various tools you can use to achieve it.

*Chapter 6, Cropping and Composition*

In this chapter, you will learn about cropping and composition in Affinity Photo.

*Chapter 7, Basics of Workflows and Balancing Dark and Light*

In this chapter, you will discover the most common methods to balance the brightness in your edit.

*Chapter 8, Blend Mode Fundamentals*

In this chapter, you will discover the basics of blend modes and how to use them in your layers and edits.

*Chapter 9, Basics of Stock Brushes in Affinity Photo*

In this chapter, we will explore the fundamental use of brushes in Affinity Photo.

*Chapter 10, Working with Color in Affinity Photo*

In this chapter, you will explore the basics of color and how Affinity represents it in the interface.

*Chapter 11, Compositing in Affinity Photo*

In this chapter, we will perform a full, multi-image composition in Affinity Photo.

*Chapter 12, Photo Restoration and Portrait Retouching in Affinity Photo*

In this chapter, we will learn how to restore and retouch your photos using Affinity Photo tools.

*Chapter 13, Advanced Color Concepts and Grading*

In this chapter, we will learn how to color grade like a pro using a variety of tools.

*Chapter 14, Destructive Filters and Tools in Affinity Photo*

In this chapter, we will identify the methods available for destructive editing in Affinity.

*Chapter 15, Creative Effects and Specialty Brushes in Affinity Photo*

In this chapter, we will build our own creative assets using the Affinity Photo program.

*Chapter 16, Working with Text and Shapes in Affinity Photo*

In this chapter, we will learn how to work with text and shapes to create a YouTube thumbnail image.

*Chapter 17, Editing in Other Personas in Affinity*

In this chapter, we will explore the other areas of Affinity used by professional artists.

*Chapter 18, Exporting and Artist Efficiency Tips*

Take your efficiency up a level with these efficiency tips to get you working quicker.

# To get the most out of this book

Basic knowledge of computers is all that is required. It is expected that you have the program installed and you can open the program....other than that, I will handle the rest.

# Download the example project files

It is recommended that you download the project files prior to reading each chapter, as there are projects for most (not all) of the chapters: `https://packt.link/npONj`

# Download the color images

We also provide a PDF file that has color images of the screenshots and diagrams used in this book. You can download it here: `https://packt.link/Flykq`

# CiA videos

I have also included **Code in Action** (**CiA**) videos for some chapters. You can access them at `https://packt.link/Sie8u`.

# Conventions used

There are a number of text conventions used throughout this book.

`Code in text`: Indicates code words in text, database table names, folder names, filenames, file extensions, pathnames, dummy URLs, user input, and Twitter handles. Here is an example: "To illustrate this, I have also included a downloadable file (see the `Clipping mask practice` file)."

**Bold**: Indicates a new term, an important word, or words that you see onscreen. For instance, words in menus or dialog boxes appear in **bold**. Here is an example: "Attribute-based selection options are found in the **Select** menu at the top of the program."

> **Tips or important notes**
> Appear like this.

# Get in touch

Feedback from our readers is always welcome.

**General feedback**: If you have questions about any aspect of this book, email us at `customercare@packtpub.com` and mention the book title in the subject of your message.

**Errata**: Although we have taken every care to ensure the accuracy of our content, mistakes do happen. If you have found a mistake in this book, we would be grateful if you would report this to us. Please visit `www.packtpub.com/support/errata` and fill in the form.

**Piracy**: If you come across any illegal copies of our works in any form on the internet, we would be grateful if you would provide us with the location address or website name. Please contact us at copyright@packt.com with a link to the material.

**If you are interested in becoming an author**: If there is a topic that you have expertise in and you are interested in either writing or contributing to a book, please visit authors.packtpub.com.

## Share Your Thoughts

Once you've read *Professional Image Editing Made Easy with Affinity Photo*, we'd love to hear your thoughts! Scan the QR code below to go straight to the Amazon review page for this book and share your feedback.

https://packt.link/r/1800560788

Your review is important to us and the tech community and will help us make sure we're delivering excellent quality content.

# Download a free PDF copy of this book

Thanks for purchasing this book!

Do you like to read on the go but are unable to carry your print books everywhere? Is your eBook purchase not compatible with the device of your choice?

Don't worry, now with every Packt book you get a DRM-free PDF version of that book at no cost.

Read anywhere, any place, on any device. Search, copy, and paste code from your favorite technical books directly into your application.

The perks don't stop there, you can get exclusive access to discounts, newsletters, and great free content in your inbox daily

Follow these simple steps to get the benefits:

1.  Scan the QR code or visit the link below

https://packt.link/free-ebook/9781800560789

2.  Submit your proof of purchase
3.  That's it! We'll send your free PDF and other benefits to your email directly

# Part 1: Foundational and Navigation Basics for Affinity Photo

This first part will take you through universal topics in the areas of photo editing and digital art in general. You will find that these core techniques will serve you throughout your entire career, and while some of the chapters may seem rather mechanical, developing the muscle memory around what the program does and how to edit will be fundamental as the work gets more and more artistic later. Remembering mechanical drills early on helps you to create what is in your head later.

This part comprises the following chapters:

- *Chapter 1, Fundamentals of Vector versus Raster Art and Basics of the Interface*
- *Chapter 2, Opening Your First Document*
- *Chapter 3, Layer Fundamentals – The Heart of Affinity Photo*
- *Chapter 4, The Basics of Masking in Affinity Photo*
- *Chapter 5, Selection and How to Achieve It in Affinity Photo*

# 1

# Fundamentals of Vector versus Raster Art and Basics of the Interface

This chapter combines the basics you need to navigate the digital art space and the details of the Affinity Photo program. It is combined in terms of topics because one cannot be separated from the other in terms of functionality.

In the first part of this chapter, we will get the nomenclature right, and by the end, you will understand how photo editing in a raster (or pixel-based) program is different from a vector-based program (such as Affinity Designer). We will cover the basic terms and how to set the size and resolution of your image up for success. This is critical in the first steps because, without the right parameters, the edits you make will never print sharply, nor look the way you want them to.

Once we get through the introductory materials, we will then dive into the interface and tour the major sections. Knowing how to navigate *any* software is a prerequisite for success, and so we want you to know where to find things as we get deeper and deeper into the content.

In this chapter, we will cover the following topics:

- Differentiating between raster (pixel) and vector-based images
- Why you would use one over the other, and how to choose the correct type of art
- Understanding DPI and why it matters
- Understanding the role image size plays in image quality
- Understanding color profile and how it factors into your image

## Technical requirements

The technical requirements for Affinity Photo are not as significant as they are for other programs. As taken from their website, the technical requirements for the machine are shown here. I have included the requirements for the desktop versions, as this is the focus of this particular manual:

- Windows-based PC (64-bit) with a mouse or equivalent input device
- Hardware GPU acceleration
- DirectX 10-compatible graphics cards and above
- 8 GB RAM recommended
- 1 GB of available hard drive space; more during installation
- 1280x768 display size or larger
- Windows® 11
- Windows® 10 May 2020 Update (2004, 20H1, build 19041) or later

These are the requirements for Mac:

- Mac Pro, iMac, iMac Pro, MacBook, MacBook Pro, MacBook Air, or Mac mini
- Mac with Apple silicon (M1/M2) chip or Intel processor
- 8 GB RAM recommended
- Up to 2.8 GB of available hard drive space; more during installation
- 1280x768 display size or larger
- macOS Ventura 13
- macOS Monterey 12
- macOS Big Sur 11
- macOS Catalina 10.15

## Differentiating between raster (pixel) and vector-based images

There are two different types of formats that exist in Affinity Photo – **raster** and **vector** elements – and it is important to understand both. Objects such as a digital photo are raster-based (composed of pixels), and objects that are drawn with the Pen tool are vector-based (formed from mathematical functions).

Raster style (commonly called **pixel-based art**) is based on the pixel, which is the building block of all digital art. A **pixel** can be thought of as the unit from which all art is produced:

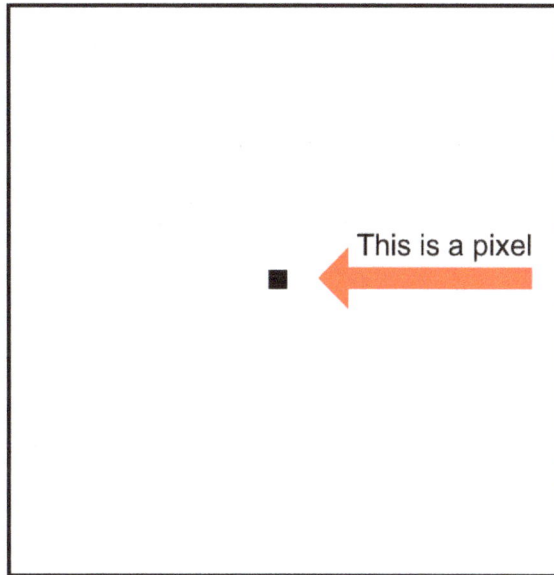

This is a pixel

Figure 1.1 – The pixel as the building block of digital art

Pixel images typically have two forms of information: **size** and **color**. When pixels are placed into position next to one another, they begin to form pictures, and subtle variation in the colors forms the detail. This is how your monitors, television sets, and other electronic devices work. The images you see on a monitor are simply created by pixels displaying their information in a way that allows us to see shapes, shades, and text. In *Figure 1.2*, the smooth red sphere is actually just a collection of pixels, and when we magnify it, we can see the different pixels that form the picture:

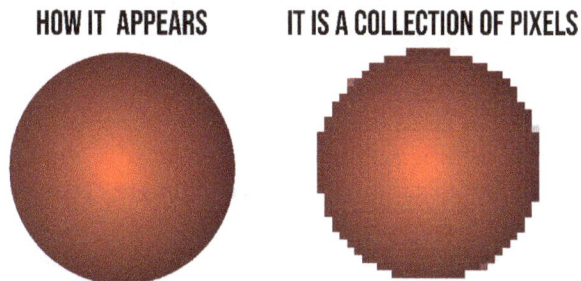

HOW IT APPEARS          IT IS A COLLECTION OF PIXELS

Figure 1.2 – The effect of magnification on pixelation

On the other side of digital art, there is vector-based art, and as the name implies, vector-based art is based on mathematical vectors. It is still drawn in pixels, but the mathematical equation done by the program is the primary driver of what is displayed on the screen (see *Figure 1.3*):

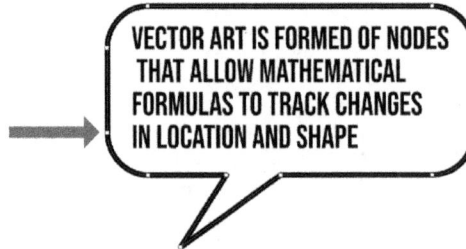

VECTOR ART IS FORMED OF NODES THAT ALLOW MATHEMATICAL FORMULAS TO TRACK CHANGES IN LOCATION AND SHAPE

Figure 1.3 – Vector art composed of nodes

## Why you would use one over the other, and how to choose the correct type of art

The simple answer of what format to choose comes down to one important factor: **scalability**. Scalability is the ability to go through various sizes without losing detail and is the most important factor in choosing what style of art to go with for a project. Other factors may also exist, such as the following:

- Timeframe
- Type of client
- Purpose of the art
- Budget

Pixel-based art does not scale well. In professional design shops, typically, logos are printed on something as small as a business card and can be as large as a vehicle graphic. For this reason, a majority of corporate logo-type art is created in a vector format.

On the other side of the coin, pixel art is based on a collection of pixels within a defined area, and when expanded to extreme values far from their original size, the pixels are spread (sort of like stretching a molecule), and then the program has to try to interpret what is in the spaces created. This interpretation leads to fuzzy borders and poor-quality images. The same is true when compressing an image: if the image was originally made on an A4 sheet of paper, there are a certain number of pixels in it, and if you compress it downward, the result is the same number of pixels fighting for visibility, leading to a poorly detailed image.

The following figures show an example of an image constructed to be printed at 16x20", which was then shrunk down to be added to this book (so a 80% reduction in size). Notice the limited quality:

Figure 1.4 – Original image at original size

Figure 1.5 – The effect of compression on detail

This is the primary difference between pixel-based (raster) art and vector art; vector art does not have any scalability issues because, as you grow and shrink, the mathematical formula automatically changes the lines and makes adjustments, leading to no loss of detail.

As we will learn in this book, Affinity Photo has tools that work in both pixel and vector-based art, and we will learn how to wield these tools to create our images as we go along, but right from the outset of this journey, I wanted to give you a rule to guide you from here until forever when it comes to working in digital art:

**Work in a size close to your intended output; image quality suffers as scaling occurs, so always be thinking ahead to the final application of your art.**

## Understanding DPI and why it matters

As previously mentioned, pixels form the images, and how many there are in an area creates the detail or the **resolution**. So, the typical way this is communicated in both the print and electronic art world is in **dots per inch** (**DPI**). This represents how many pixels are present in an area of one square inch, and quite simply, the more dots per inch, the higher the amount of detail.

Typically, if you are making a piece of art for a digital screen, the general DPI is 72, as most monitors cannot go past that. However, if you are doing print work, the minimum DPI is 300, as anything less will show fewer and fewer details.

Setting the DPI affects the file size because, obviously, the more information (in this case, DPI), the larger the file is going to be. In the following figure, simply changing the DPI of the file from 72 to 300 increased the size of the file substantially:

300 DPI=1.41 MB
72 DPI=964 KB

Figure 1.6 – The effect of DPI on file size

## Understanding the role image size plays in image quality

In addition to being the building blocks of images, pixels are also a unit of measure for **size**. In digital screens, screen resolution is typically given in the unit of pixels in an area. As an example, a typical monitor has a size (in this example, given in inches) of 24 wide by 14 high and boasts 1080p.

Now, what is 1080p? This is the **resolution**. The "**p**" in 1080p refers to pixels, and it is saying that this monitor is 1080 pixels tall. Simply put, the more **p**, the higher the resolution because there are more pixels, and more detail is possible on the monitor. And as a result, the more pixels, the higher the resolution, and the sharper your image is going to look.

Now, don't get crazy on your DPI; the human eye can only see and comprehend a certain amount of detail, so booting the DPI up to an insane value will not make your images look any better, so remember: *moderation in all things*.

## Understanding color profile and how it factors into your image

The last topic we need to cover to get you up and running is the topic of the color profile. Believe it or not, there is no absolute version of "red." If you doubt this, stop by your local paint counter and look at the vast number of swatches available to cover a wall. For this introduction, I will share with you the two most common color profiles, and then tell you how they are different. Aside from that, we will not talk much more about them until later chapters, because I want to keep it simple.

The two most common color profiles are as follows:

- RGB
- CMYK

Notice the differences in the exact same image in the following comparison; it is the same image, only with different color profiles:

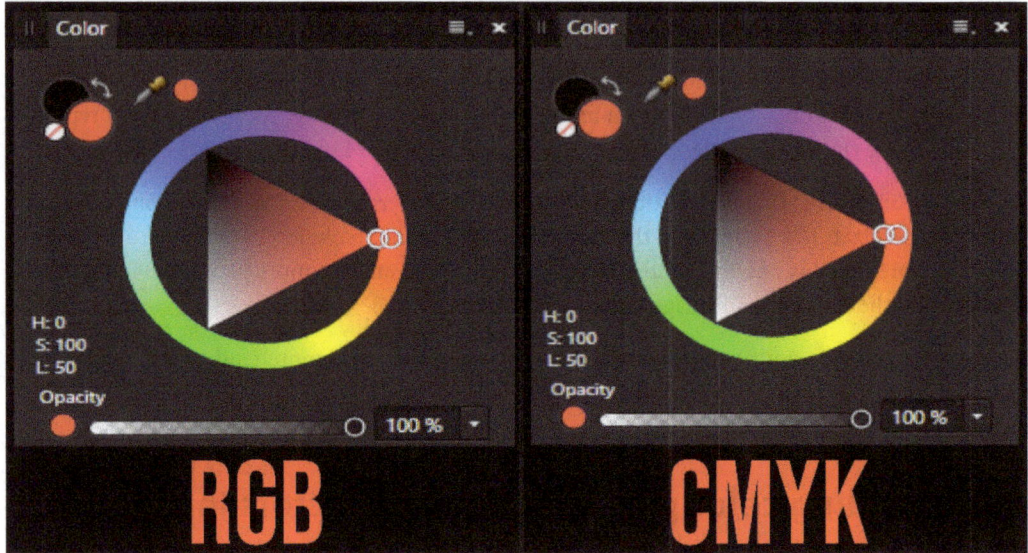

Figure 1.7 – The most commonly used color profiles in digital art

Let's define them a bit more specifically.

## RGB

**RGB** stands for **red, green, and blue**. This color profile is used for artwork for digital use, as this is how your monitor displays colors. However, today, most printers have programs that will convert any color to an RGB color, so it is quickly becoming the standard across the board. All the colors that you see are composed of these three colors in various combinations (we will deal with these combinations in *Chapter 10*) but for the purpose of this introduction, all you need to know is that it is the one we will be working with most frequently.

## CMYK

**CMYK** stands for **cyan, magenta, yellow, and black**, and these are the colors of the printer cartridges in your home printer. All the various colors you see are made by varying combinations of these four colors.

For the beginner, we will be dealing with only one color space, which is the traditional RGB8 color space. (We will discuss variants in the more advanced sections, but for now, just stick to the normal RGB8 color space for this conversation.)

The interface is your control panel for all the wonderful things you can accomplish in Affinity, so your understanding of the interface will be crucial to your success. To make you as familiar as possible as quickly as possible, I have broken it down into multiple sections and given a short explanation of how to get back to the default in the event you make a mistake with the interface…because as a beginner, you will.

Let's start by opening the program and clicking on **File | New**. It does not matter what size document you create, just click on **Create** so you can explore the interface along with us. In the next section, we will look at main areas of the Affinity Photo interface.

## The six areas of the Affinity Photo Interface

When you open the program for the first time, you will be greeted with this default interface (or some version of it – slight modifications may exist based on developer changes, but it has never significantly changed in all the years I have been using it). For simplicity's sake, we are going to divide the interface into six areas so you can navigate to the areas I discuss during the explanations later.

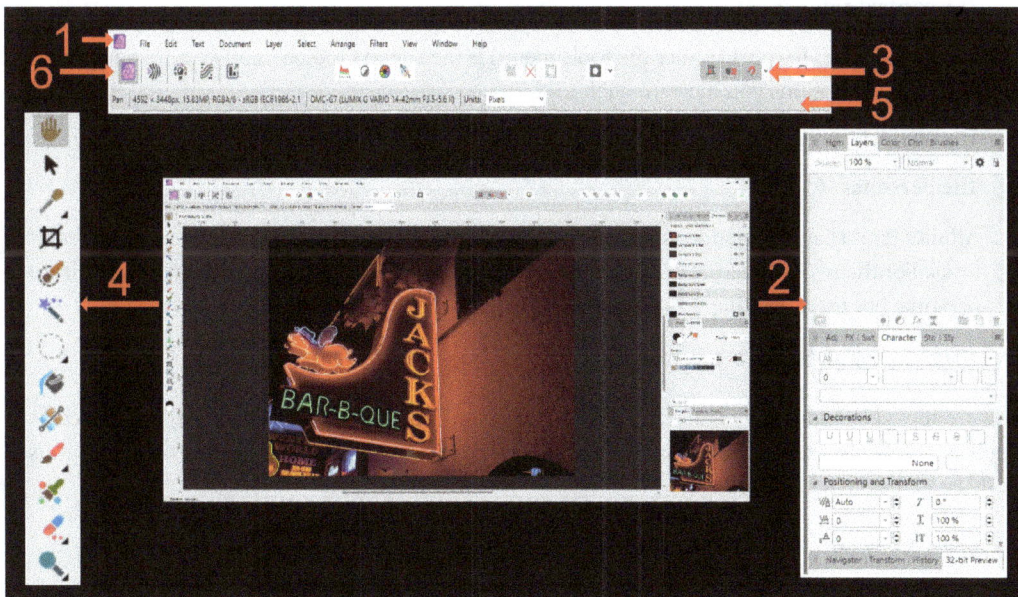

Figure 1.8 – Major sections of the Affinity interface

Let's go over these areas one by one:

- **The menu bar (1)**

    The menu bar is in the upper-left corner at the very top of the window. This is where most of the basic functionality is stored; things like **Open**, **Place**, **Save**, and **Export** are all on the **File** tab.

- **The studio area (2)**

  On the right side, there is a set of panels called Studio panels. These are groups of commands dedicated to certain things, such as **Color**, **Layers**, and **Channels**. The panels form the studio and can be hidden or retrieved (it is not uncommon to click a panel out of the interface…but do not worry, I will show you how to get it back). There can be a left and right studio in the default layout (in this figure, we only have the right studio shown).

- **The toolbar (3)**

  Above the canvas space, there is a section referred to as the toolbar. These are common tools such as **Flip**, **Rotate**, **Quick mask**, and so on.

- **The tools (4)**

  Separate from the toolbar is the **tools** area, which is along the left-hand side of the studio by default, and it contains the actual tools we will use to edit. Tools such as brushes, gradients, and selection tools are all in this area.

- **The context toolbar (5)**

  Right above the viewport (under the tools menu) is the context toolbar, and I want you to think of this as the options you have for each tool.

  The context toolbar will change with the tool.

- **The personas (6)**

  Affinity Photo has a tremendous amount of versatility and multiple personas to perform different tasks. For the majority of the book (all chapters except *17*, *18*, and *20*), we will be in the **Photo** persona, but there are chapters on each of the other personas.

## Customizing your Studio panels

Panels are movable and can be rearranged by clicking and dragging them up and down, as well as brought out closer to the workspace where you are editing. We will be working with this in various projects as we begin actually editing.

You can customize what panels are shown by clicking on **Window | Studio**, and then checking the panels you want to have showing based on your workflow:

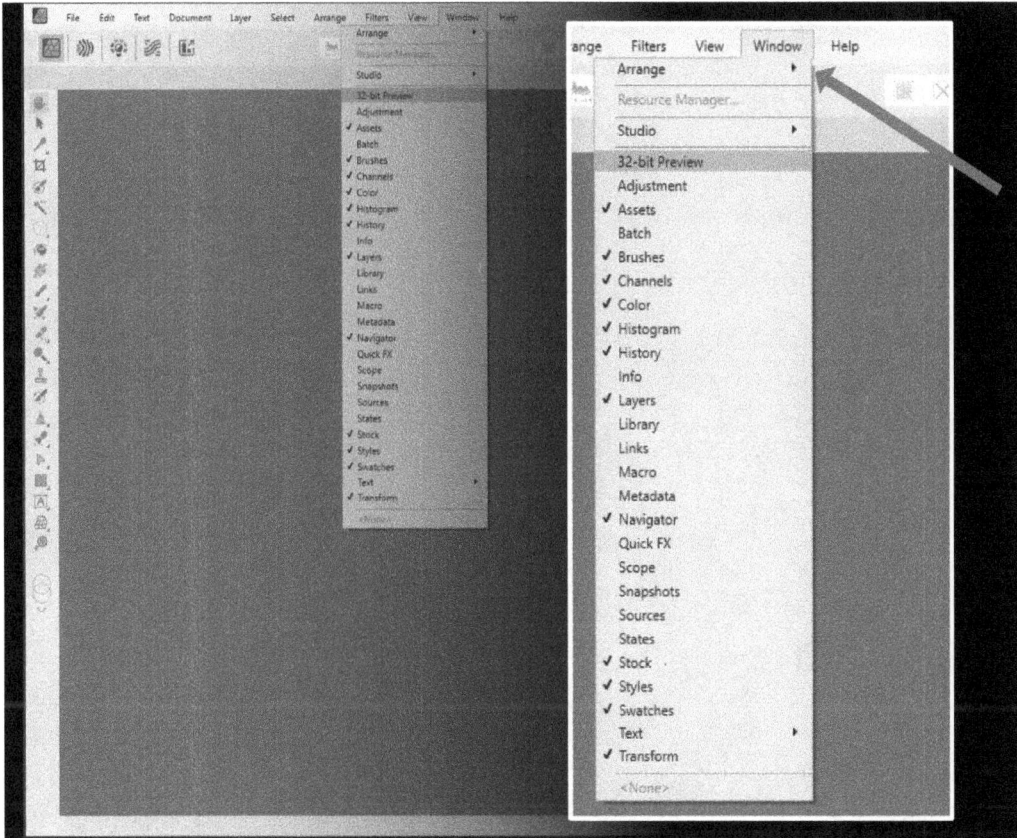

Figure 1.9 – Customizing Studio panels

## Customizing your toolbar

This toolbar can be customized by going to the **View** menu and clicking on **Customize Toolbar…**:

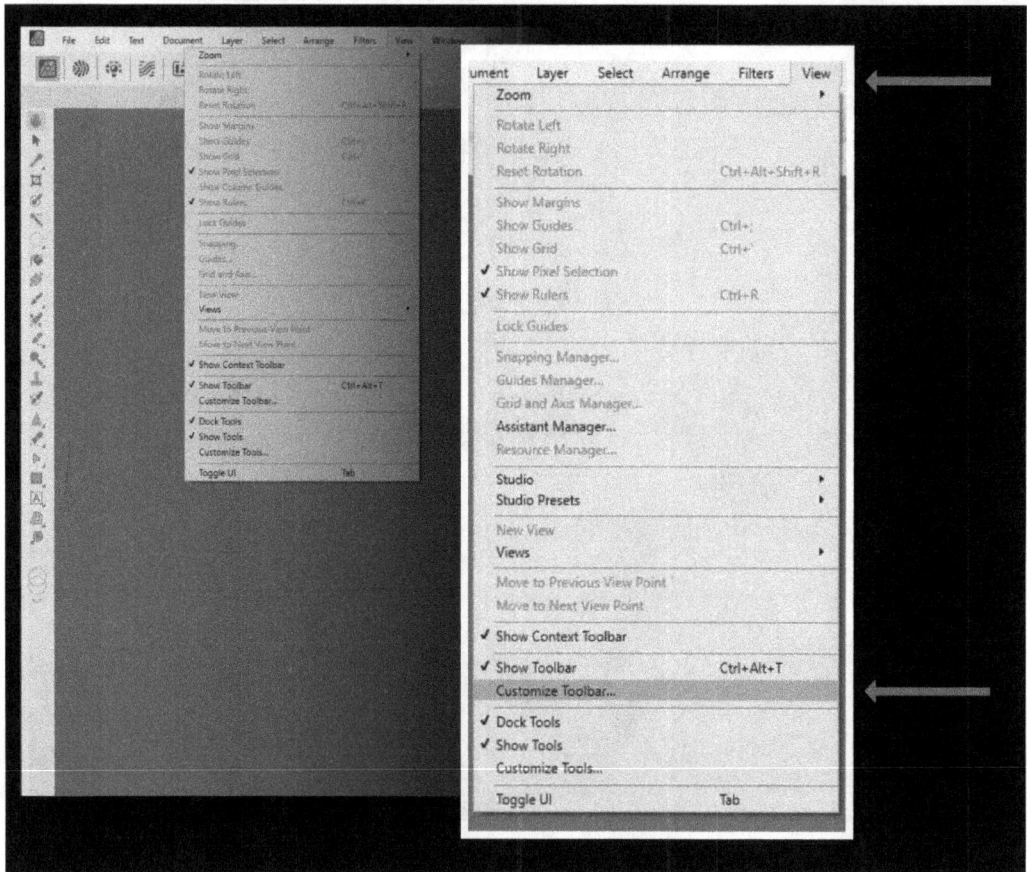

Figure 1.10 – Customizing the toolbar

To add a tool to the toolbar, simply drag the tool into the open space (noted by the dotted lines in the toolbar), and do the opposite to remove them:

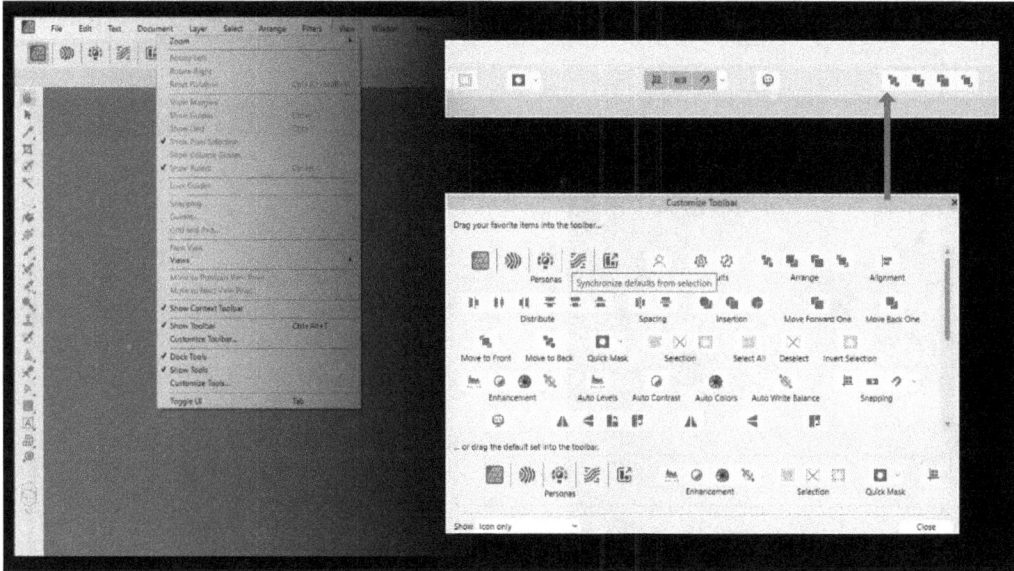

Figure 1.11 – Dragging images into the toolbar

## Customizing your tool menu

You can customize the tools by clicking on **View** | **Customize Tools…**:

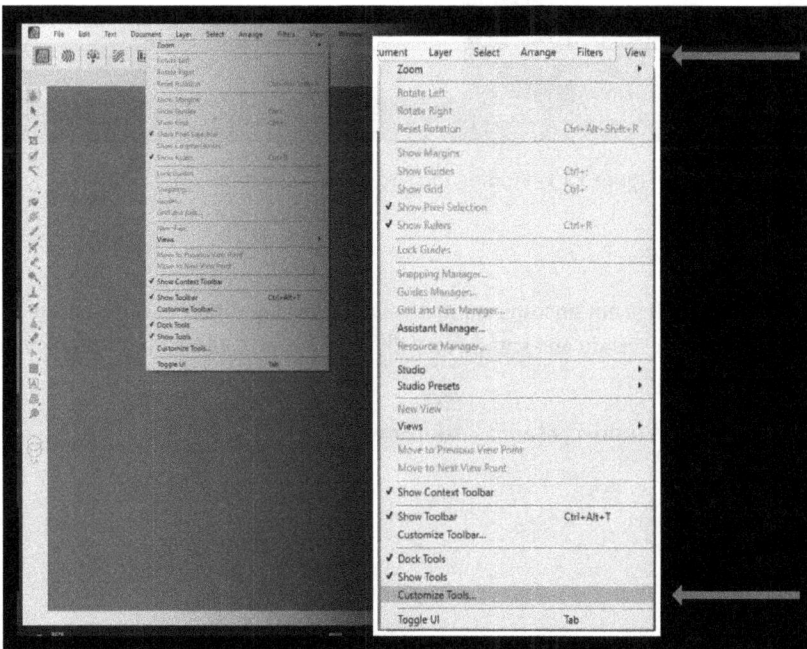

Figure 1.12 – Customizing the tools

To add a tool, simply drag it over to the tools area and it will be added; reverse the operation to remove tools that you do not want:

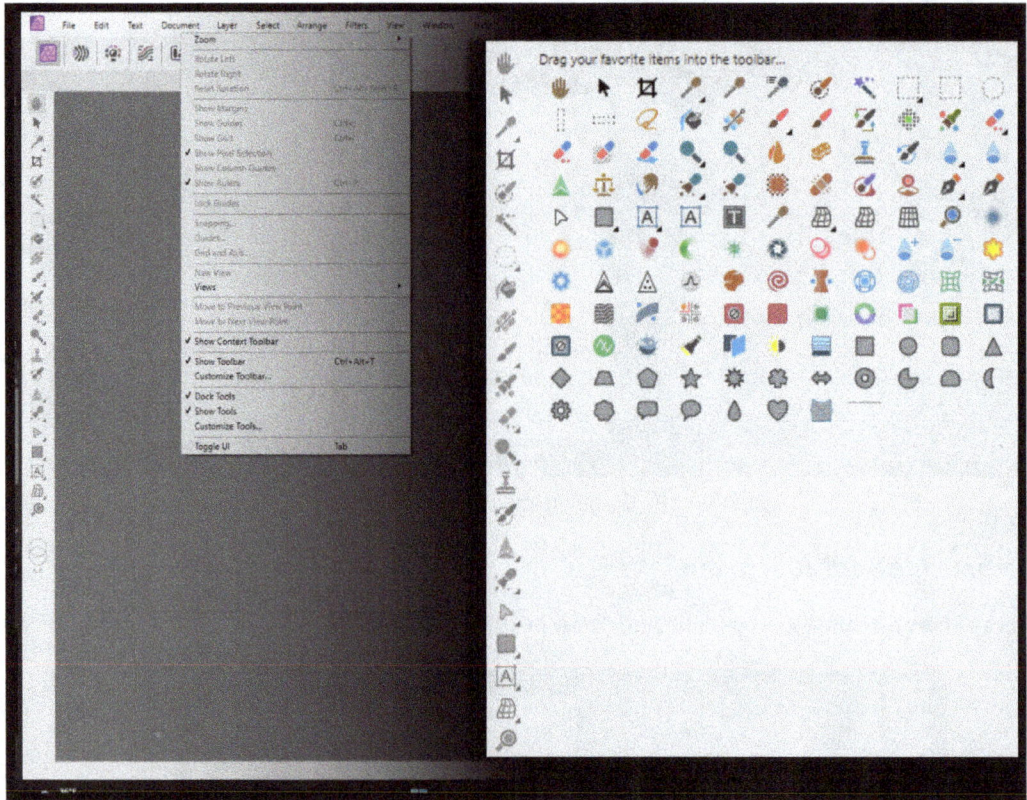

Figure 1.13 – Adding tools to the tools section

## Saving your workspace

As you develop as an editor, it is not uncommon that you will develop different workspaces, or layouts, where you keep your tools in certain areas, and you will favor certain tools. Affinity allows you to save workspaces for just this purpose.

To save a workspace, go to **Window | Studio | Add Preset…**:

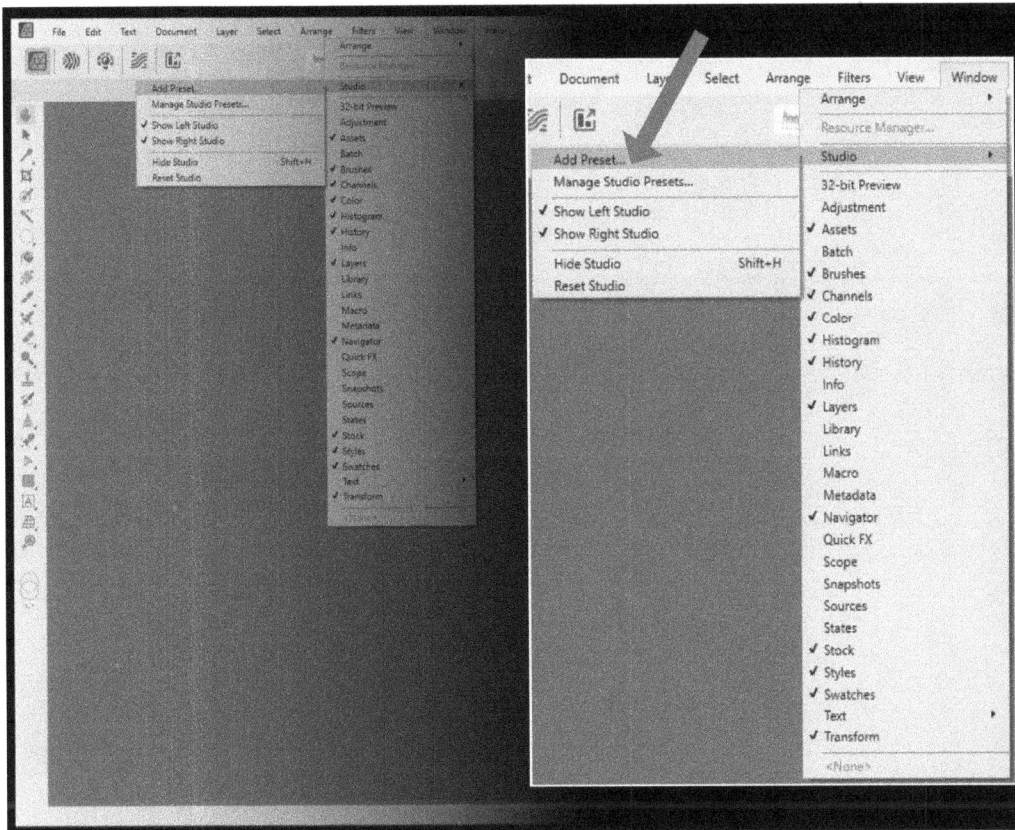

Figure 1.14 – Adding studio presets

You can manage presets as well if you decide to change (add or remove) them later.

## Summary

In this chapter, we have covered the fundamentals of all digital art programs, so if you are new to the digital art space, you have an idea of how one little dot, one little pixel, can change the entire digital landscape. Also, you learned some of the terms we will be dealing with in the later sections of the book, such as vector and raster-based images. You can at least now navigate the major sections of the interface (don't worry, there is a lot more to come, we do not expect you to be an expert in the program's deep functionality at this point).

In the next chapter, we will open up our first document and cover some more fundamental terms to make sure your art comes out the way you want it to. We will be talking about the difference between canvas and document size, as well as some professional workflow shortcuts such as the creation of presets and templates.

# 2

# Opening Your First Document

In this section, we will cover the creation of your first document in Affinity Photo, bringing in the terms we learned about in the previous chapter. We will begin by exploring the differences between opening an existing photo, creating a new document, and placing an image in a new project. Then, we will learn about the fundamental differences between terms such as "document size" and "canvas size" and how to create a preset. Decisions made at this early stage relating to the document are 100% correctable; however, setting yourself up for success right from the jump is always preferred to correcting something later in the process

In this chapter, we will cover the following topics:

- Bringing in an existing photo into Affinity Photo for editing
- Creating a new document for a project
- Placing an image into an existing project
- Resizing the document you just created
- Creating presets and templates for use in other projects
- Professional tips, tricks, and important points

## Bringing in an existing photo into Affinity Photo for editing

If you want to bring an existing photo into Affinity Photo to edit it, all you have to do is carry out the following steps:

1. Open the Affinity Photo program.
2. Close the splash screen (see *Figure 2.1*).

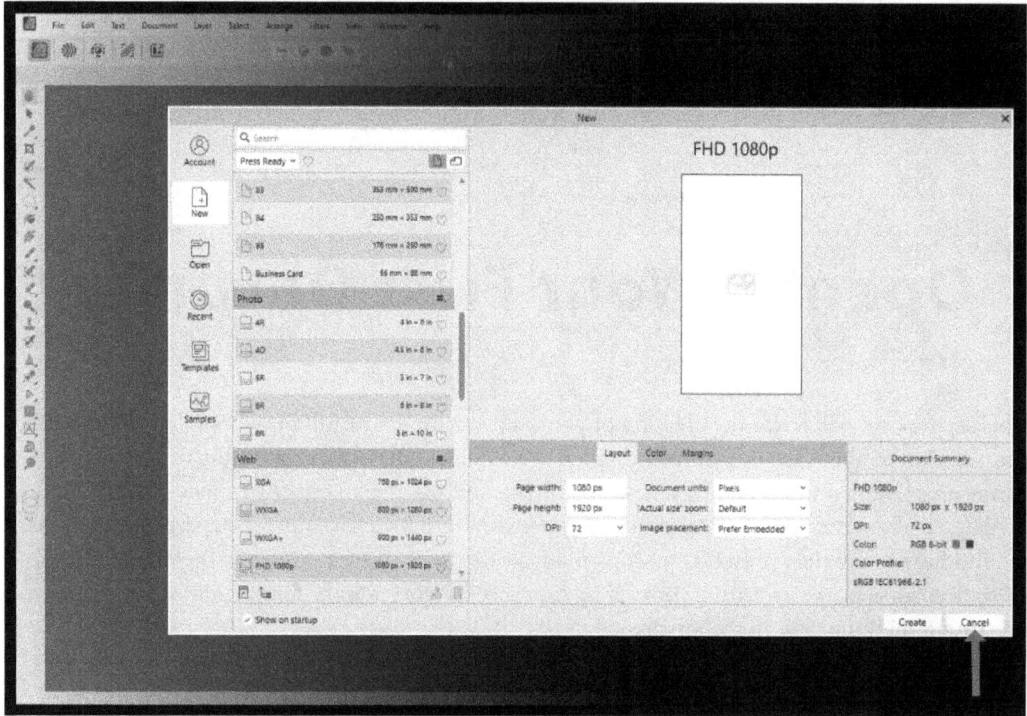

Figure 2.1 – Closing the splash page

3.  Go to the **File** tab in the menu bar.

4.  Click on the **Open…** command.

5.  Search for the image or file you want to open and click **Open**:

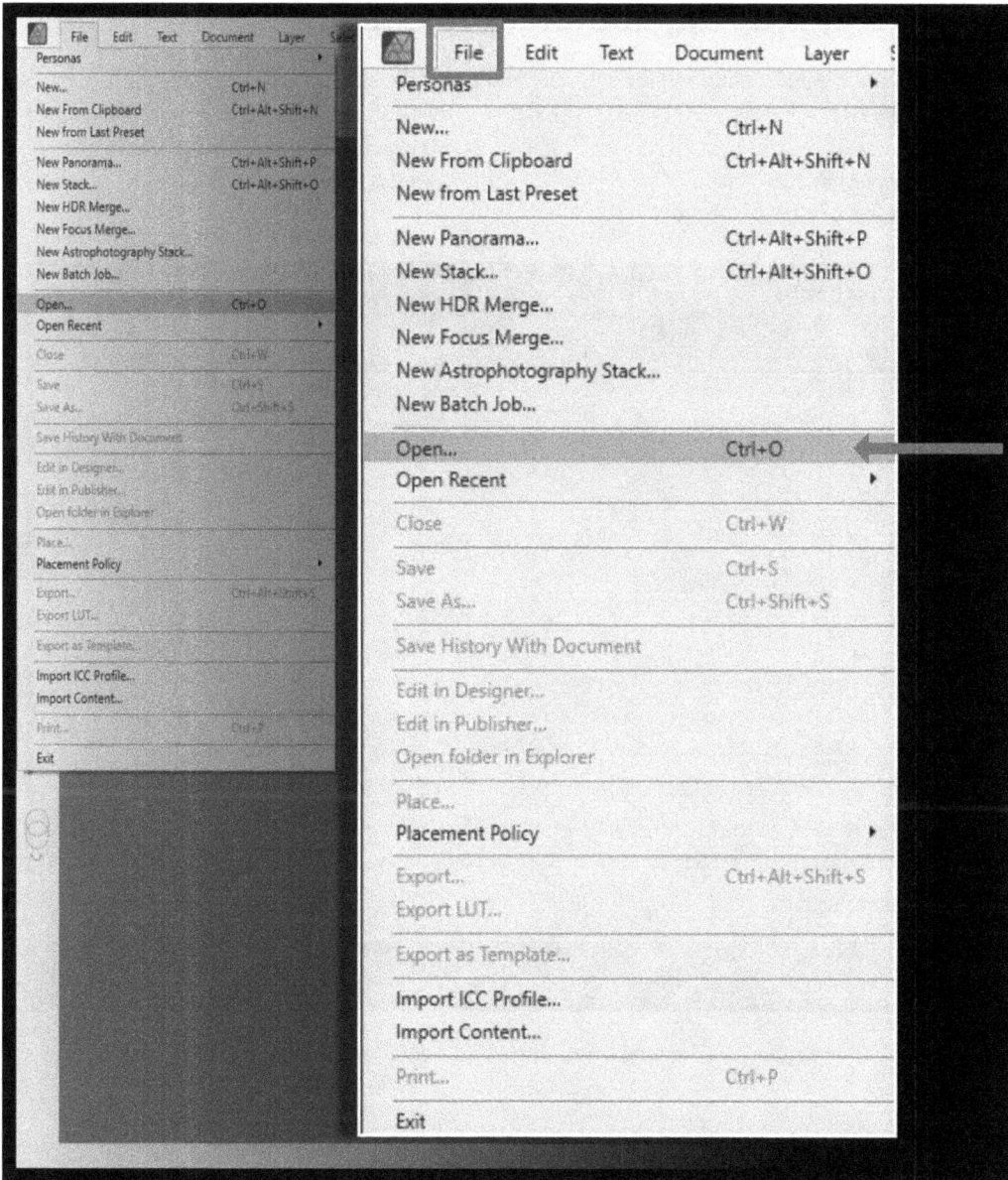

Figure 2.2 – Opening a document

From here, you are free to edit the image.

# Creating a new document for a project

When you want to create a document for a project and know the size, such as a banner for a website (1,920x1,080 px), or maybe you want to make a 16x20" canvas print, then you need to make the project document the size of your project. To do that, click on **New** either on the splash screen or in the file menu in the upper left-hand corner (see *Figure 2.3* and *Figure 2.4* for examples of where they are in the current version of Affinity Photo):

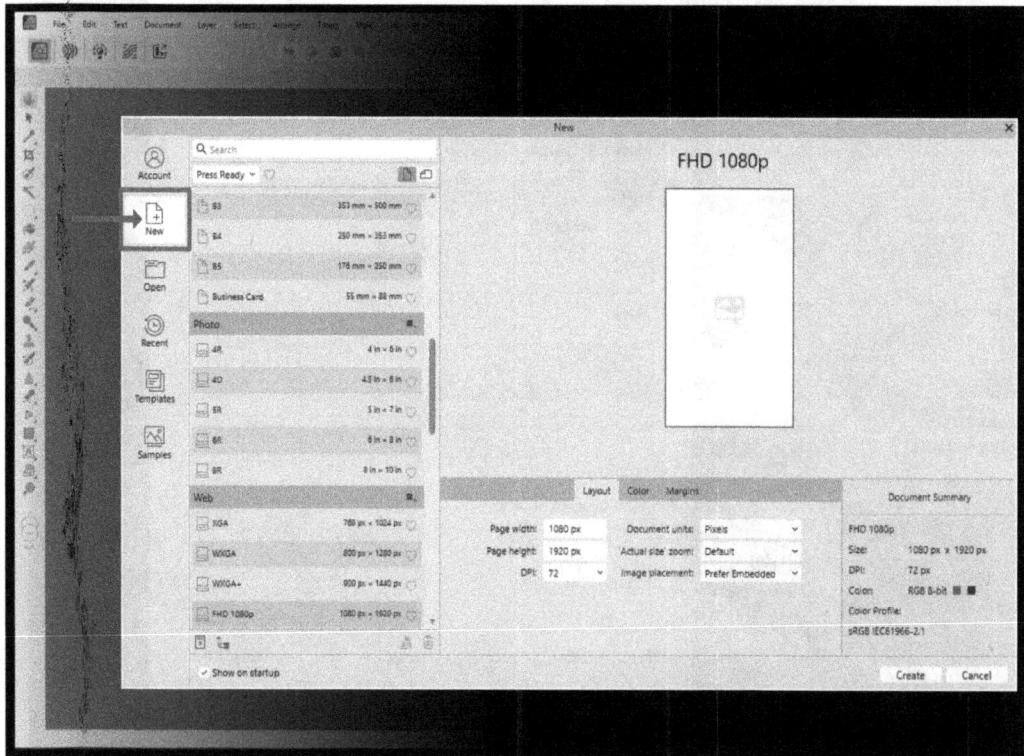

Figure 2.3 – Creating a new document from the splash page

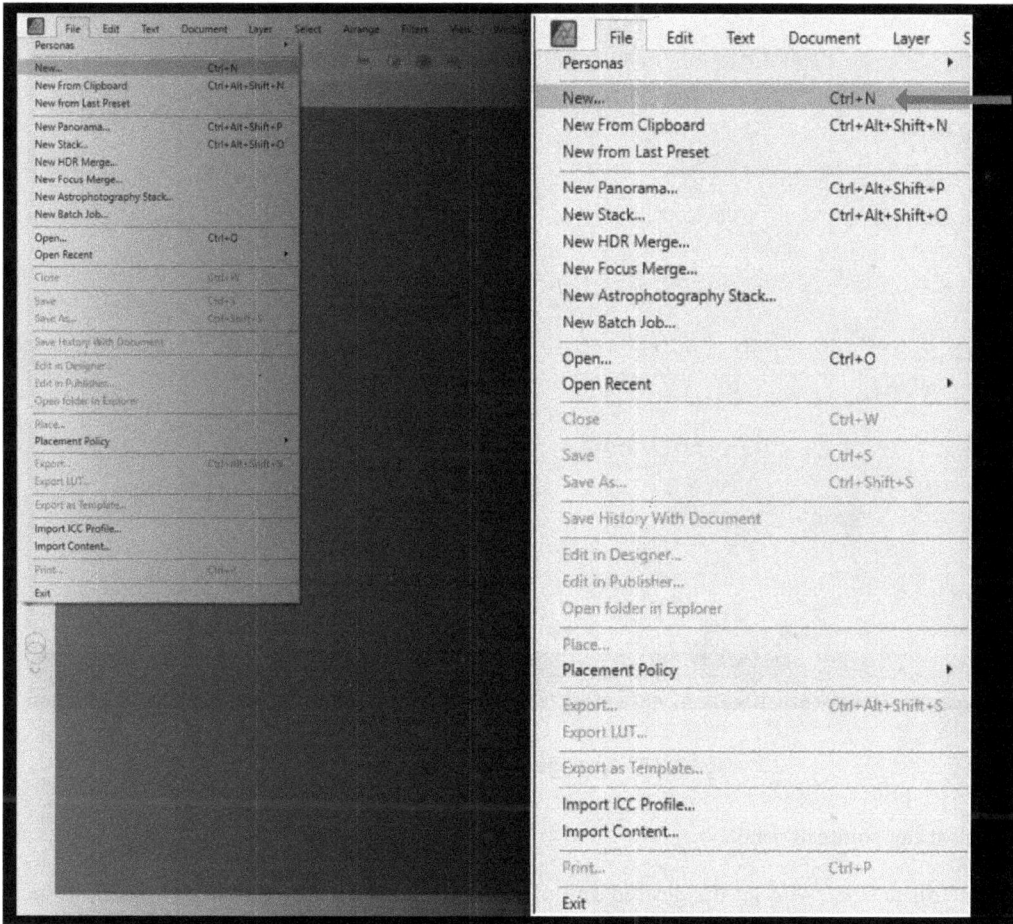

Figure 2.4 – Making a new document from the file menu

To read the new document screen, we have attached a keyed photo of the important portions of the screen so that you can follow along (see *Figure 2.5*):

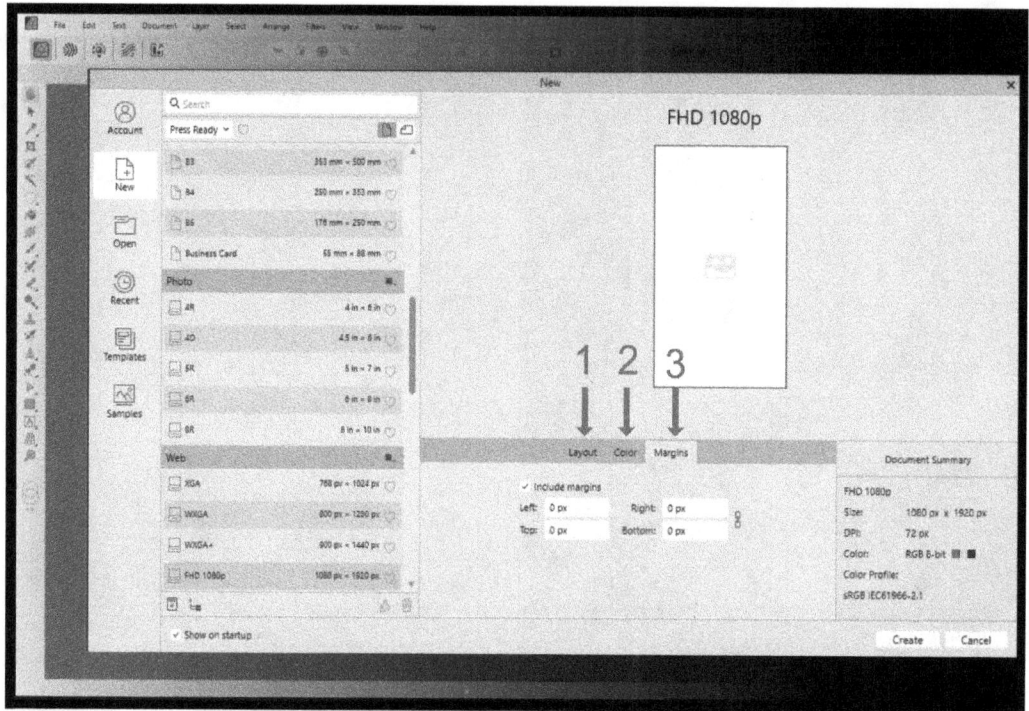

Figure 2.5 – Settings for new documents

Let's look at the points in detail:

- **1**: This is where you set the desired unit of measure, size, and the desired DPI (recall these terms from *Chapter 1*) based on your project. Remember if you are printing, the DPI should be a minimum of **300**, and for digital work, it should be **72**. The best practice I follow is setting it to **300** DPI initially and then saving a 72-DPI copy: this way, I can always start out with the higher DPI.

- **2**: In this section, you will set the desired color profile for the project (recall the term from *Chapter 1*) ). Here, and for *most* things, you will work in **RGB8**; however, if you are printing the image, make sure you consult your printer's needs in terms of the color profile. Some printers will ask you to convert it into CMYK, so make sure if you desire to print the image that you know your printer's color requirements.

- **3**: In the event that you are creating a document for printing and margins are required, you can set them here. Again, this is not essential for digital work, but most printers do not go all the way to the edge, so a certain margin is common.

In this example, we are going to create a 16x20" piece of art for print at 300 DPI with an RGB8 color profile and without margins (see *Figure 2.6*):

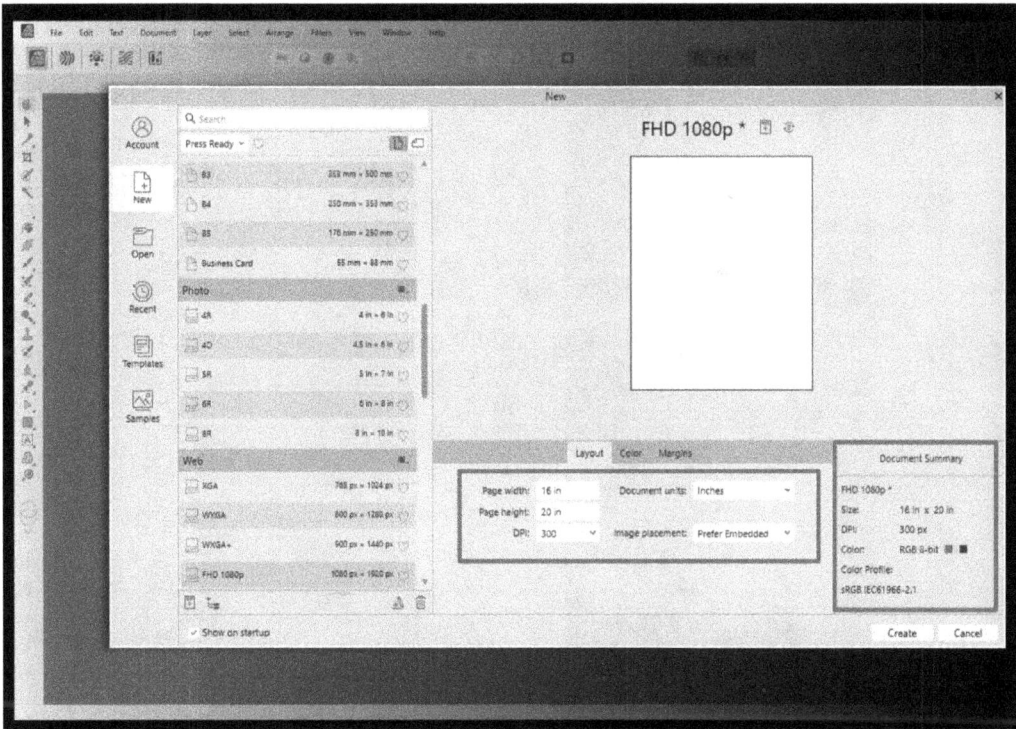

Figure 2.6 – Example of setting the document for a project

Next, we'll look at how we can place an image into an existing project.

## Placing an image into an existing project

Let's look at the steps to do just that:

1.  Open a **New** document as shown in the previous section with the settings you want it to have.

2.  Go to the **File** menu in the upper left-hand corner and click on **Place…**.

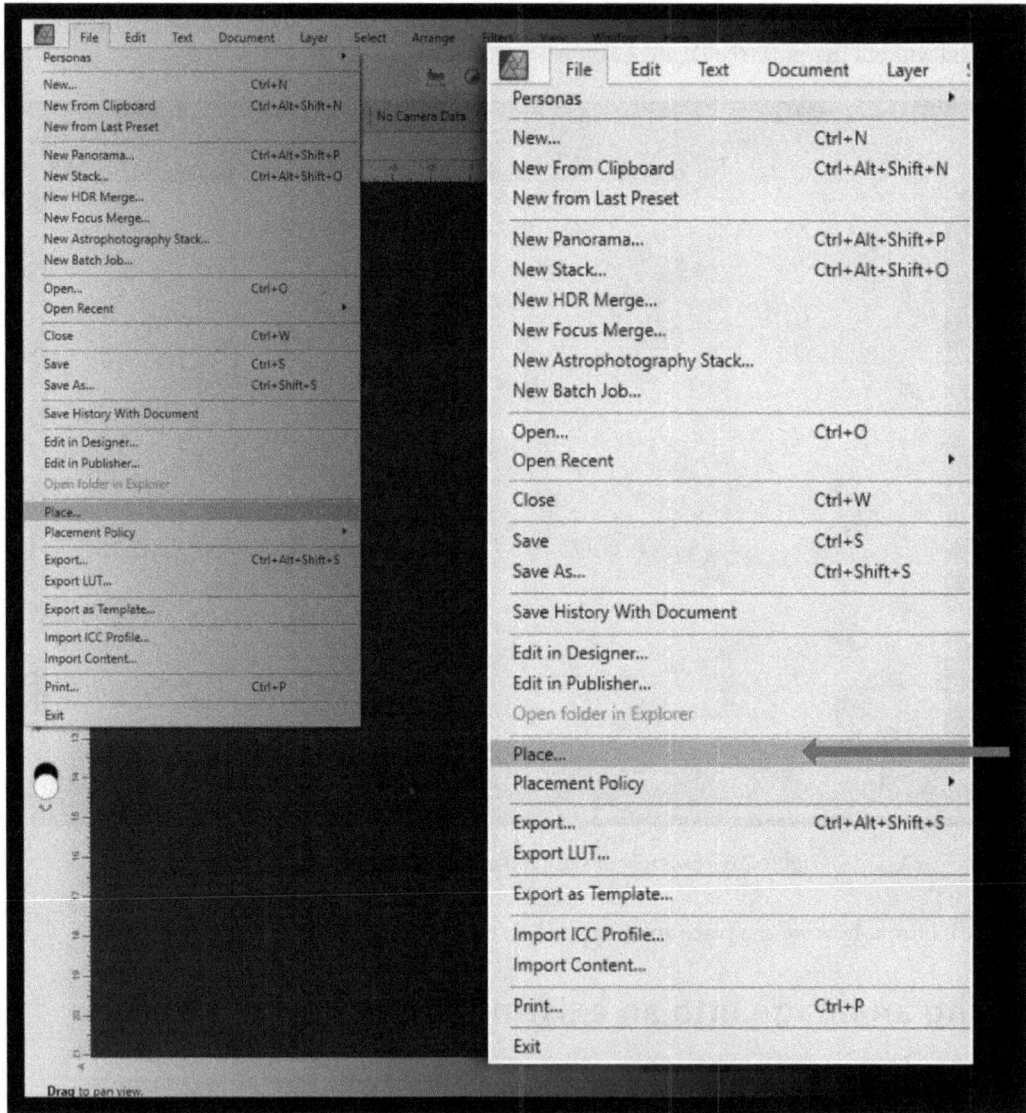

Figure 2.7 – Placing an image in an existing document

3.  Locate the file you want to place.

4.  Click and drag the file to the size you want in the image:

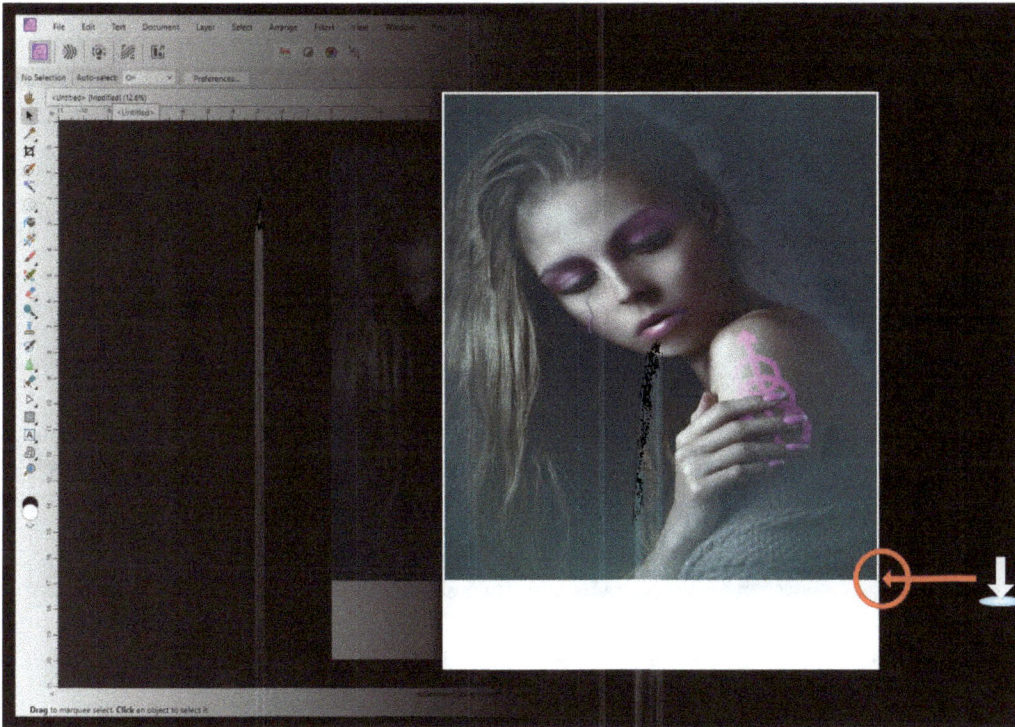

Figure 2.8 – The icon for placing the image

Now that we have planned the image, let's resize it next.

## Resizing the document you just created

Now that you have an image, either bringing in or creating and placing it, you are eventually going to want to resize it, and so we have to explore the basic terms and how exactly we do this. We are going to cover the difference between the **document** and the **canvas**, and how to adjust the parameters of the image during editing.

The options for resizing the project are under the **Document** option in the menu (see *Figure 2.9*):

Figure 2.9 – Resizing the document using the Document menu

The document is the actual image you are working with, but much like you can place a document on a Matte, the Matte can be thought of as the canvas. So, if you want to adjust the size of the image *without changing the size of the canvas*, select **Resize document....**

## When you want to change the document size

When you go to change the document size, this window pops up (see *Figure 2.10*):

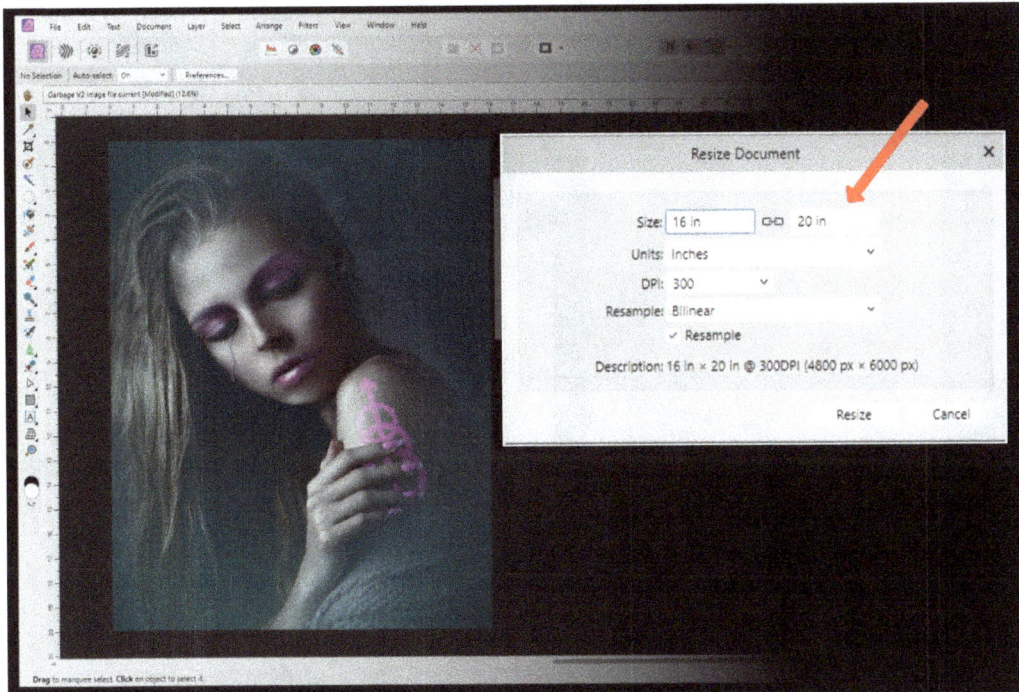

Figure 2.10 – Options for resizing including "aspect ratio lock"

This allows you to adjust the size of the document, change the DPI, and so on. The most important thing to keep in mind in this window is the aspect ratio lock. This lock keeps the width and height consistent; if you want to change this ratio, click on the lock to unlock it, and feel free to adjust it.

## When you want to change the canvas size

If you want to change the canvas size, this window comes up (see *Figure 2.11*):

Figure 2.11 – Anchor points

In this window, the important features include the aspect ratio lock, but there is a diagram showing nine positions; this is where you want to center the document on the new canvas. Clicking on the squares places the document in these positions.

As an example in this illustration, I have chosen to put the image in the lower right-hand corner of the image to show you where the program placed the image during resizing. I changed the size of the canvas to **1599.9mm x 2262 mm** as well to show you the effect. Notice that because the canvas is larger than the document and I specified the lower-right corner, this is the new placement.

This is why it is important to know whether you want to change the dimensions of the document or change the dimensions of the canvas, as they are drastically different items.

## Creating presets and templates for use in other projects

Creating the same document over and over can be tedious, so it is better to make presets or templates. Affinity comes loaded with various presets and templates for common print and digital jobs; these come up when you open a new document (see *Figure 2.12*):

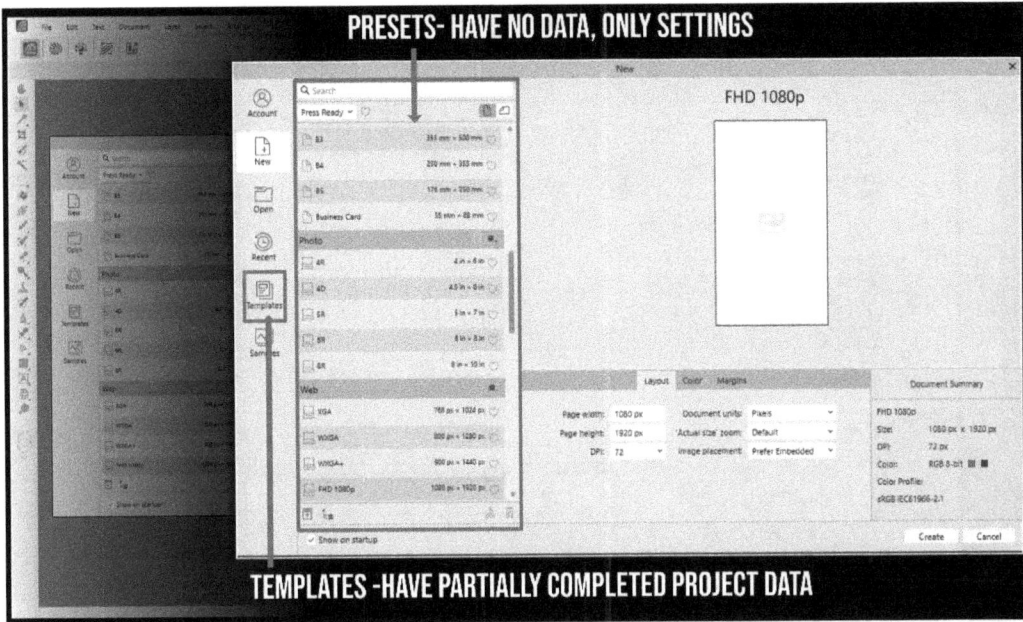

Figure 2.12 – Creating presets in Affinity Photo

## Creating presets

Now if you have a job you are working on or a certain image size you like to use, you can save your own presets; this will save things such as the size, the DPI, the color profile, and so on.

In my work as a canvas printer and T-shirt maker, a common size for me is 16x20" and I know my printer has the capability to take my RGB color profile and convert it into its software, so I will make a preset to make sure I do not have to redo this every time.

To save your own presets, use the following steps:

1.  Create a document but *before* you hit create, click on the + sign (see the lower left arrow in *Figure 2.13*). This will add the preset to your presets category.

2.  Right-click on the preset and then rename it to whatever you like (see the middle item in *Figure 2.13*):

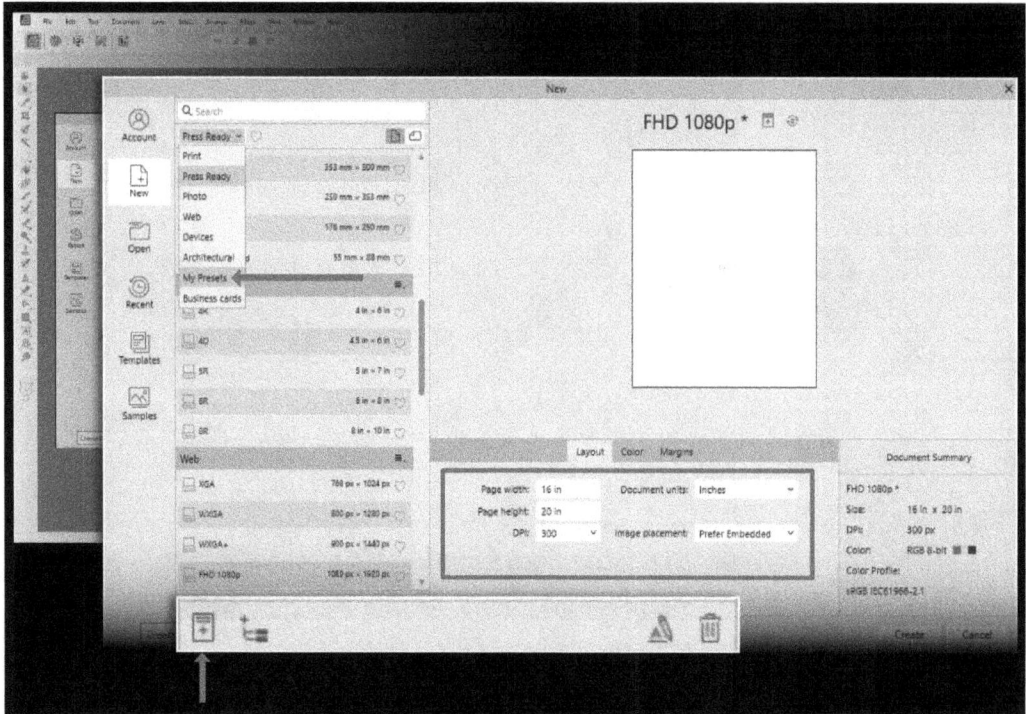

Figure 2.13 – Saving presets

## Creating templates

Templates are a special type of preset and a template includes elements that you have already added. Think of templates as partially finished projects that you use as a starting point for new projects.

Templates are on the same new screen as presets; however, they are arranged into folders, and when you think about it, projects have to exist in folders, right? So, selecting a template is the same process as selecting a preset. Meanwhile, creating templates is a bit more complicated.

An example of when to use templates could be YouTube thumbnails. For what I do as a creator, I frequently have to make YouTube thumbnails, and I am not interested in recreating them each time just to change text and images. Therefore, I would create a template.

To create a template, use the following steps:

1.  First, set up the folder where you want to save your templates. Once it is set, you do not want to move it, so make sure it is somewhere consistent.

2.  Create the document just like you would any other time and edit it to a spot where you want the template to take over.

3.  Go to **File** | **Save as template**, and place it in that folder from *step 1*.

4. Open a new document and where it says **Templates**, add the folder you created (see *Figure 2.14*).

5. Affinity will pull up the documents in that folder in the template window:

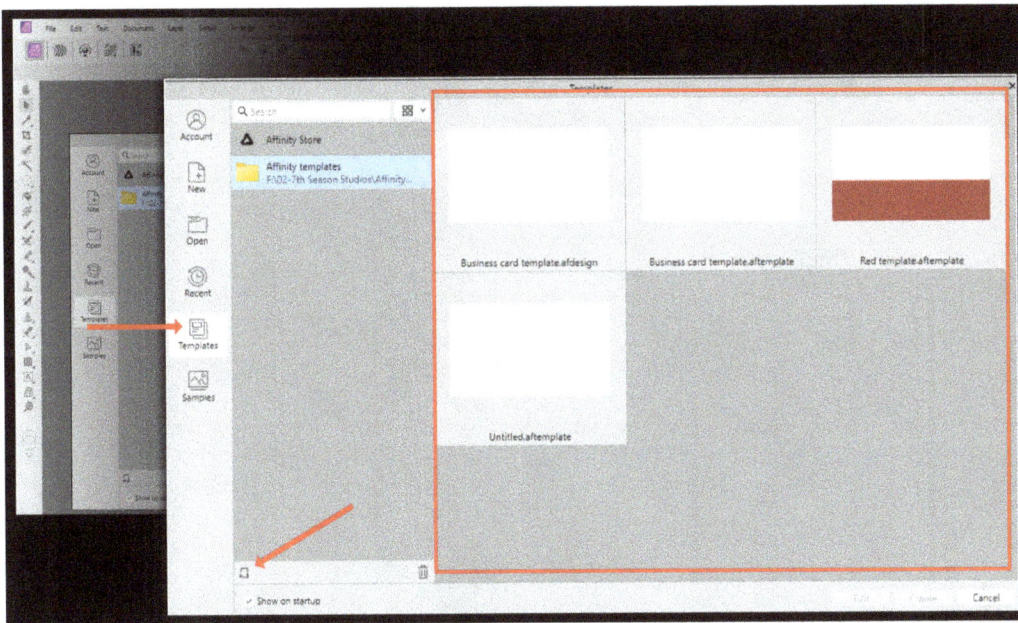

Figure 2.14 – Saving templates in Affinity Photo

## Professional tips, tricks, and important points

Here are some points that are good to keep in mind:

- Work as close to the document size you are trying to produce as you possibly can, as excessive resizing or stretching can ruin the image.

- Create templates for common projects and working sizes to save yourself time. Many people create popular templates for meme generation or branded layouts (such as YouTube thumbnails).

## Summary

To conclude this chapter, we have the fundamentals of setting up a document down, and this is important to understand because having to resize a document post-editing can be a tremendous headache as we start building masking layers and modifying effects.

Now that we have the basics of the document down, it is time to create your first image; in the next chapter, we will look at the concept of layers. Sticking with the concept of building from the ground up, pixels create layers, which are then stacked to form images. So, to understand editing, you must first understand the layer structure by opening your first image and starting to edit it – see you in the next chapter!

Now that we have the basics of the document layout, it is time to create our first single-image chapter, we will look at three ways of putting together an actual document. From the production...

# 3

# Layer Fundamentals – The Heart of Affinity Photo

In this chapter, we will tackle the single largest fundamental concept of *any* photo editing program… the concept of layers. **Layers** come in a variety of types, so we will first discuss what layers are and how we can add them, delete them, and change them. After that, we will look at various types of layers, including adjustment layers, a concept called live filter layers, and lastly, we will touch on text style layers.

In this chapter, we will cover the following topics:

- Understanding layers
- What are adjustment layers?
- Live filter layers
- Text layers
- Layers – practical editing practice
- Professional tips, tricks, and important points

## Technical requirements

You will need the Chapter 03 project to follow along with me, the link to which can be found in the Preface.

## Understanding layers

All digital art programs are based on a concept called **layers**, and all digital art comes from a unique combination of layers. Think of layers like sheets of paper, so these sheets are stacked on top of one another to form the image.

Layers are handled in Affinity Photo in the **Layers** studio tab (see *Figure 3.1*). Remember, if you do not have this panel, you can turn it on by going to **Window | Studio** and then checking the box next to the **Layers** panel. This will most likely place it on the right-hand side of the screen (as shown in *Figure 3.1*). I am going to click and drag my **Layers** panel out of the studio and place it in the workspace for ease of viewing during this book (if you want to follow along, just click and drag it out into the workspace):

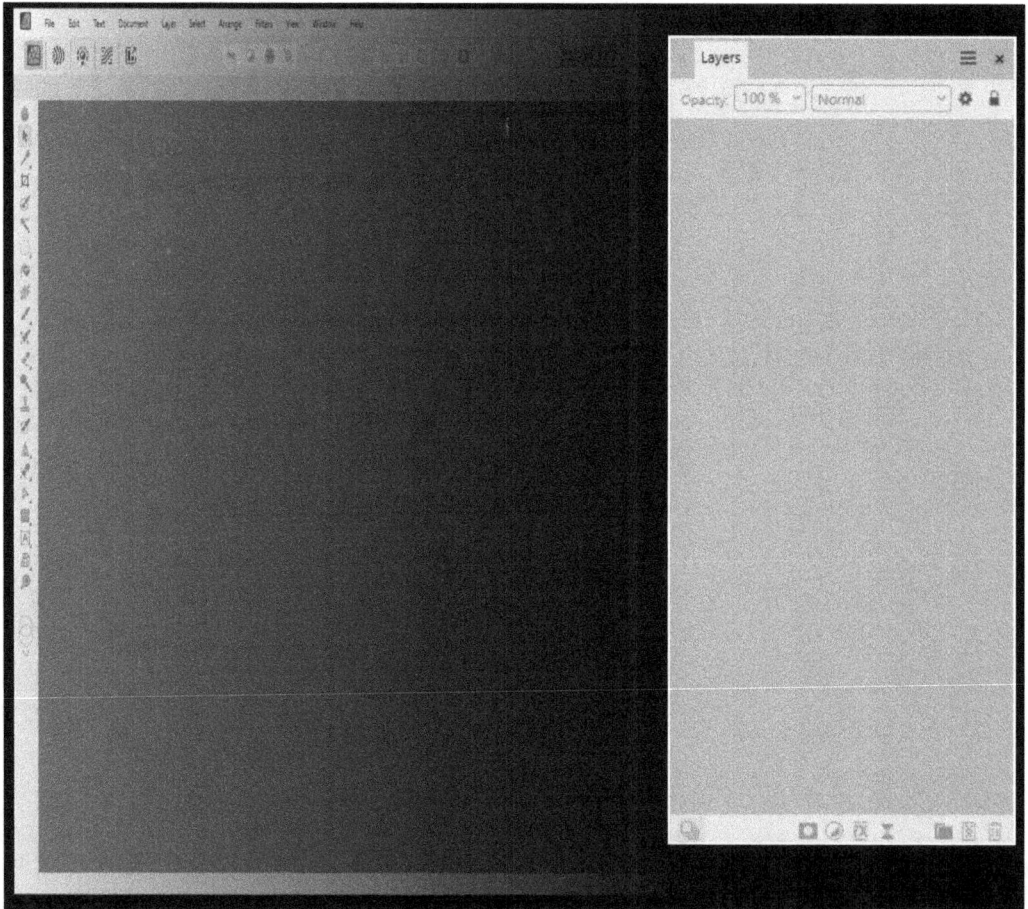

Figure 3.1 – Studio overview

Let's begin by opening a new document. I am going to open up our first downloadable image (refer to *Figure 3.1*). I am going to go to **File | Open** and find the location where I saved it, and this will open the image in the workspace, as shown in *Figure 3.2*:

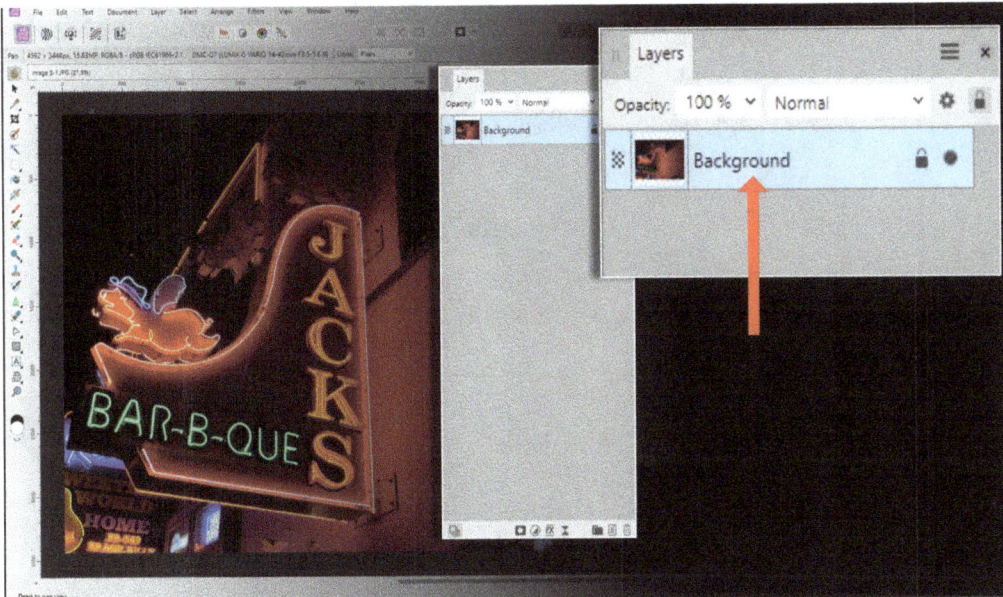

Figure 3.2 – Workspace containing the layer

Notice there is now a layer in the **Layers** panel – in this case, it is titled **Background** and is referred to as a **Pixel** layer. This is an important designation, as there are different types of layers. Let's define a few of them now for simplicity:

- **Pixel layers** – When Affinity identifies a layer as a pixel layer, it means that the program understands that what it is looking at is a collection of pixels that can be adjusted and manipulated for editing photos, drawing, painting, and so on. This means it can be selected and adjusted. For most photo editing, you will be working with a pixel layer.

- **Image layers** – Sometimes when you download from sites, Affinity Photo will not recognize the image as a pixel layer, but instead, will use the word **Image**. You will not be able to draw or use any tools or effects on the image layer until it is rasterized as a raster layer. The word **Image** shows up on the side of the thumbnail (see *Figure 3.3*) for the difference in here.

  Image layers are not recognized by Affinity as being composed of pixels, so if you try and select parts of the layer, nothing happens.

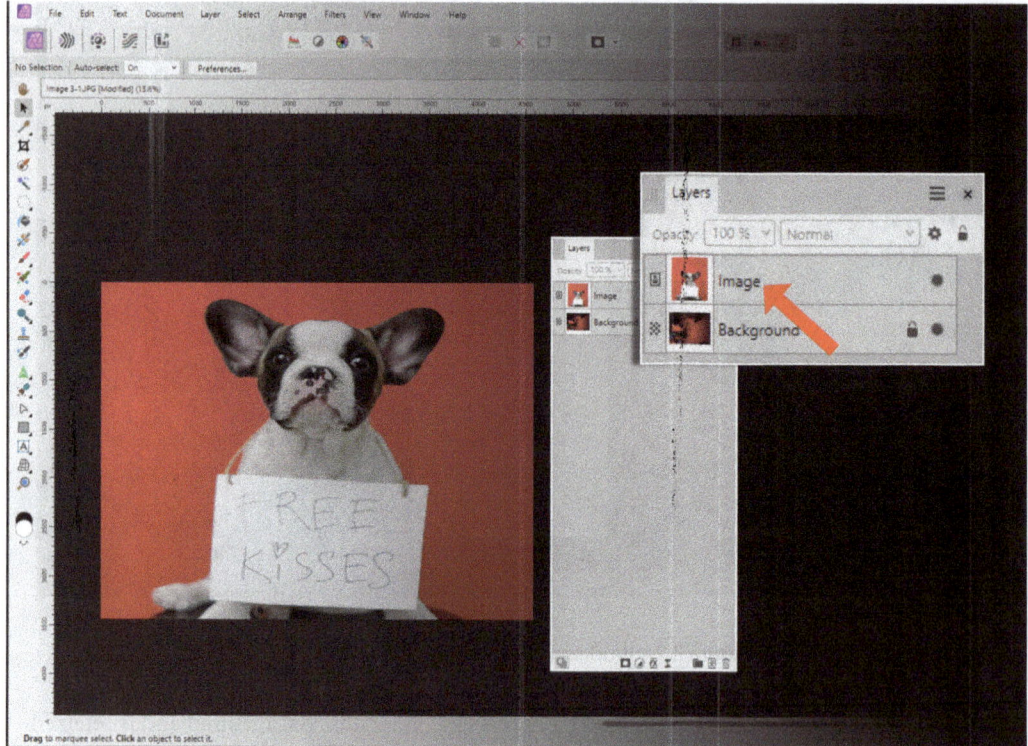

In *Figure 3.3*, I added an image of a dog into the mix; notice the differences in how they are labeled. This **Image** layer cannot be selected and needs to be what we call **rasterized**.

## What is rasterizing a layer?

In order to make image layers selectable, as a best practice, we do what is called **rasterizing the layer**. The act of rasterizing the layer turns an image into a pixel layer. To rasterize an image, hover the cursor over the layer in the panel and right-click. Select **Rasterize** and you will notice that the type of image now says **Pixel**. Now, it is selectable and ready to work with (see *Figure 3.4*):

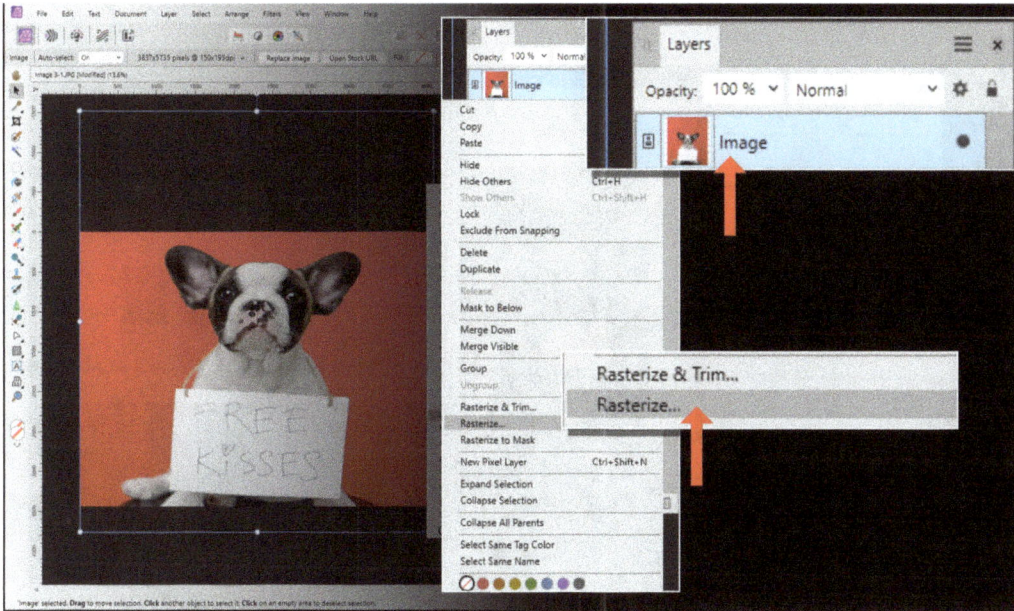

Figure 3.4 – Shows the rasterized designation

## Working with layers – the demographics

Layers are fundamentals of all digital art programs, and whether you go into vector art, Photoshop, 3D art, and so on…layers are the method these programs use to tell where things are in the world. Think of layers like locations on a road map; it is up to you as the editor to plan the trip and get from A to B, but unless we know where the city is that we are visiting, we have no idea how to plan the trip, how long it will take, or whether we are moving in the right direction.

Layers provide these guideposts by telling Affinity *the dog layer is on top of the swimming pool layer, but underneath the layer that contains the child.* In short, layers form the relative location of things.

### Layers have order

As mentioned earlier, layers are like sheets of paper, and as such, if you stack one piece of paper on top of another, you cannot see the piece of paper that is underneath. The same is true with layers. The **Layers** panel shows the order of layers with the lowest layer at the bottom of the stack. To illustrate this, follow these steps:

1.  Go to the **File** menu and click on **Place…**. We are going to place an image into this project from somewhere on your computer.

2.  Go to the downloadable files for this book and locate **Image 3-2**.

3.    Click on **Open**:

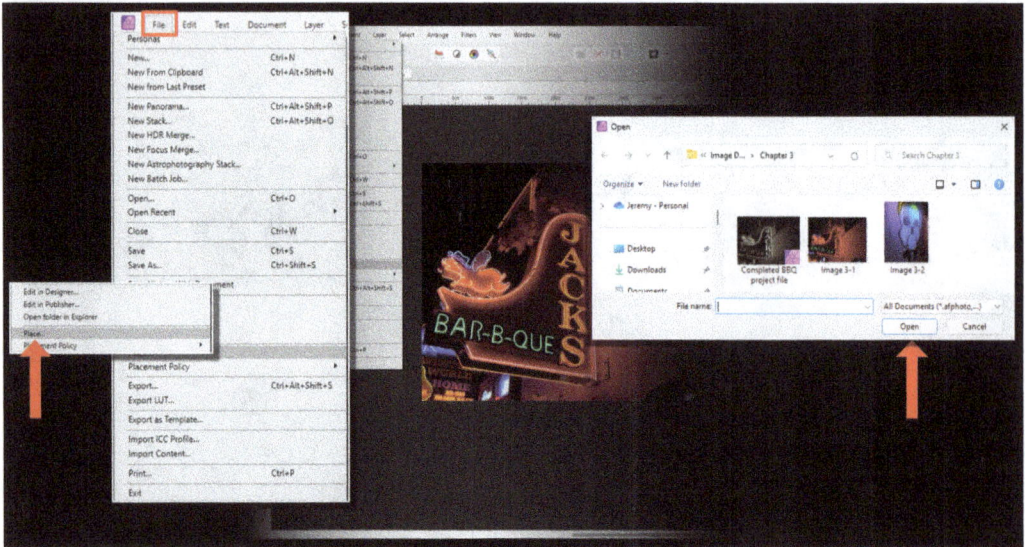

Figure 3.5 – How to open a layer

4.    Now, the icon will change to an arrow with a little circle (see *Figure 3.6*); click and drag the image until you get to the size you want (don't worry, you can always resize it later):

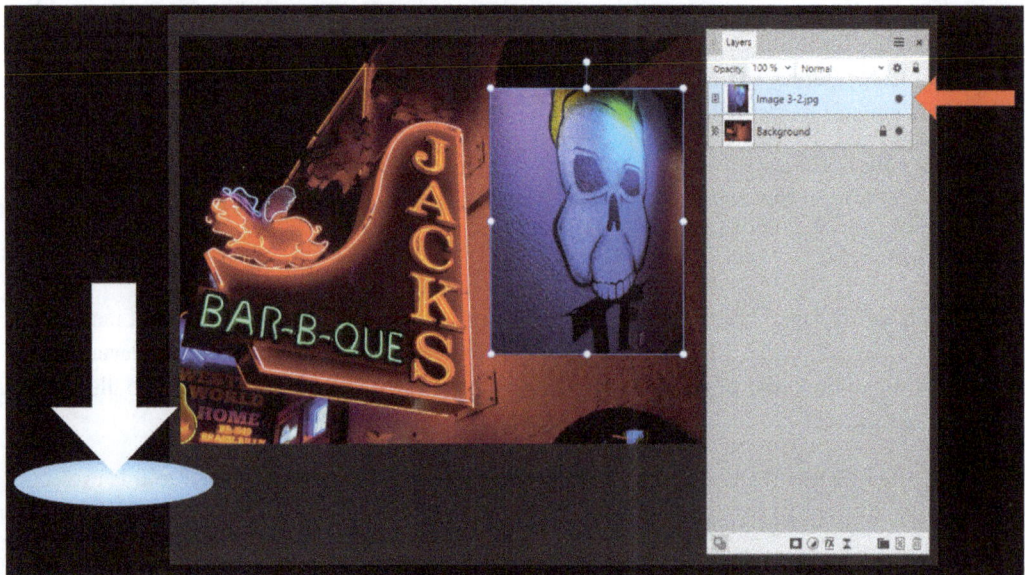

Figure 3.6 – Image 3-2 has been added

If you mess up with the layer selected, just hit the *Delete* key on your keyboard and try again. Notice the order of the layers; in this case, **Image 3-2** is on top of **Image 3-1** (which is the **Background** layer). See *Figure 3.7*.

Also, notice the image was loaded as an image file, so feel free to rasterize it as mentioned in the previous section.

### Reordering layers

Now that we know there is a hierarchy of layers, we will need to be able to reorder them as we edit.

To do this, simply left-click on the layer to select it (the layer will turn blue in color), and with the button clicked, drag it where you want it in the stack (see *Figure 3.8*). Notice that, in this example, we can no longer see **Image 3-2**; why is that? It is because the **Background** layer is on the top of the stack, so you cannot see the layer underneath it.

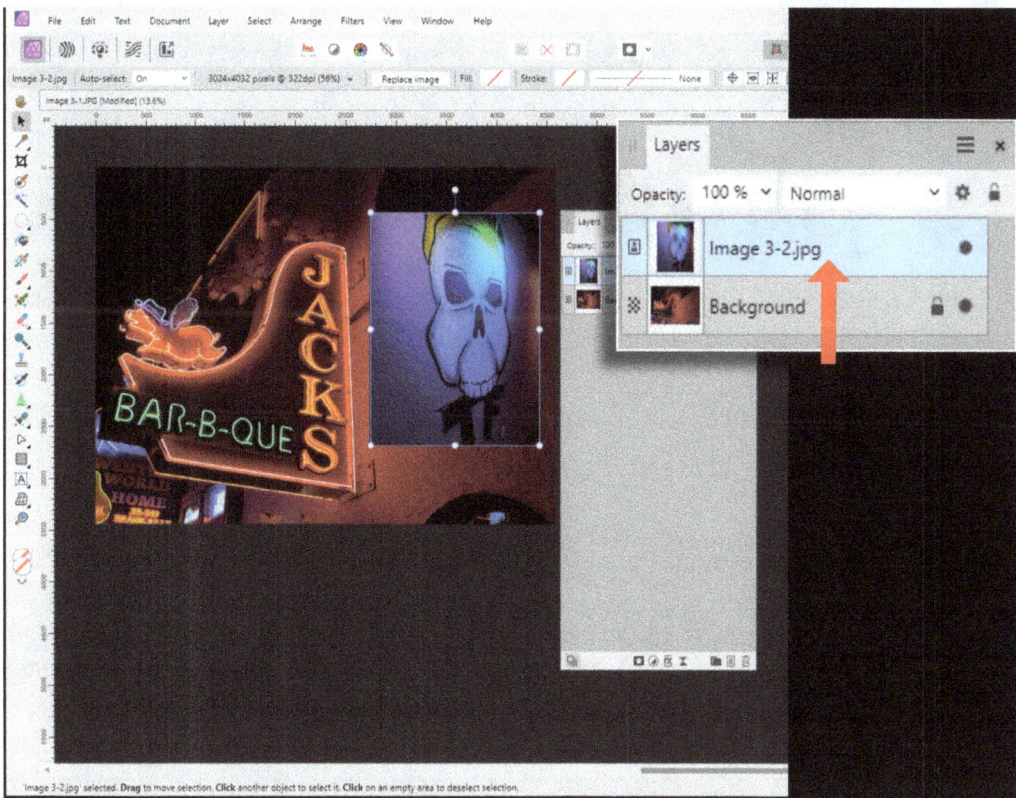

Figure 3.7 – Image 3-2 on top of Image 3-1

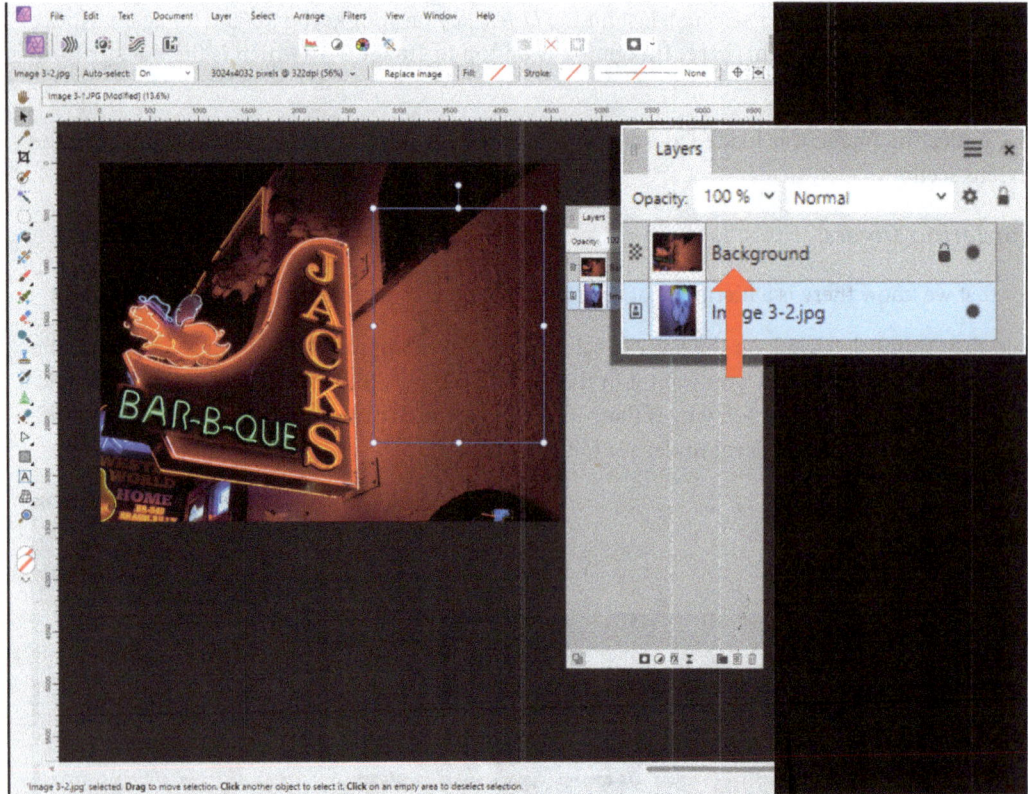

Figure 3.8 – Repositioned background

I am going to drag **Background** back to its original position at this point for the rest of the chapter.

> **Pro tip**
>
> 99% of the time, the reason you have problems in your composition is your layer structure is wrong; it is always the simple things that cause you issues.

## Deleting layers

Sometimes, you don't need a layer any longer and you want to remove it from the project. There are two ways to do this.

The first way is to right-click on the layer (notice it turns blue) and then choose **Delete** (see *Figure 3.9* for the visual):

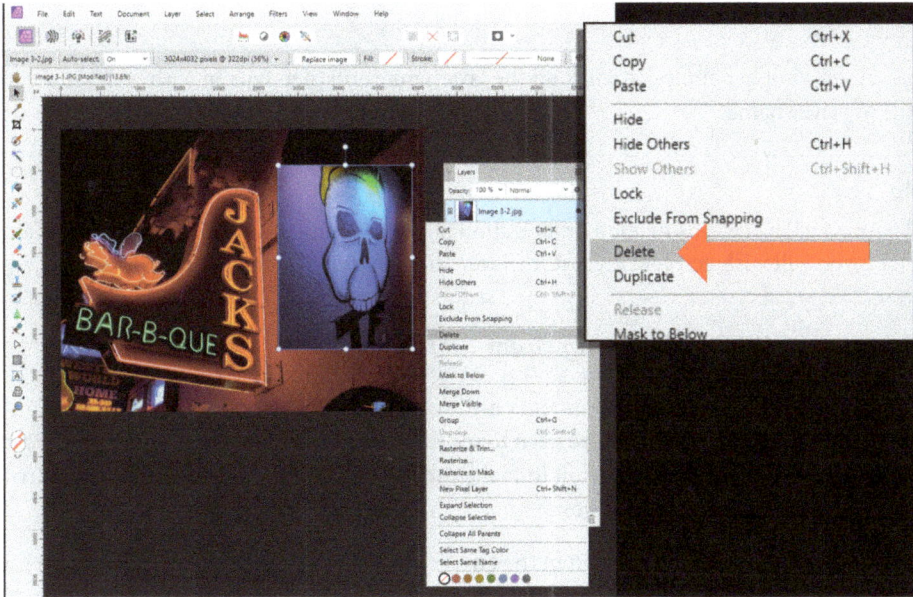

Figure 3.9 – Delete option for layers

The second way is to click and drag the layer to the trash can in the lower-right corner of the **Layers** studio panel (See *Figure 3.10*):

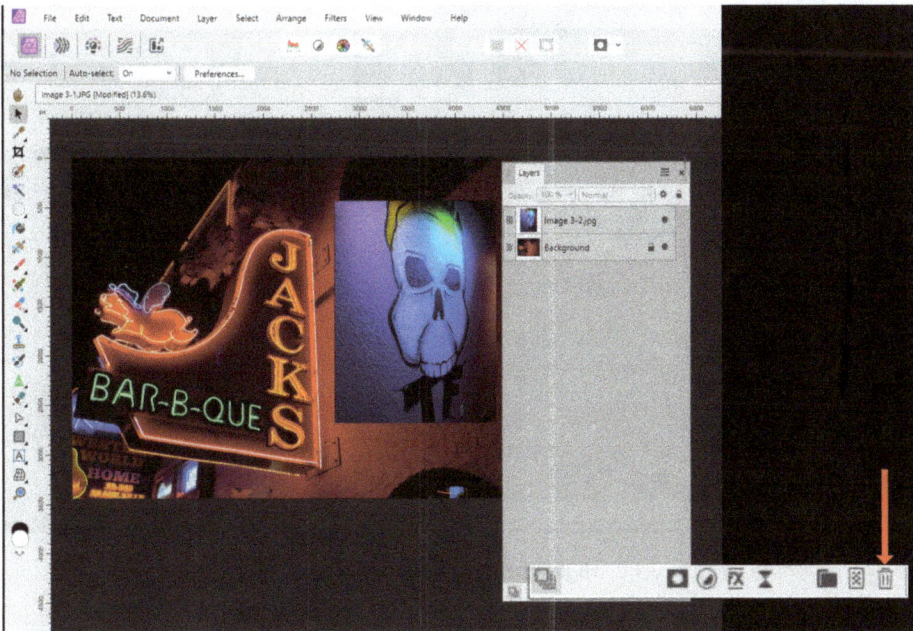

Figure 3.10 – Clicking and dragging to the trash can

## Renaming layers

To rename a layer, simply double-left-click on the layer name and then type in with your cursor; this will change the layer name.

> **Pro tip**
>
> Good naming of layers leads to good organization of projects, and when you are 54 layers deep in an edit, you will wish you had named the layers so you know where you need to adjust.

## Making layers visible and locking them

Another critical aspect of layer management is **hiding** and **locking** layers. Sometimes, we want to hide a layer, as it may be in position but we need to see the object underneath, and sometimes we may have gotten a layer just perfect and we want to make sure we do not accidentally click it and throw it out of order and make unforeseen changes.

To hide a layer, simply click on the *eye* icon in the **Layers** panel (showing in *Figure 3.11* as item **1**), and to lock a layer, simply click on the *lock* icon in the **Layers** panel (showing in *Figure 3.11* as item **2**):

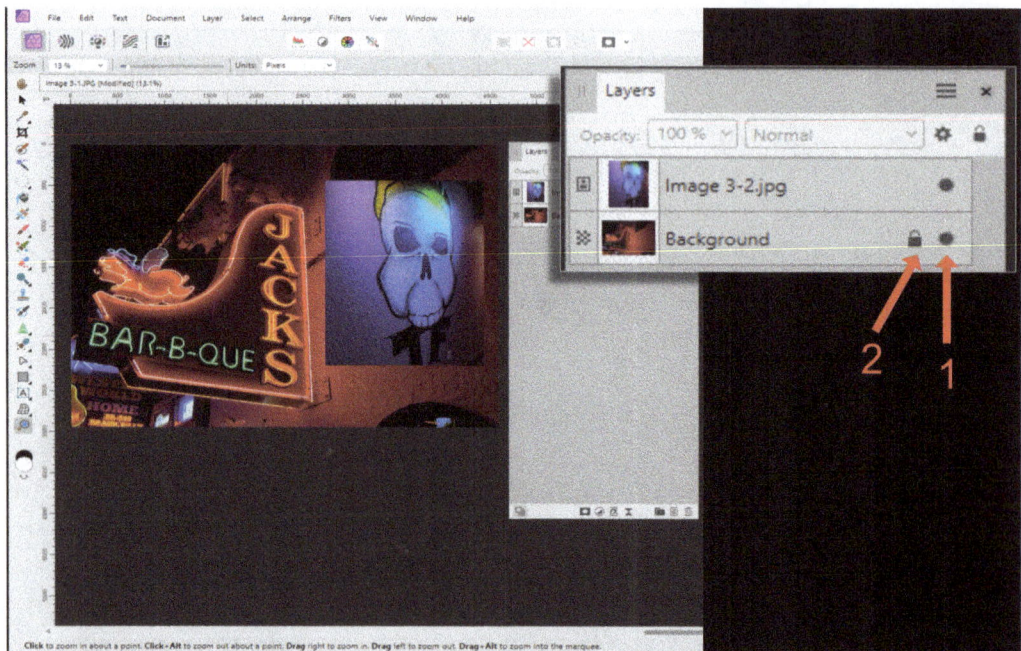

Figure 3.11 – Highlights hiding and locking layers

## Adjusting blend modes and opacity for layers

While there is an entire other chapter dedicated to blend modes in this discussion on layers, I just want to show you how to adjust them, as well as the opacity of the layer, let's take a moment to define the **opacity** and **blend** modes.

**Opacity** is how transparent or solid a layer appears; **100%** means it is solid, whereas **50%** means that the layer is somewhat see-through. A value of zero opacity means the layer is invisible (see *Figure 3.12*, item **1**)

Blend modes tell the layer *how* to interact with the layer underneath it, and they can be used to create some amazing effects, which you will use in the projects in this book. While it is outside the scope of this section to discuss the intricacies of blend modes, in the **Layers** panel, this is where you adjust them (see *Figure 3.12*, item **2**):

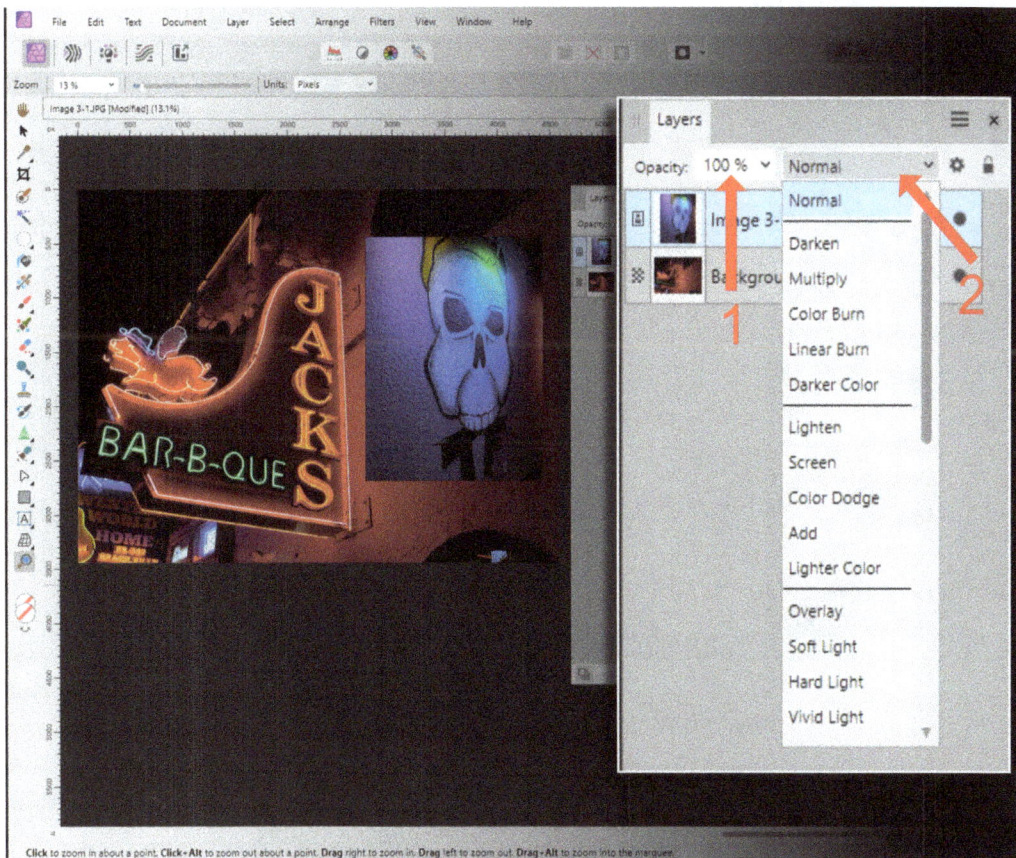

Figure 3.12 – Shows the location of blend modes and Opacity

## *Grouping layers*

The last concept around the basic demographics of layers is the discussion on grouping layers. Grouping layers is sort of like creating a folder, and in that folder, you put the individual pieces of paper (or layers).

So, why would you want to group layers? Simply put, to make adjustments to *all* the layers at once, or to move all the layers at once. Think of grouping as your best way to *organize* your work, and when combined with the naming conventions, it allows you to pull off very complicated edits, all while remaining organized.

To group layers, simply use the following steps:

1.  Hold *Ctrl* and left-click the layers you would like to group. They will turn blue as they are selected.

2.  Right-click and choose **Group** (see *Figure 3.13*, item **1**) and this will create a group (see *Figure 3.13*, item **2**):

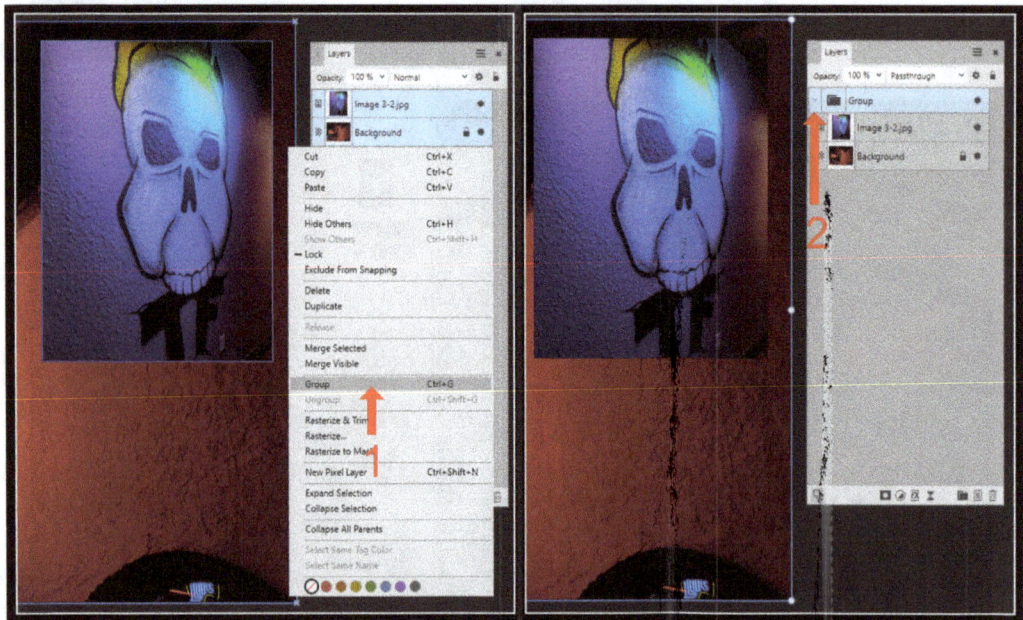

Figure 3.13 – Shows the position of nested layers

Notice the layers are inside the group; they are nested. Layers not only have an order but sometimes there is a linkage between layers that guides other layers; this is called **nesting**.

When you create a group (see *Figure 3.14*, item **1**), you will see a little twirly down arrow (see *Figure 3.14*, item **2**). If you click on the twirly down arrow (that is now a technical term), you see the layers underneath it are indented. This implies these layers are nested inside the group:

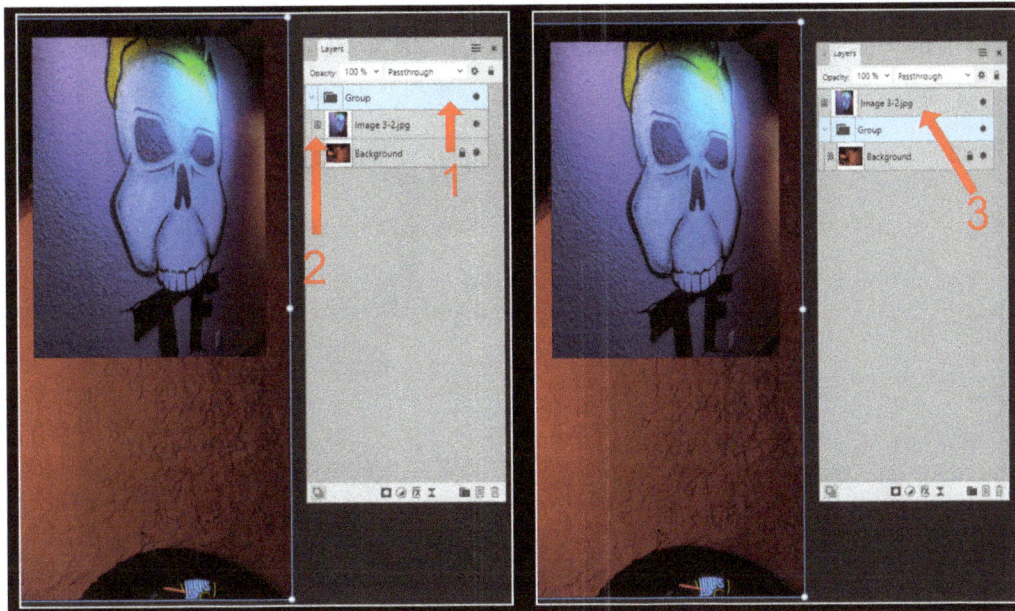

Figure 3.14 – Shows the removal of a nested layer back outside the stack

So, if you move, change, or adjust the group, the layers move with it. You can always remove layers from a group by clicking and dragging the image outside the group. Notice in *Figure 3.14*, item **3**, the blue layer is now outside the group, not indented, and therefore no longer nested.

We will use this technique all the time in our edits.

## What are adjustment layers?

Inside the **Layers** panel, there is a type of layer called an **adjustment** layer, and as the name implies, this allows you to adjust the image. They are located in the lower areas of the **Layers** studio panel (see *Figure 3.15*, item **1**)

To create an adjustment layer, simply click on the type of layer and you will see it appear in the layer stack. In *Figure 3.15*, notice item **2** after we added a black-and-white adjustment layer. Notice the layer's position in the stack. This means that **Black & White Adjustment** is applied to every layer *below* the adjustment layer.

Every adjustment layer has adjustments you can make, and each layer is different. We will be exploring the most popular adjustment types in our projects through the book. We will not be covering each one, but we will cover those that we use most frequently in photo editing. In the **Black & White Adjustment** layer, depending on the color or hue of the underlying images, you can adjust the sliders to your taste based on what you are trying to do with your image (see *Figure 3.15*, item **3**):

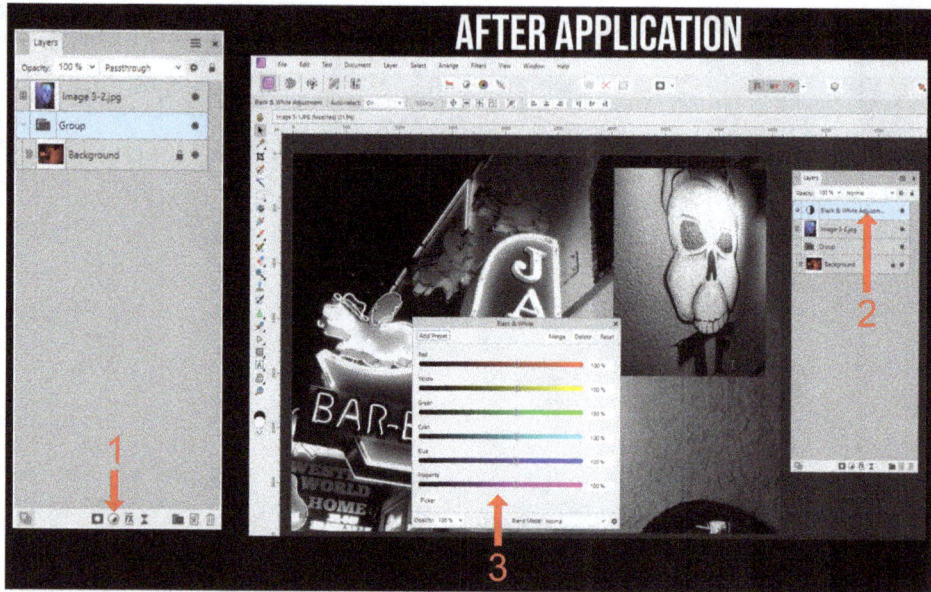

Figure 3.15 – Shows the application of a black-and-white adjustment layer

Adjustment layers can be stacked and combined. Notice in *Figure 3.16* that after we added **Black & White Adjustment**, we added a **Lens Filter Adjustment** layer to give it an aged sepia tone.

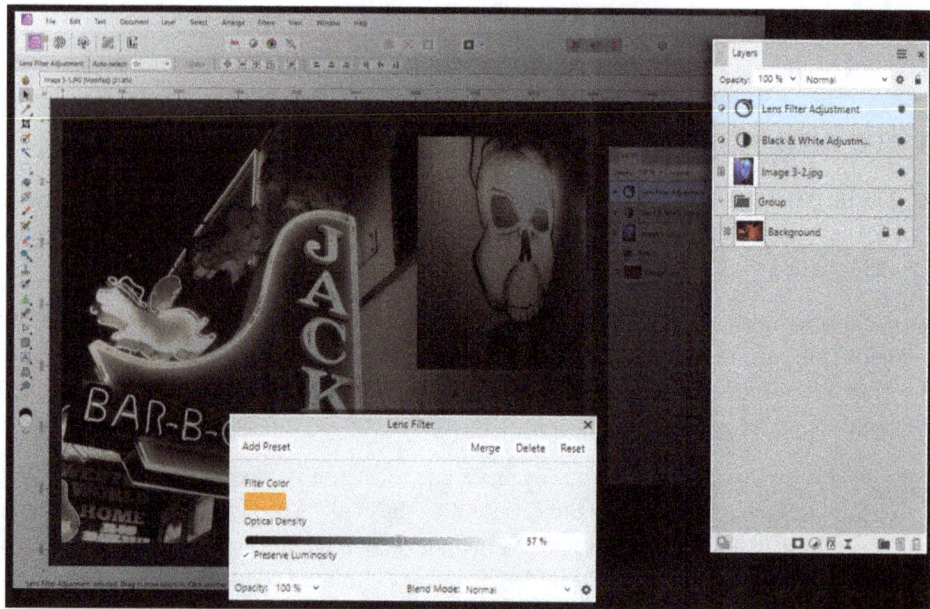

Figure 3.16 – Shows the addition of a lens filter adjustment

Now that we have covered adjustment layers, let's move on to live filter layers.

## What are live filter layers

In a later section of the book, we will be covering destructive filters (see *Chapter 14*), but in this section, we want to cover a form of filter that *creates* a layer; these are called **live filters**. Live filters are a special type of feature in Affinity Photo that has given it a substantial advantage.

So, what makes a live filter so awesome? Well, the short answer is that they can be adjusted at *any* time. A destructive filter (which we will talk about later in the book) is applied to a layer, and then it is *not* adjustable, thus you are stuck with the adjustment you made. A live filter, on the other hand, can be constantly adjusted.

Live filters are found in the **Layer** menu, under **New Live Filters Layer** (see *Figure 3.17*, item **1**).

There are many types of live filters, but the common unifying property is that you can click on the layer and come back to readjust. In the following figure, we have the live filter layer of **Vignette** (a very popular effect in editing). Notice the adjustments can be modified through the slider to change the exposure, hardness, and so on (see *Figure 3.17*, item **2**).

Live filters are attached to layers. Notice in *Figure 3.17*, item **3**, the layer is indented, which means it is nested to that layer. So, if you added another layer on top of that base layer, the vignette would not be present on that new layer because it is nested to the layer it is tied to.

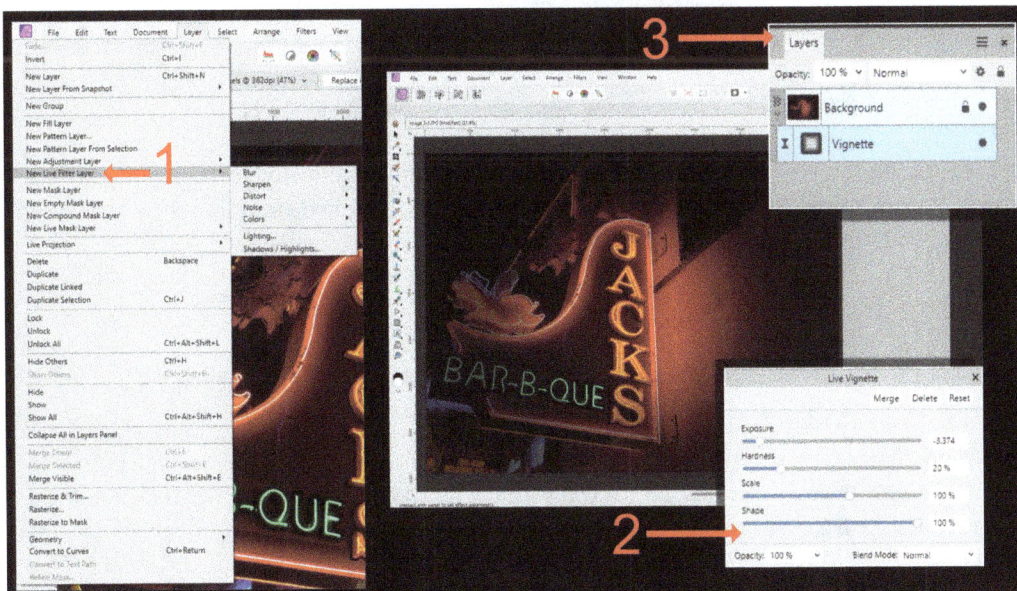

Figure 3.17 – Attachment of a vignette live filter layer to an image

If you wanted to add a vignette to the finished image, above all the layers in that image, then you can simply click it and drag it to the top of the stack. Notice the location of the **Vignette** layer in *Figure 3.18*; it is above all the layers, so it is applied to the topmost position of the composition:

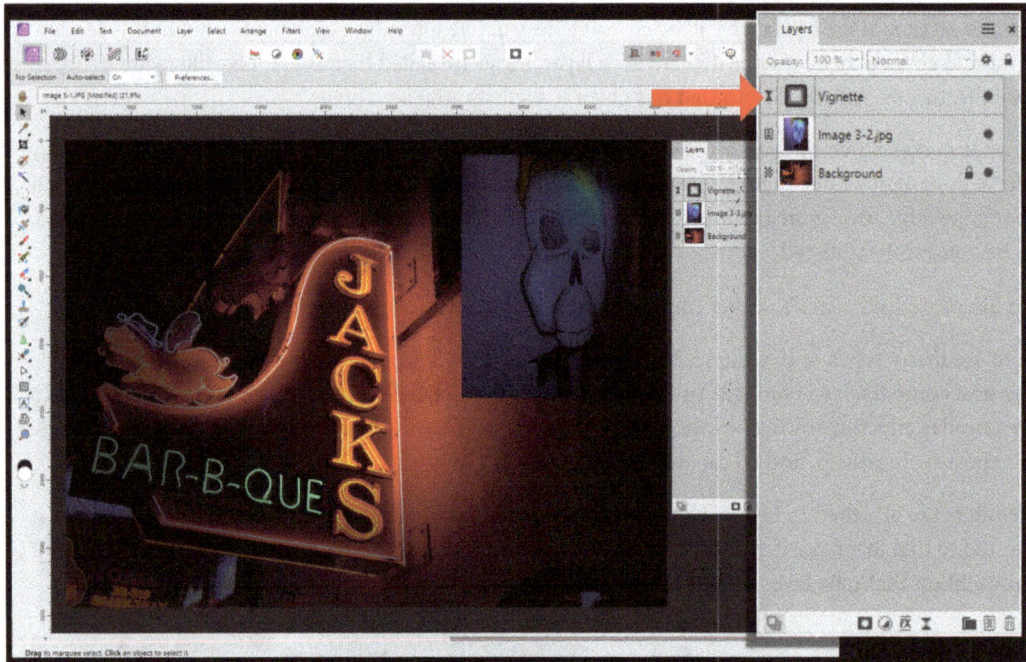

Figure 3.18 – Shows the Vignette layer above all layers in the edit

Text layers are the last type of layers we are going to cover in this introductory section. We will deal with text adjustment and specifics in later chapters, but for now, I need you to realize these are a type of layer that you will encounter, and they behave *exactly* the same as other layers.

## What are text layers?

Text layers (for now) are created by the **Text** tool, and in this case, we will be creating the text using the **Artistic text** tool (see *Figure 3.19*, item **1**).

To use the **Artistic text** tool, simply select the tool and type on the image. Notice the type of layer it creates. To modify the text, just click on the text with the tool *still selected* and adjust it as necessary (see *Figure 3.19*, item **2**):

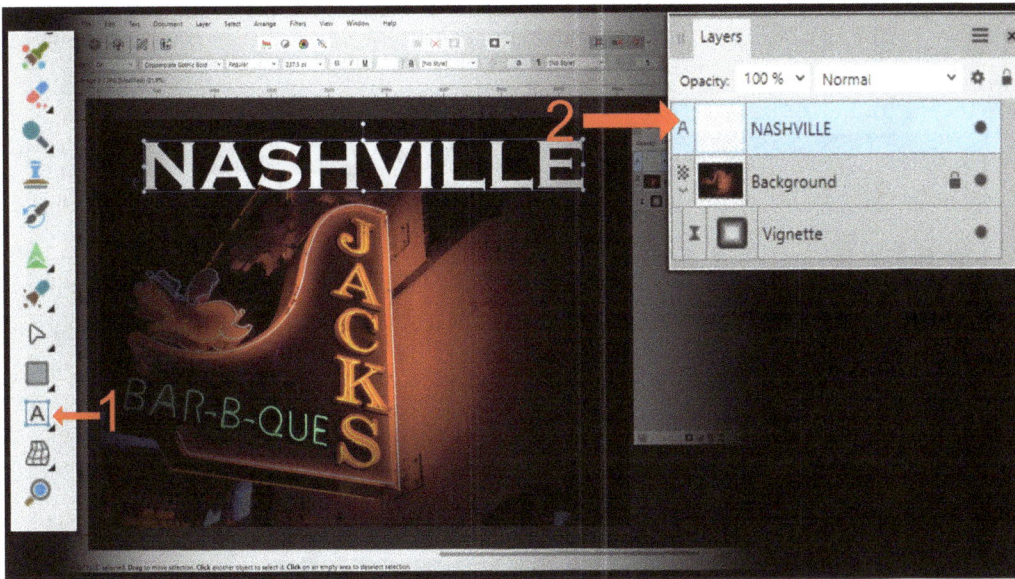

Figure 3.19 – Shows the existence of a text layer

The text has a position in the composition, a blend mode, and opacity. The reason we wanted to cover the text layer early in this book is that many people will want to create memes and pieces of content collateral, and that all involves text.

## Layers – practical editing practice

Okay, it is time for your practical application project, to combine everything we learned in this chapter and previous chapters. For this one, we are sticking with the BBQ image since the adjustments can be used on the entire image without masking, and we are going to create an old world sepia-style image, adding a good amount of wear and interest.

In the images that came with this course, open `Img. 3-1` (**File | Open, then** locate the file path).

Rename the layer from `Background` to `BBQ` (double-click on the layer and type in the name).

Add **Black & White Adjustment** (remember, adjustment layers are found at the bottom of the **Layer** panel with the half circle). Adjust the color specifics as shown in *Figure 3.20*, item **2**). Notice that as you adjust the reds, there is a profound change, as the neon red color is cast over the entire image. What we are doing by creating the black and white layer is desaturating the image, or removing the color.

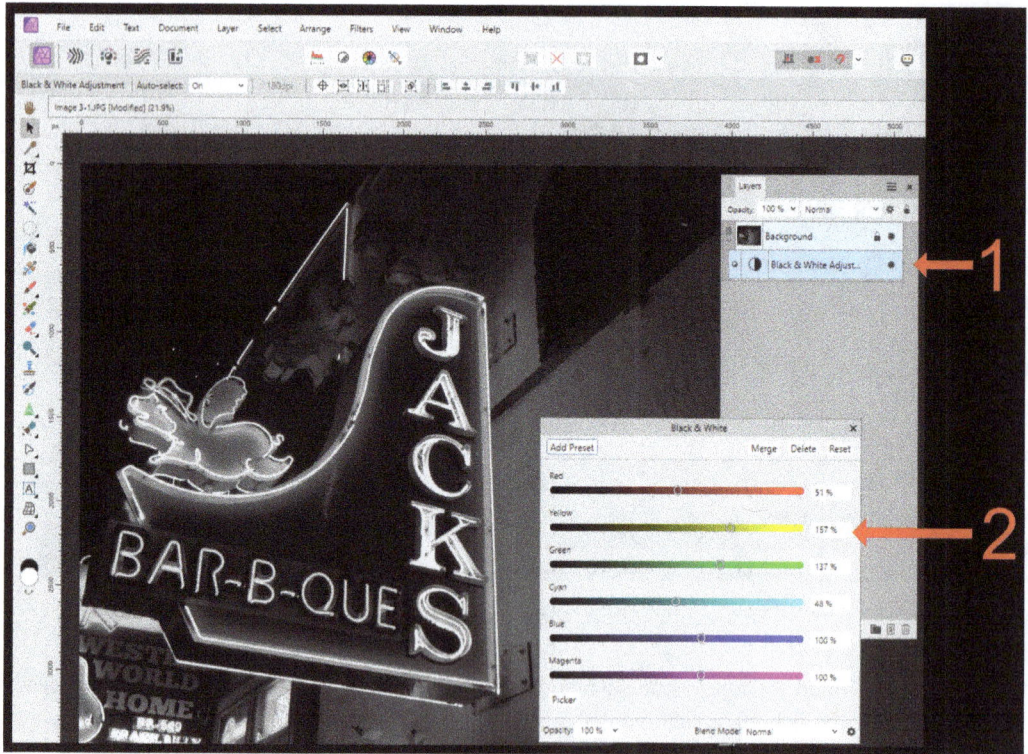

Figure 3.20 – Shows the addition of the black-and-white layer

Now we will add a **Lens Filter Adjustment** layer *above* the **Black & White Adjustment** layer. This order is important; if we add the layer *below* the **Black & White Adjustment** layer, then the **Black & White Adjustment** layer will not allow **Lens Filter Adjustment** to show (see *Figure 3.21*, item **1**). Notice the adjustments in item **2** of *Figure 3.21* to the optical density and the color we chose (to adjust the color, simply click on the colored box).

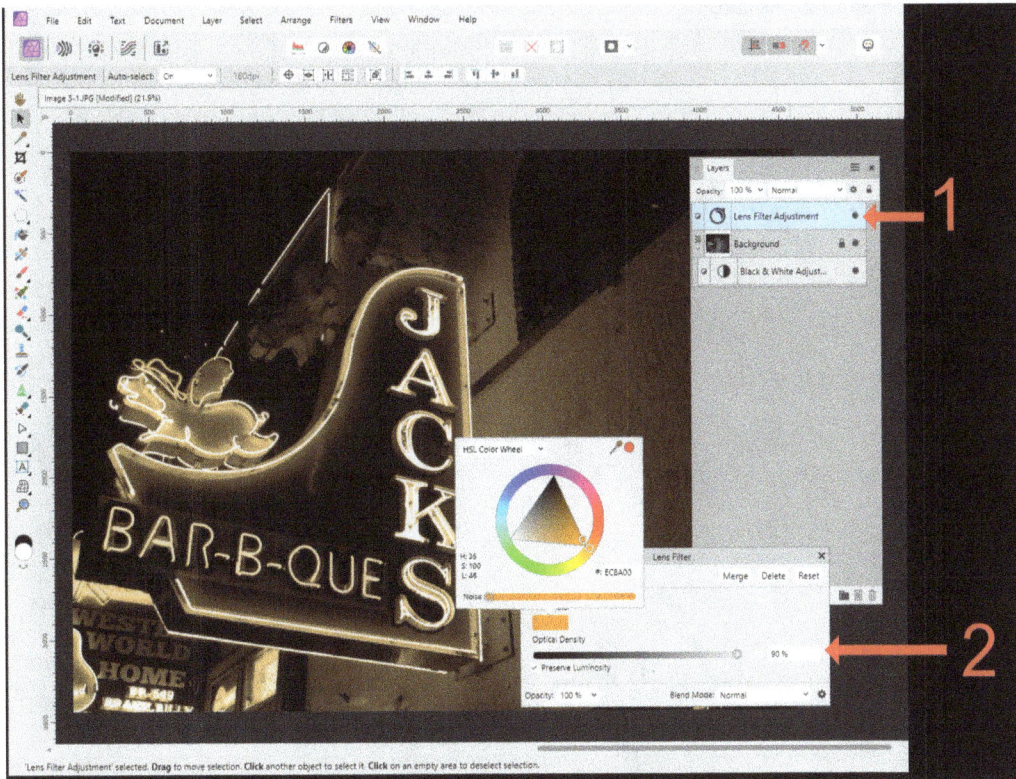

Figure 3.21 – Shows the addition of a lens filter layer

Next, it is time to weather it; for this, we will use a **Live Filter** layer, and create what is called **noise**, specifically **Add Noise**. To accomplish this, refer to *Figure 3.22*. Notice the location of the **Live Filter** layer (above it all), and I changed the type of noise to **Uniform**. Do not forget to adjust the slider to taste:

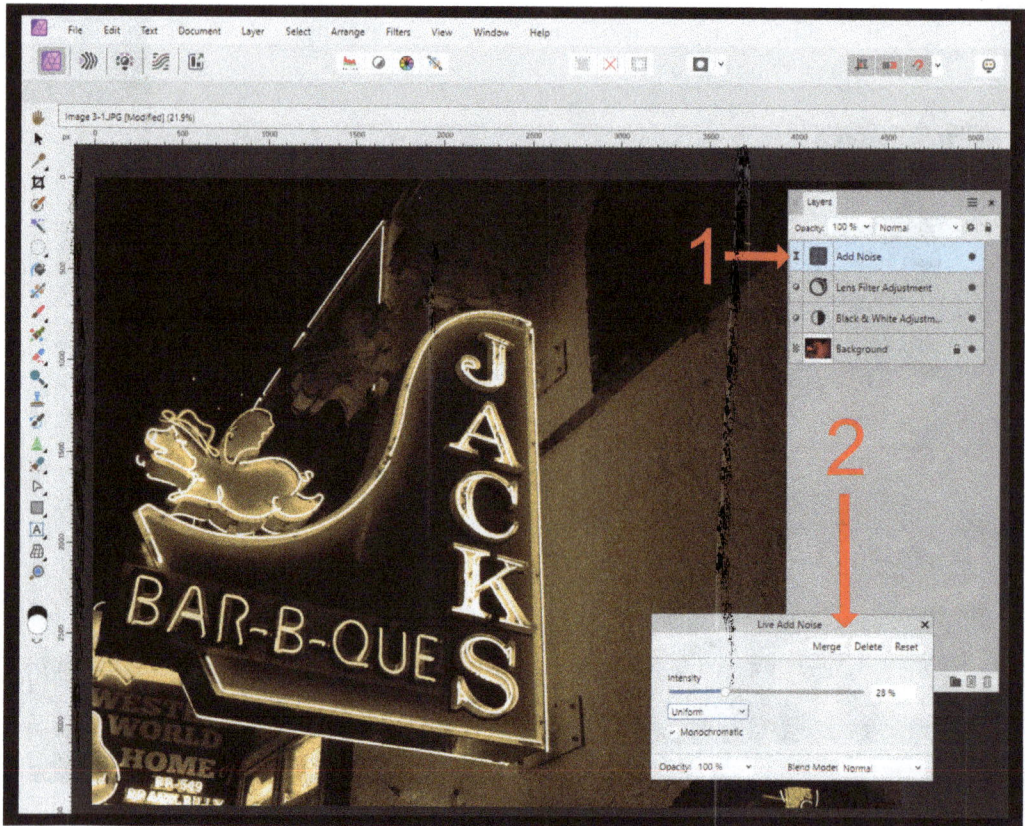

Figure 3.22 – Addition of a noise live filter layer

Lastly, we will add a **Vignette** layer (you saw this before), which is located under **Layers | Live Filter layer color | Vignette** (see *Figure 3.23*). Notice the location of the **Vignette** layer and the settings (see items **1** and **2**):

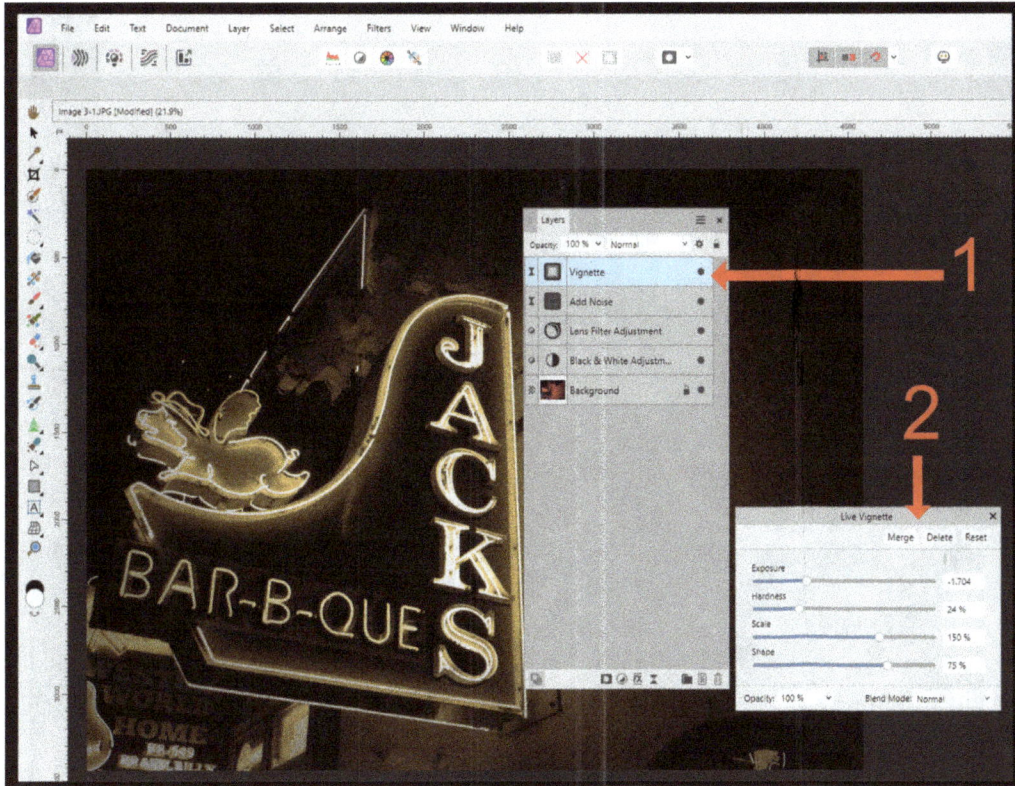

Figure 3.23 – Addition of the Vignette layer

Now, once we get done, let's go in one last time and adjust it (there are always adjustments in editing).

I want to adjust the intensity of the **Lens Filter** layer to get a more desaturated look. To do this, double-click on the **Lens Filter** layer and make adjustments to **Optical Density** (see *Figure 3.24*):

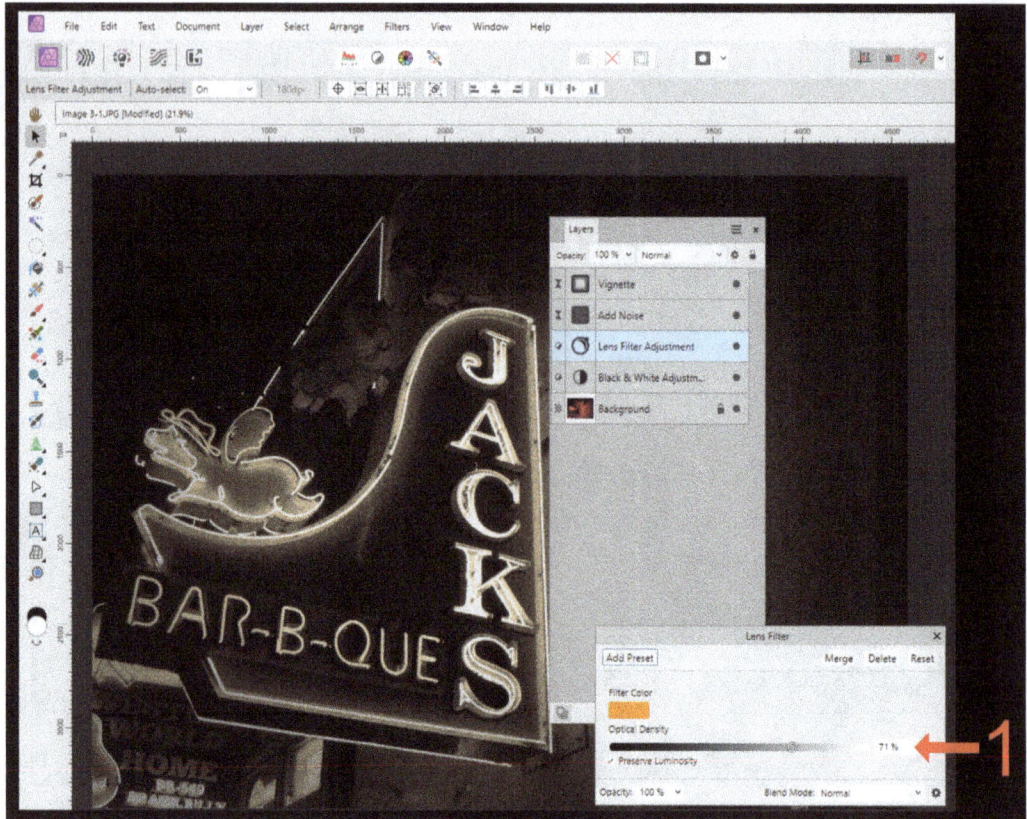

Figure 3.24 – Revision of the lens filter

## Professional tips, tricks, and important points

In this chapter, we have talked about layers, and I cannot stress enough the importance of messing around with layers and structures and just getting to know how layers behave. Set yourself two images, and then just move things around, nest them, rename them, add an adjustment layer, and see what happens. Remember, you got into this because it is fun!

Remember to name your layers – good organization makes for good projects.

I typically lay out all my layers *before* adding adjustments, text, or live filters, as I want to make sure my composition is right.

In both adjustment layers and live filter layers, you can adjust the blend mode; playing with these different modes does completely new and interesting things to your edits, so play around. I call this *blend mode roulette*.

# Summary

This was a long chapter, but an essential one. I will tell you without a doubt that 90% of the issues you will encounter with editing photos will be layer issues, either wrong groups, messed up orders, or some version of the two.

On the plus side, in this chapter, you completed your first edit. Congratulations!

The adjustment layers and the live filter layers in this chapter were applied to the entire image and so, in the next chapter, we will discuss masking, which is another essential skill that allows us to apply these powerful layers to only parts of an image.

# 4

# The Basics of Masking in Affinity Photo

Masking is an essential skill in any editing program, and so in this chapter, we are going to cover the principles of masking, the methods you can use to mask different types of masks (such as paint-on masks and vector-based masks), when you can use masking (practical uses), and lastly, we will apply these techniques into a full edit. By the end of this chapter, you will have a firm awareness of what masking is used for in the workflow at the basic level. Later in the book, when discussing selection, we will revisit masking based on things such as selection. So, for right now, let's keep it simple.

There are a variety of masks, such as the gradient, mask, layer mask, clipping mask, group mask, and so on. But for this section, we'll just cover the basics, the two most utilized styles of masking in the work that I do. It is important to note that masking is an art, and as such, every editor develops their own versions, workflows, and methods. It is impossible to cover every conceivable style of masking, and a quick search on YouTube will deliver countless variations. This manual covers the most applicable universal ways that I have successfully taught to over 100,000 students.

In this chapter, we will cover the following topics:

- Basics of masking and making your first mask
- Making a gradient mask
- Utilizing a clipping mask – vector
- Practical masking application
- Professional tips, tricks and important points

## Technical requirements

The CiA video of this chapter can be found on `https://packt.link/1UOJw`.

You will also need the `Chapter 04` project to follow along with me.

# Basics of masking and making your first mask

In this introductory section, we will learn the basics of masking, starting first with the principles because while techniques frequently change, principles never do. If you understand the principles and come to terms with the idea that there is no one perfect way to mask, then we can move on to the most basic type of mask you can make, a brush-on mask.

We will cover the steps needed to get you up and running immediately using a brush-on-mask approach, no selection is needed, and no advanced techniques are required, just a simple immediate repeatable solution, so let's get started.

## Masking principles

When I teach masking, I find it best to break it down into three fundamental principles because principles never fail; only techniques do:

- **Principle #1**: Masking is simply a way of saying *what we are hiding and what we are showing*.

  When you mask an object, you *hide* part of it, obscuring some of it from view. So, when I am editing, the first question I ask is, what do you want to show, and what do you want to hide?

- **Principle #2**: Masks are separate layers you add to other layers.

  In *Chapter 3*, we discussed the concept of layers; masking layers only deal in black and white, and there is no color information. You also attach (nest) them with other layers.

- **Principle #3**: Black conceals and white reveals.

  Masking works on Luminosity, and Luminosity is a fancy way of saying "light and dark," so if your mask is bright white, the information is 100% visible (revealed), and if the layer is completely 100% black then *nothing* shows (concealed).

So, now that we know the principles, let's look at how we do it. It is worth noting that while we will primarily use layers, all the techniques can also be used on groups of layers so that you can apply masks to entire groups.

## Making a simple mask on a Pixel layer

At the simplest core, we need a repeatable, easy way for you to mask without waiting to learn everything before doing something. So, this method requires three things:

- Understanding what a mask layer is
- Knowing where to find the **Color** studio tab
- Knowing where to find the digital brush in the toolbar

A few pages down the road, we do have a video to show you how to find the previous items, so for right now, keep reading, and we will make it all flow.

To illustrate the principles and align them to what we have discussed previously, we will open `Masking practice 1` from the downloads and then set up to follow along. This file is not very interesting; there are two contrasting colors set up to allow you to see exactly what happens with the mask function. This is best illustrated in the following screenshot:

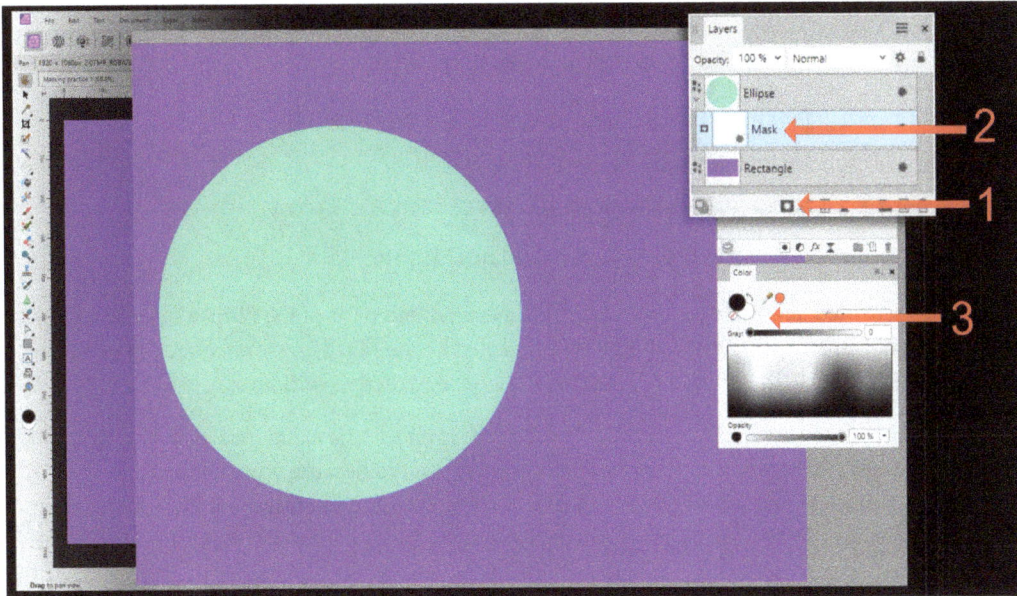

Figure 4.1 – Practice file for blend modes showing the location

Let's go over the elements of the preceding screenshot:

- Item 1 shows the location of a **Mask** layer in the **Layers** tab.

- Item 2 shows the **Mask** layer being nested into the **Ellipse** layer. Remember that this means the **Mask** layer will only affect the **Ellipse** layer (see *Chapter 3* for the section on nesting).

> **Important note**
>
> I have pulled the **Color** studio tab out and placed it into the workspace to make the visual easier. To follow along, if you want your color space to match mine, follow the CiA video, the link to which can be found in the *Technical requirements* section, or refer to *Chapter 10, Working with Color in Affinity Photo – Working with the Color Panel and the Swatches Panel in Affinity Photo* section, to see how to manipulate the **Color** studio panel.

- In Item 3, notice that the black circle is in front of the white circle. We will be painting the black color onto the mask layer, so make sure it is the color selected (refer to the CiA video where I demonstrate how to select the color).

Now, notice in *Figure 4.1* that **Ellipse** is the layer we will mask out, that is the teal color. If we paint on the **Mask** layer and not on the **Ellipse** layer, we should see the ellipse start to disappear from the image.

Watch the CiA video (see the *Technical requirements* section for the link) to select a brush (or refer to *Chapter 8* for the basics of the **Brushes** panel). Notice the important points that are highlighted in the video.

To paint on the **Mask** layer similar to the video, follow these steps:

1.  Select the brush tool from the left-hand side of the screen.

2.  Make sure **Opacity** is at **100%** in the context toolbar.

3.  Make sure black is the color in **Color** studio tab.

4.  Ensure you are painting on the **Mask** layer and not **Ellipse**.

Notice in the CiA video how as I paint black on the **Mask** layer and parts of the **Ellipse** layer disappear. This ellipse is not actually going away, it is being concealed by the **Mask** layer (refer back to principle #3 earlier in the chapter). The process of masking is a non-destructive workflow.

This is the first time in the book that the term non-destructive workflow is mentioned. A non-destructive workflow is a workflow that allows you to easily undo what you did because you did not destroy it. As an example, a professional editor will always mask an object rather than erase what they do not want because if they change their mind, the image can be restored by simply painting white on the **Mask** layer.

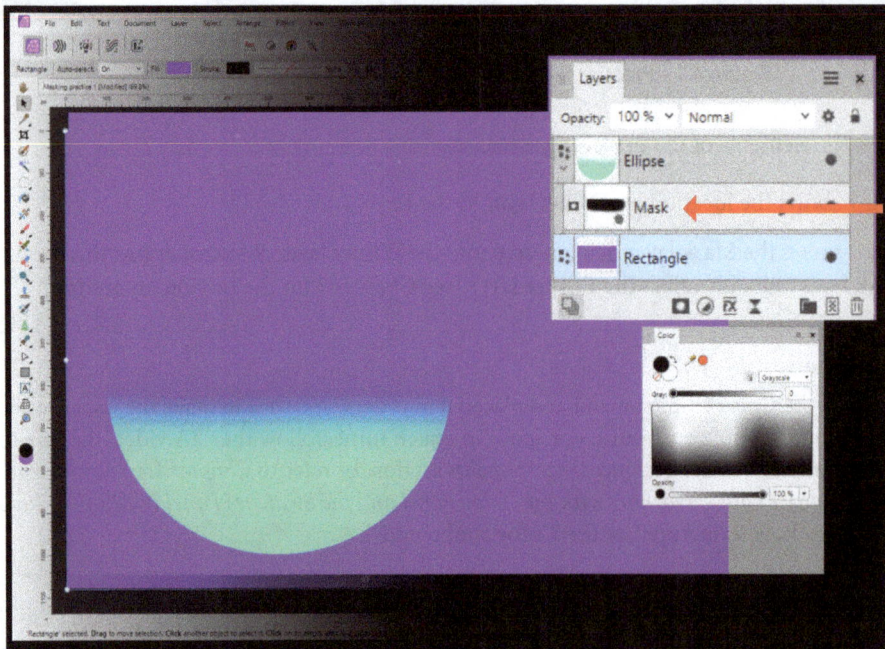

Figure 4.2 – The preview of the mask on the Ellipse layer

Notice in *Figure 4.2* that the **Mask** layer now contains some black markings on the layer representing the area of the **Ellipse** layer that is concealed. Previously, the entire last layer was white, telling us that the entire ellipse layer was revealed.

Next, we will learn to undo a mask.

## How to undo a mask that you placed on an image

If you placed a mask on an image and applied black coloring (thus concealing part of the image) and you want to undo it, there are several options:

- Hit **Edit | Undo** until you get back to where you painted.

- Move the mask layer to the trash and create another one.

- Switch to white paint and paint over the mask. You will see the image you masked begin to reappear (refer to the CiA video, the link to which is mentioned in the *Technical requirements* section). Notice the important places where the video pauses. The following are the steps to undo the mask and make the image visible once again:

    I.      Select the brush.

    II.     Check **Opacity** in the **Context** toolbar.

    III.    Make certain your color is white in the color studio.

    IV.     Make certain you are painting the white on the **Mask** layer.

Notice in the following figure that the **Mask** layer previously had black on it, and now there is substantially less black color. This is because we painted over with white, and the fact the layer is largely white means that there is very little left (if anything) being hidden on the layer:

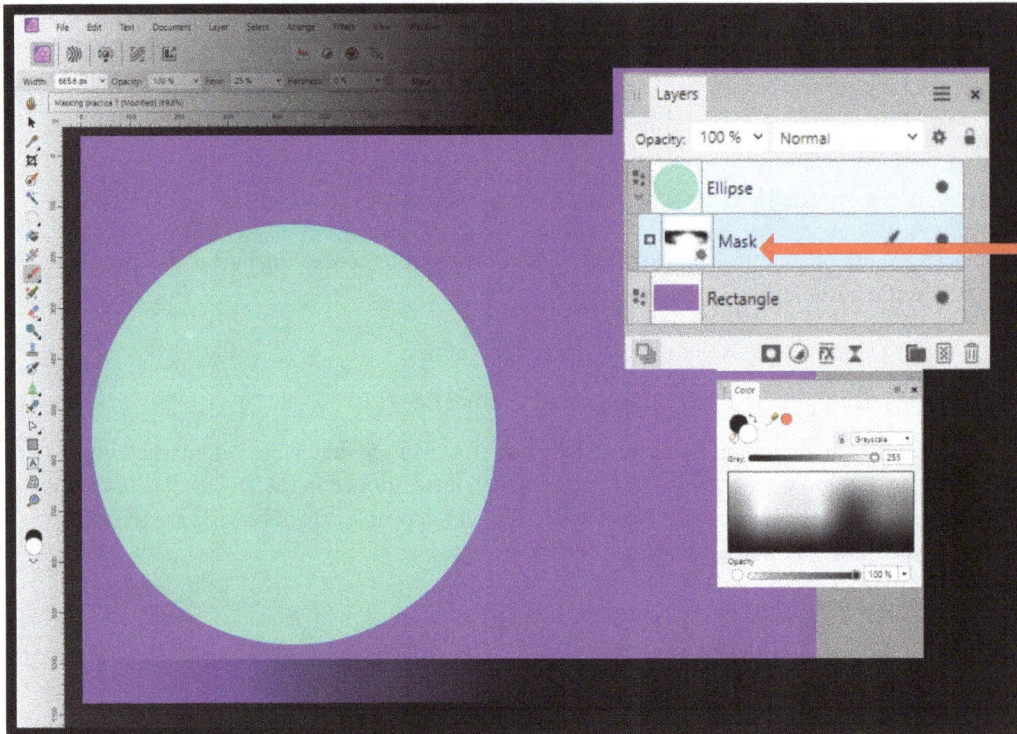

Figure 4.3 – A preview of the Mask adjustment after revealing the image again

This skill of being able to create and revise the mask layers is one of the most frequently used skills I engage in during an edit, and while we have been applying them over the **Mask** layers, they can also be used when dealing with other layer types, such as Adjustment and Live Filter layers.

## Masking Adjustment layers or Live Filter layers

In the previous section, we looked at how we can add a **Mask** layer to a common **Pixel** layer, and it was a simple technique where we select the **Mask** layer, nest it into the **Pixel** layer, and finally paint on it.

In the previous chapter, we covered many types of layers, and one type of layer is called an Adjustment layer, and the other is called a Live Filter layer (see *Chapter 3* for the details on adjustment and Live Filter layers). We can mask a **Pixel** layer with an adjustment or Live Filter layer, but the process is slightly different. Notice in *Figure 4.4* that there is a Pixel layer for this piece of graffiti, and an adjustment layer has been applied to it (in this case, a **Black & White Adjustment** layer) is added above it.

Also, notice that the square representing the Adjustment layer is 100% white, and we know from earlier discussions on masking that white reveals. So, if the square is white, we need to read it as *all the adjustment layer is showing*, or *all is revealed*.

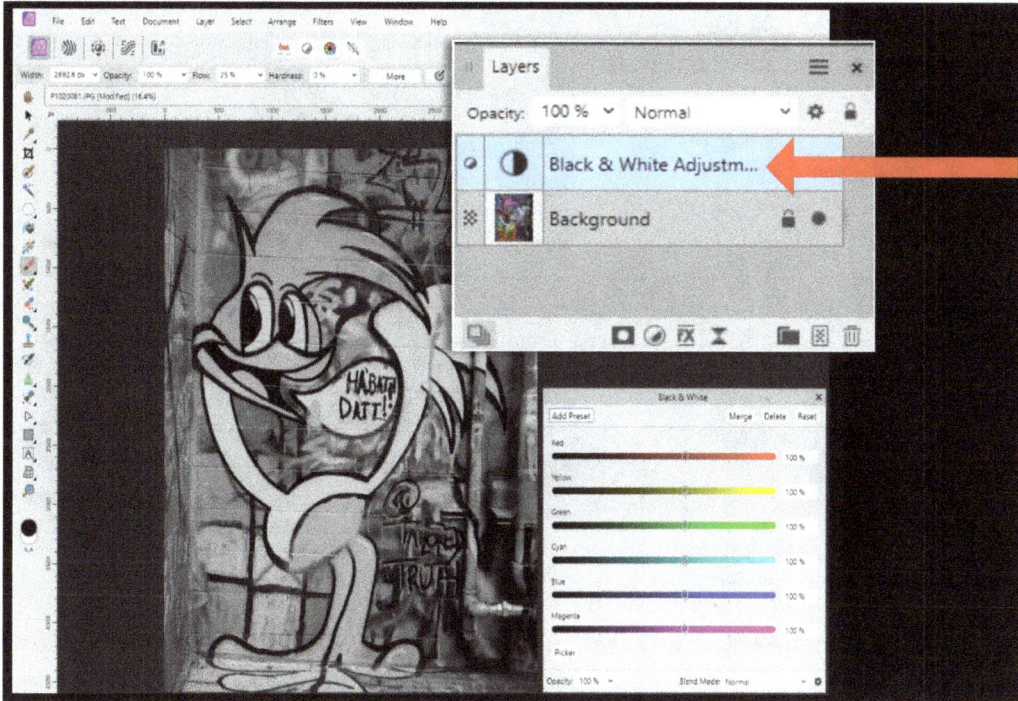

Figure 4.4 – The white square denotes a built-in masking layer

Now, this is an important point— the Adjustment and Live Filter layers have built-in masks, meaning we do not have to add a **Mask** layer to them. We just simply need to paint the layer in black and white, and it will mask the layer. This is explained in *Figure 4.5*:

- In **IMAGE 1**, we have masked everything. Notice that the square is completely black.
- In **IMAGE 2**, we have masked only half of it.

The square represents what was masked, and the image represents what is shown:

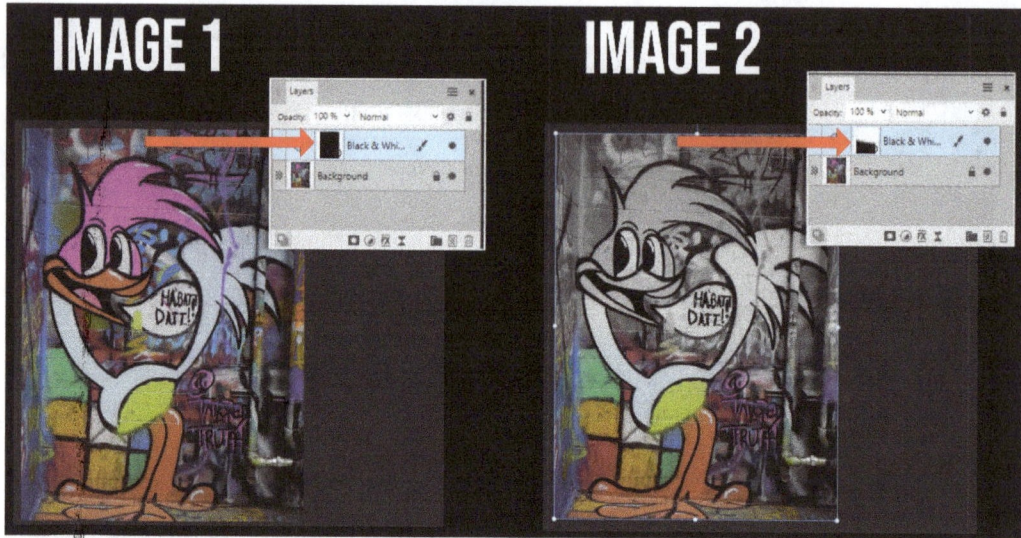

Figure 4.5 – Partial coverage of a masking layer concealing part of the black-and-white adjustment

Refer to the CiA video (the link is mentioned in the *Technical requirements* section) where I show how I masked the two images so you can see the process.

## When to use this technique

As a professional, we mask adjustment layers when we only want to apply the Adjustment or Live Filter layer to each part of the image. An example of this editing is shown in the following screenshot:

Figure 4.6 – A project breakdown of a more advanced edit showing adjustment layers with masks

Let's analyze the preceding screenshot:

- Both live Adjustment layers and Live Filters are represented here. The **Gaussian Blur** layer is a Live Filter layer) and **Exposure Adjustment** is a Live Adjustment layer.

- The stacking is important—notice that the image (or pixel) is at the base, which means every adjustment and Live Filter layer is acting on it. Plus, notice there are no nested layers.

- Each layer is masked by hand using the same quick brush method we have been teaching.

- For some masks, such as the **Unsharp Mask** layer, notice how we only applied it to the foreground and blacked the foreground out.

This simple landscape edit shows how to apply a masking technique to the Live Filter and Adjustment layers. We have included the working file for this in the downloads for the book if you would like to follow along and play with the file (see the `Masking landscape project file`).

Now that we have covered the fundamentals of the simple brush-on masks, it is time to dig deeper into gradient masks (this builds on the same fundamental principles of black and white but utilizes another tool).

# Making a gradient mask

In the previous sections, we learned how to apply a simple painted mask to an image, but sometimes you want that smooth transition from one image to another, or a smooth transition from the adjustment into the natural image, such as a color layer into the sky. This is where gradient masks come in. Gradient masks are commonly used in double exposures.

To understand the gradient mask, you have to understand the **Fill** tool. We will use this tool later in the course for color grading, but now is the time to explore this tool for masks, so let's get into it.

## Where is the gradient tool, and how do we read it

The gradient tool is found in the tools area on the far left and looks as shown in *Figure 4.7*, along the left-hand side of the interface (remember, if you do not have it, go to **View tools** | **Customize tools**, and drag it into your menu):

Figure 4.7 – The fill tool location, context, and the layer setup

The context toolbar will change to reveal the menu as shown at the top of *Figure 4.7*. Let's take a look at the options in the context toolbar:

- **Context**: We will be applying the gradient to **Fill** of the object, not the stroke of the object.
- **Type**: There are numerous types. For photo editing, we will focus on three types of gradients: **Linear**, **Radial**, and **Elliptical**.
- **Color**: This will be the color of the gradient, and in this case, we will be exclusively working in black and white. It currently shows a red line, implying no color.
- **Rotation button**: This rotates the gradient by 90 degrees
- **Reversal button**: This changes the direction of the gradient.
- **Aspect lock button**: Locks the aspect ratio to 1:1. Make sure this is unchecked to make elliptical gradients.

In the next subsections, we will learn how to use the **Linear**, **Radial**, and **Elliptical** gradients.

### Creating a new Linear gradient

The **Fill** tool is used by clicking and dragging across the image and then releasing it when you get it as far as you want it (refer to the CiA video for an example).

The context toolbar for the **Linear** gradient now looks as follows:

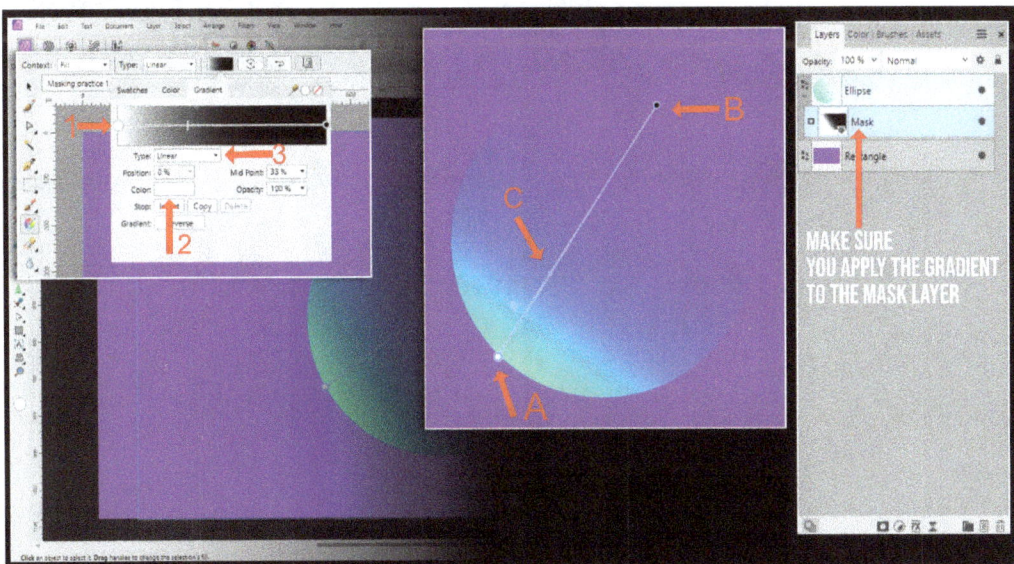

Figure 4.8 – The mechanics of the tool, including points and sliders

Let's explore the tool by referring to the preceding screenshot:

- (**1**) in the preceding screenshot is the original point of the gradient: notice that the circle is bigger than the black circle; that is the point that is selected. To change the black point, simply click there, and it will be the selected point.

- (**2**) represents the color of the origin point of the color. This shows the color of the gradient's origin. You can click here to change the color. Remember that changing the color of a mask does not matter because you are working in black and white and not color values.

- (**3**) is the type of gradient. Use the drop-down menu to select the type of gradient, in this case, **Linear**.

- (**A**) is the origin point in the working window. This is where we started the click and drag.

- (**B**) represents the endpoint. This is where we ended the click and drag. To reposition, simply click again and move it.

- (**C**) is the balance slider. This slider shows the balance of dark to light. Adjust it where you need it to get the desired mask strength.

### Creating Radial and Elliptical gradients

The **Radial** and **Elliptical** gradients are created the same way as **Linear** with two notable exceptions:

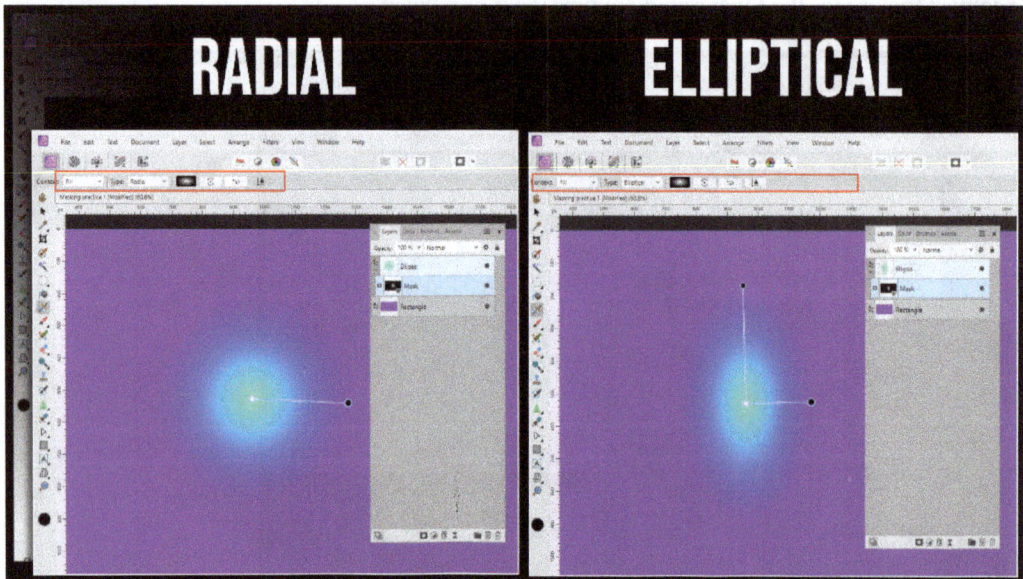

Figure 4.9 – Illustrating the difference between Radial and Elliptical gradient

Let's see what is going on in the preceding screenshot:

- The **Radial** and **Elliptical** gradients have an origin point, and then they fan outward.
- You must turn the aspect ratio button off with **Elliptical**; if not, it will always be **Radial**.

See the CiA video for how to make both the **Radial** and **Elliptical** gradients. Now that we know about gradients and their types, let's understand how to use them in photo editing.

### Using gradient masks in editing

When editing, a common use of this technique is applying a recolor adjustment to an image. In this case, we have included a practice file (`Gradient mask practice`) for you to follow along. The end result of this edit is to make it appear that there is a purple light source off-camera, such as a purple neon glow.

Take a look at *Figure 4.10*:

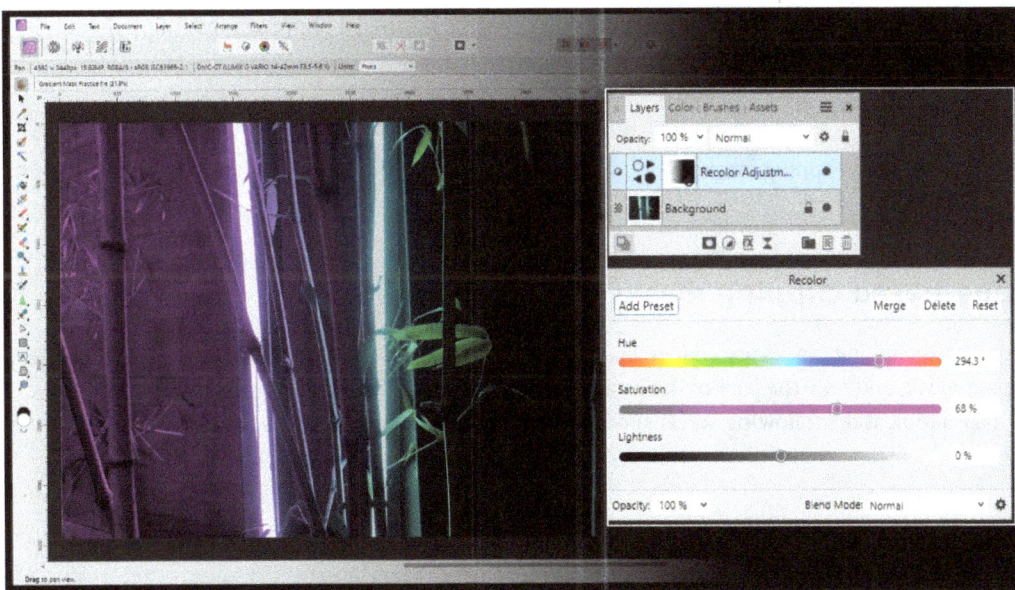

Figure 4.10 – The application of the hue, saturation, and luminosity (HSL) adjustment with a gradient mask

Note the following points in the preceding screenshot:

- The layer with the mask applied is an Adjustment layer
- We have applied a **Linear** gradient running from the left of the page to the right
- Notice the black mark in the white square for the layer, denoting the presence of the mask

To see how the layer was applied and the mask was created, refer to the CiA video (see the *Technical requirements* section for the link).

So, in summary, when it comes to gradient fills, be it **Linear** or **Radial**, the mechanics we have learned earlier in the chapter are always the same: the details and the tools change, but the effect and the intent will be consistent.

Now let's move on to another form of masking known as a clipping mask; this one uses a technique known as a vector shape to achieve a result.

# Utilizing a clipping mask – vector

As covered in *Chapter 1*, the world of digital art is divided up into two distinct categories, pixel (or raster) based art and vector (or math-based) art. So far in this chapter, we have dealt 100% with pixel-based art, that is, painting on mask layers using pixel-based brushes in colors of black and white.

However, another type of masking is utilized involving vector-based art, and the most common term for this style of mask is called a **clipping mask**. So, why is it called a clipping mask? Simply put, it clips the image to only the shape you want to see.

There are two types of clipping masks:

- Shape-based clipping masks
- Curve-based clipping masks

## Shape-based clipping masks

Shape-based clipping masks use shapes, such as squares, triangles, hexagons, and so on to create the clipping mask, and *only* the part of the image *inside* the clipping mask is shown. For an example of this, take a look at the following screenshot:

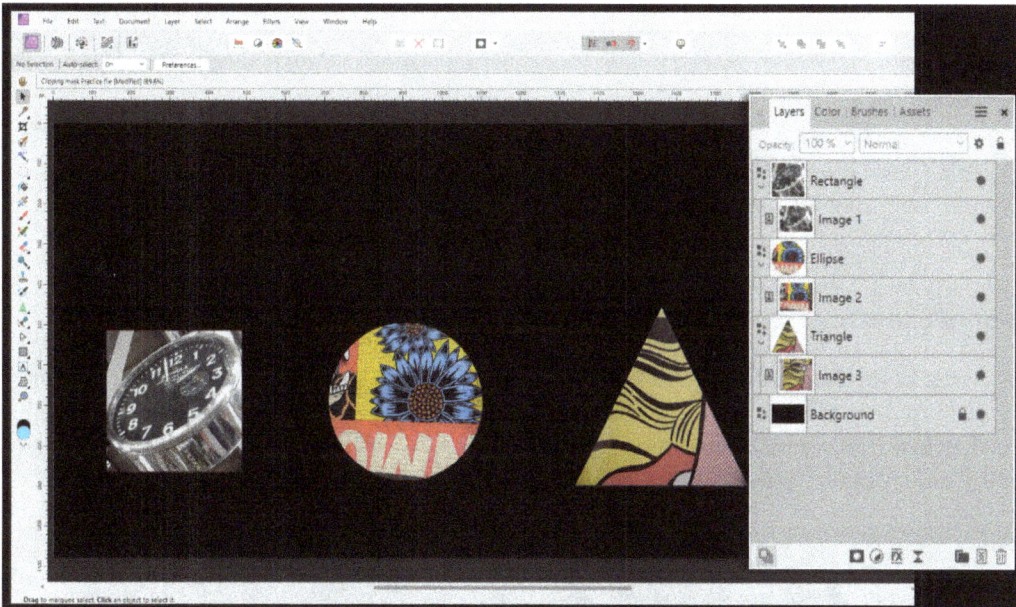

Figure 4.11 – The application of three shape-based masks on images

Notice the flowing items in the **Layers** tab:

- The shapes are the layers (see the names **Rectangle**, **Ellipse**, and **Triangle**)

- Notice that the image files are nested inside the shapes (for example, **Image 1** is nested inside the square)

- Only the part of the image inside the shape is visible; that is, the image is clipped to the shape, which is why it is called a clipping mask

Let's understand how to create a shape using a clipping mask in the next section.

### Creating a shape-based clipping mask

To create a shape-based clipping mask, follow these steps:

1. Draw a shape from the tools section in the size you want to use as the mask.
2. Place the image in the project.
3. Drag the image into the shape, effectively nesting it.

> **Tip**
>
> If you cannot see your image, it is still there; you need to move it to where the shape is. This is not uncommon (check out the CiA video where I did this to the square).

To illustrate this, I have also included a downloadable file (see the `Clipping mask practice` file), and I have captured it in a CiA video.

In the next subsection, we will see when we need to use this shape-based clipping mask.

### When to use a shape-based clipping mask

An example of this is in print work or web design when you are looking to utilize a circle to crop someone's headshot. See *Figure 4.12* for an example of this technique in a poster. This can also be used in modern art. Observe the image on the right side; every part is the same image. Just a rectangle clipping mask has been added, and then the size and color have been changed:

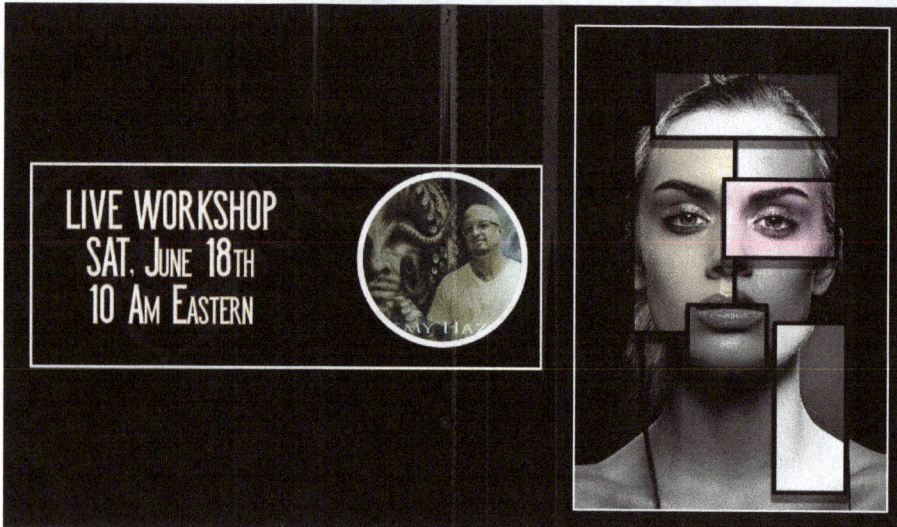

Figure 4.12 – The application of shape-based clipping masks

## Curve-based clipping masks

> **Note**
>
> Working with the pen tool is one of the most confusing elements of digital art programs, and hence a ton of practice will be required to operate these tools effectively, so plan accordingly and manage your expectations. These sorts of masks are not recommended for those looking for an immediately successful mask, but rather a mask that, when mastered, is a high-demand skill.

A curve-based clipping mask is created using the pen tool, and then the curve created is adjusted utilizing the node tool (see **1** for the location of the tools):

Figure 4.13 – Using the pen tool to create clipping masks – showing nodes

Notice that the bold red line in the preceding screenshot is present around the main subject. The figure was traced with a pen tool by applying several nodes. Notice that some of them are square and some are round:

- Square nodes are shaped like squares (see *Figure 4.13*, **2**)
  - Square nodes form sharp corners, so we use them for intersections
- Round nodes are denoted as circles (see *Figure 4.13*, **3**)
  - Round nodes create smooth-flowing shapes, and so you will notice them in *most* of the mask shown in the preceding screenshot

Now that we know what a curve-based mask is, in the next subsection, we will create one.

### Creating a curve-based mask

Curves are created by selecting the pen tool and beginning at a point on the drawing to place the first node, and then depending on the type of node you want, you either click and release or click and drag. A curve is completed when the last node connects with the very first node (shown when a yellow line appears). See the CiA video for a demonstration of a simple mask being created. Note the following:

- Use click and release to create a square node
- Use click and drag to create a round node

Now that we know how to create nodes in the mask, we need to be able to adjust the nodes we have made to get a good mask.

### Adjusting the nodes

You can adjust nodes using four methods (check out the CiA video to see these methods utilized in the adjustment):

- **Adjusting the position or location**: To move a node, simply left-click and drag the node to the new desired point. See the CiA video showing the node tool being selected and moving the nodes around the left side of the apple.

- **Adjusting the number**: To add a node simply click on the curve, and a new node will be added to the curve. To delete a node, select the node (the node will turn blue) and then hit the *Delete* key.

  As a general rule, fewer nodes are always better for a smooth mask.

- **Adjusting the type**: Click the node and go to the context toolbar and click on the round or square node; this will convert the selected node to the other type.

  See the CiA video for an example toward the end where the round node is turned into a square and then back to round using the context toolbar.

- **Adjusting the handles**: Adjusting the handles will give different behaviors between square nodes and round nodes. Holding down the *Alt* key and swinging the node will convert it to a sharp node. See the CiA video where the handles are adjusted through the process to line up with the apple.

### Creating the clipping mask

Once the curve is drawn simply, nest the image inside the mask. In *Figure 4.14*, the apple is nested inside the curve; however, the curve can still be adjusted after this happens:

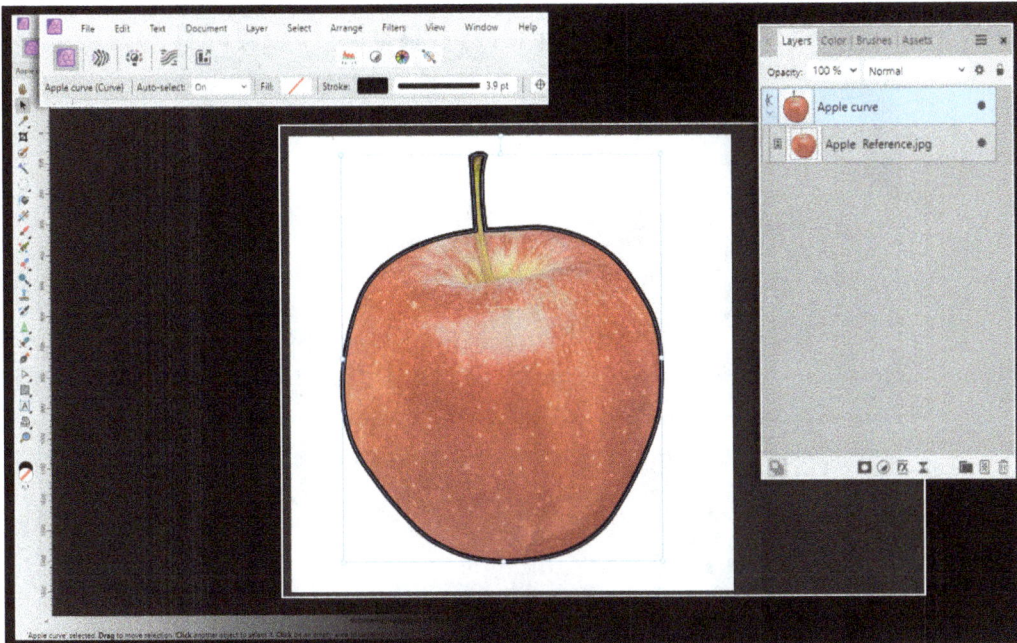

Figure 4.14 – Example of a clipping mask being used on an apple

Notice that in *Figure 4.14*, **Stroke** is set to no color. It is still there, but we only use **Stroke** for creating the mask; you don't want the color on it once you add your image.

## When to use a curve-based clipping mask

A curve-based clipping mask is looked at as the smoothest, highest-quality mask you can make, but it is not applicable to complex objects. Objects that are jumbled, such as trees with many leaves or hair that is blowing, cannot be masked this way—those are masked using advanced techniques, which we will show later in this book. Right now, we will say that if the subject is smooth, geometric, and can easily be traced around, then it is a candidate for a smooth vector-style mask. In *Figure 4.15*, I have included some projects I have done in Affinity Photo that utilized the smooth masking technique:

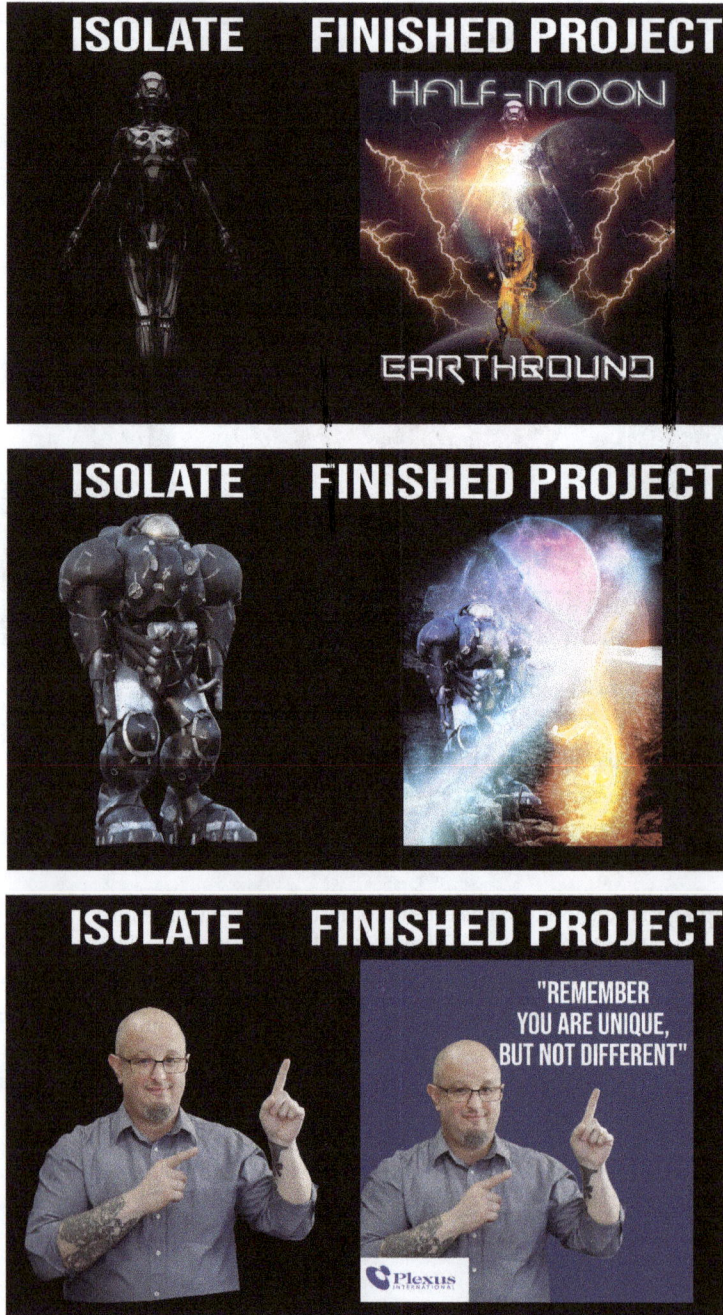

Figure 4.15 – Examples of projects done with clipping masks

Now that we have learned about clipping masks using vector, we can put them together into actual application projects.

# Practical masking application

It is now time for the application of everything we have discussed in the chapter so far. We will see how we can actually utilize these techniques in an edit. This will show you that even at this early stage, you have more skills and more power than you think.

While we introduced the existence and the masking of things such as the Adjustment and Live Filter layers in previous chapters, it is impossible in the book to go through *every* Adjustment layer and every Live Filter layer, so we will explore a few of the most frequently used in this section.

The edits in this section are cumulative, meaning they will be built on the topics covered in the previous sections, so you already have all you need to do this work.

## The edit breakdown

In this edit, we will use six layers:

- A vector mask – A **Curve** layer
- The Live Filter layer – **Gaussian Blur**
- The Adjustment layer – **Exposure**
- The Live Filter layer – **Sharpen | Unsharp Mask**
- The Adjustment layer – **Selective Color Adjustment**
- Live Filter layer – Color: **Vignette**

In the attached working file, there is a one-pixel image (see the application download).

We will be transforming this basic image into this stylized edit (see *Figure 4.18*).

Note, while the adjustment values I used are shown in *Figure 4.19* to *Figure 4.21*, the completed working file will have all of the layers and the actual final values, so always open the downloadable file for additional guidance:

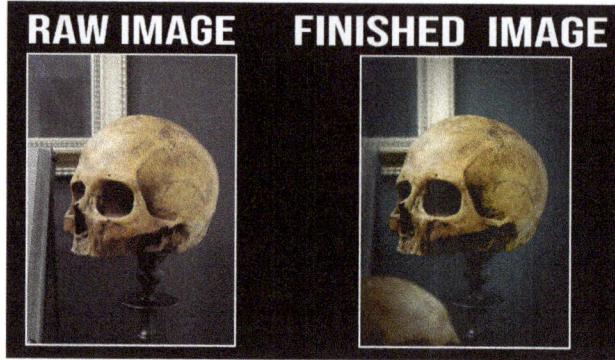

Figure 4.16 – The before and after of what we are going to make

The final layer structure (i.e, position, mask, and so on) is shown in *Figure 4.19*. Refer to this image to see where I placed it in the stack. Highlights and important points are mentioned in the *Operational steps* section for each of these layers:

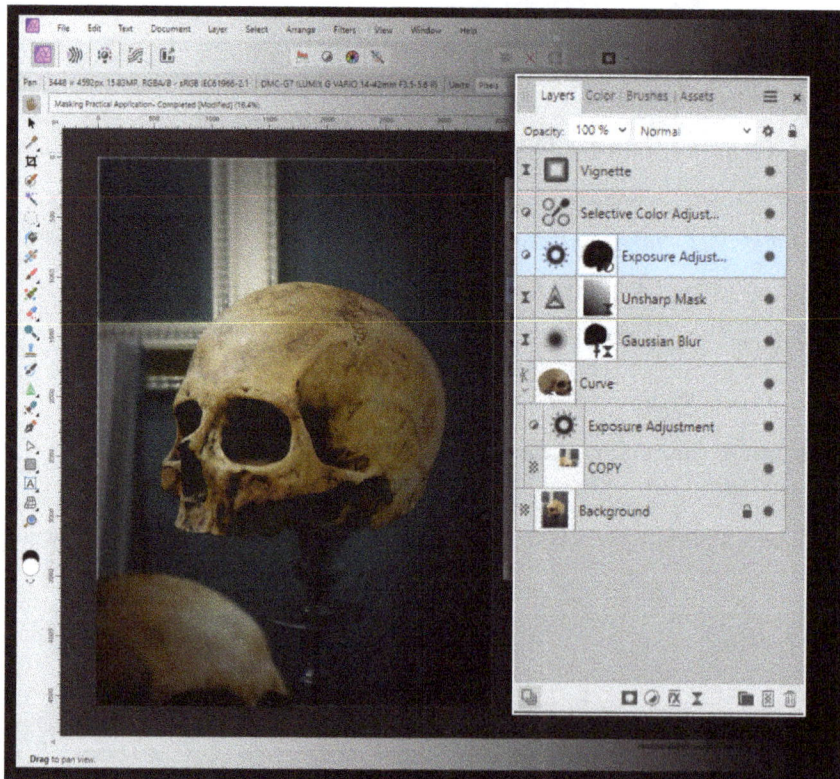

Figure 4.17 – The layer structure broken down for the edit

## Operational steps

Follow these steps to recreate the edits made in *Figure 4.18*:

1. Duplicate the Pixel layer to get a second copy of the image.

2. Label it COPY and then turn off the visibility, as we will not need it until later.

3. Generate a **Gaussian Blur** Live Filter layer (refer to *Figure 4.20* for my settings, but do what you like, after all, it is your image).

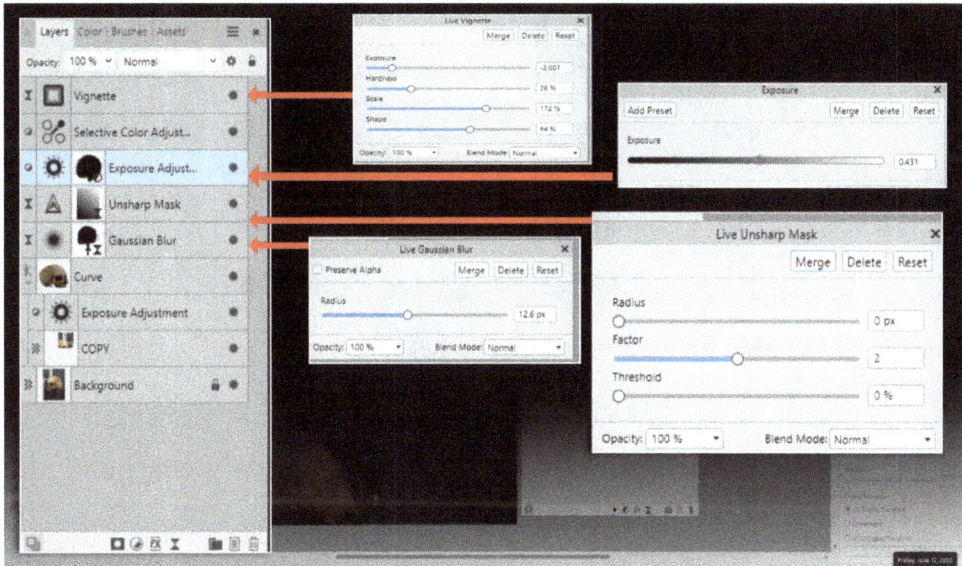

Figure 4.18 – Addition of adjustments for Unsharp Mask, Exposure, Blur, and Vignette

4. Use a simple brush painted-on mask to paint black over the skull; this conceals the blur from the main focus area.

5. Apply **Unsharp Mask** to the image, then paint around the skull using black to conceal the sharpening effect.

6. Add an exposure layer and paint black over the skull area. If you mess up, simply paint white over the top to reveal it again. You want the skull to be the brightest area.

7. Add **Selective Color Adjustment**. We will be adjusting the following categories (refer to *Figure 4.21*) for the slider position:

   · **Blacks**

   · **Neutrals**

   · **Whites**

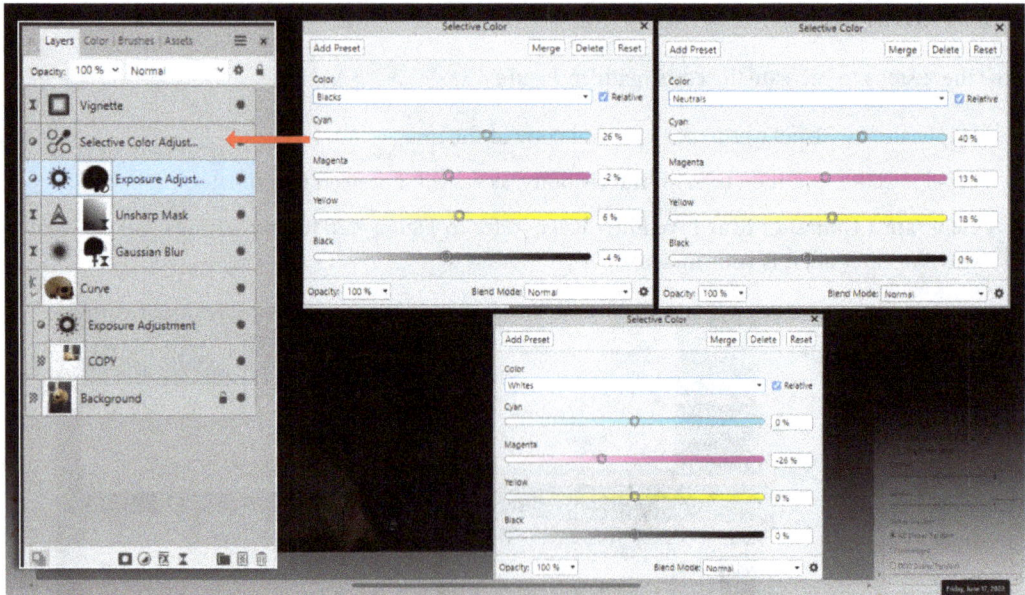

Figure 4.19 – The Selective Color Adjustment values

8.  Create a **Vignette** layer and adjust it.

9.  Unhide the **COPY** layer you made in *step 1* and trace around the skull using the pen tool to create a vector mask.

10. Insert the skull image inside the vector curve, thus isolating it (see *Figure 4.19* for the nesting position). Notice where the **COPY** layer is nested in the curve.

11. Go to **Arrange** (in the file menu under the **Arrange** layer) and with the **Curve** layer selected, flip it horizontally to face the opposite direction.

12. Move the skull and increase the size to position it using the Move tool, giving the illusion it is in the foreground.

13. Apply **Exposure Adjustment** inside the curve to apply it only to the skull, not the image (remember that applying nested adjustments only applies them to whatever is in the nest. We want this skull to have its own exposure that differs from the main subject.

The previous project is very simple and contains all of the techniques you have learned so far and some new ones such as selective color that you may never have seen before. I hope you see how you can combine the layers with the masking techniques and get some amazing results.

## Professional tips from this edit

In photo editing, three factors contribute to a subject called **atmospheric perspective**. This refers to how you are telling the story of the image for the viewer using atmosphere and the three variables that create it. The three elements are as follows:

- Lightness (luminosity)
- Focus
- Saturation

In an image, the objects in the background should have the following attributes:

- Be darker
- Be less detailed
- Have less saturation

The subjects closer to the foreground should be the following:

- Lighter
- More detailed
- More heavily saturated

Using simple adjustment layers such as **Exposure**, we darkened the background and concealed that adjustment around the skull, thus highlighting the subject.

The **Vignette** layer draws the eye to the center focal point.

**Unsharp Mask** is only seen on the skull, thus, when combined with the **Gaussian Blur** layer, it gives the balance of detail and ambiguity.

With this being said, let's move on to some of the common things that I struggled with as a new editor to make sure your journey is smoother than mine.

# Professional tips, tricks, and important points

Masking is an essential skill; much like in basketball, you have to be able to pass, dribble, and shoot the ball. So, while it may seem complicated, it is simply a matter of practice and knowing the application of the technique, usually through practice, practice, practice.

As a professional, I use brush masks constantly for compositions, and I can give you the following tips to help you along your way:

- Always use a soft brush. No matter how precise you are with the stylus, a hard-lined brush will show up as a bad mask. If you need a hard line, use a vector mask.

- You can always check the preview (remember the white square) to see how the mask is being shown.

- Take your time creating vector masks; do the fewest number of nodes to capture the detail, but realize that your viewer will not be looking as closely as you are, so a bad mask is not the end of the world, and it does not have to be perfect.

- I use a bold red line when creating my vector masks because it helps me see where I am working.

- I use the **Gradient** and **Elliptical** masks for color corrections and to add focus to my images by masking lights and detail to draw the eye.

- While we have dealt with individual layers and masking, remember you can apply the same concept to the group of layers

## Summary

In this chapter, we covered some basic principles of masking; however, in order to edit effectively, we have to be able to do one other basic skill that relates to masking, and that is the process of selection. When masking is combined with selection, we can begin to use selection as a mask, so let's say you have a balloon that you want to mask, you can select that balloon and then save it for later work.

In this chapter, we learned the three principles of masking and the two most powerful and frequently used techniques on how to mask, the simple brush-on and the vector clipping mask. This is important knowledge as we will begin to stack basic skills one on top of the other, and that is why the book is written the way it is, so keep going and combining all the new knowledge in each chapter. The masks build on the layers, which build on the idea of a pixel-based system.

But as a last comment, I just want to point out one last time before we leave the masking topic in this chapter, you have skills at this point to create some beautiful edits, and you should *never* wait until you know *everything* before you make *something*. So, edit, create, experiment, and most importantly, have fun. In the next chapter, we will start discussing selection in photo editing.

# 5

# Selection and How to Achieve It in Affinity Photo

On the short list of essential skills for any photo editor is the topic of **selection**. Selection needs to be practiced because not every selection technique will work for every situation. If I am editing an image that has trees and leaves, I will adopt a method significantly different from what I would use with a hard-edged item such as a building. The process of selection is something that can only be learned through trial and error... there is no substitute for actual editing in the program.

It is physically impossible to show you all the tools and techniques to select for every conceivable situation and so, in this book, I wanted to give you the simplest, most universal tools for selection to get you up and moving in your editing. Feel free as you grow as an editor to search out new methods and techniques that match your needs for your style of editing. What I am showing here are three or four techniques I use 95% of the time that do not rely on a specially selected photo or a certain lighting situation.

In this chapter, we will cover the following topics:

- Principles of selection
- Tools of selection
- Saving your selection

Let's begin first with a theoretical framework to get to know exactly what is going on when we select in a pixel-based image.

## Technical requirements

The CiA video for this chapter can be found at `https://packt.link/AKQrQ`.

You will also need the `Chapter 05` project to follow along with me.

# Principles of selection

We will begin the discussion with the principles of selection, as principles never change and are universal. Too many people dig directly into techniques and so they end up working on a tool-by-tool basis; in reality, they would have been far better served learning about the principle and then selecting the tools that did the job. That being said, let's dive into the principles around the selection technique in Affinity Photo.

## What is Affinity Photo doing when it selects?

As we covered earlier in the book, raster-based art is composed of pixels, and pixels are nothing but small squares placed next to one another that contain color and size information. We as humans make associations in our brains about what these things mean; this is an important distinction. Take a look at the following figure and note the things you observe:

Figure 5.1 – A sample image showing the principles of selection in application

In the preceding figure, we see many things… but remember what I said, we as humans see the following things:

- A human
- A wall
- A few squares (frames)
- Bricks

But what does Affinity Photo see? Here is a hint, it is not the object; there is no awareness that the image is a painting on a wall (although AI is improving). Here is what Affinity Photo sees:

- A value of brown (various hex code colors)

- A value of blue

- A darker shade of red next to a lighter shade (dynamics)

- A color of blue next to a value of reddish orange

Affinity Photo has no idea that the subject is a person, as it sees the world using three parameters:

- Color values

- Lightness values

- Location values

We need to retrain ourselves as editors to see the image as Affinity Photo sees it, and to accomplish this, I have developed a way to translate our wishes as editors and humans into the language of Affinity Photo using three statements.

## How to think of a selection – the three statements

This is a simple method I taught myself to help me to select when I started in photo editing, as I got tired of never having my selections turn out the way I wanted them in my head, and the reason was I was asking Affinity to interpret my needs instead of changing my language to communicate with the program. Thus, I developed three statements to help with universal selection.

Use these when faced with a selection, and it will help you choose correctly and see the world as Affinity does so you can make better selection choices.

### Statement #1 – what we want to accomplish

From our human brain, we engage in activity to achieve a result, so this is what we want to do. This expresses our end desire, but it is purely only useful to us; Affinity Photo does not care about what we want to do... it only asks what we want the program to do.

As an example, in *Figure 5.1*, you will see the blue paint running down at the bottom of the image. If we wanted to clean up the image, we may say *I want to remove the blue paint from the outside of the brick frame.*

Affinity does not recognize terms such as *brick*, *paint*, *outside*, and so on… those are words that we use. So, in short, we have to change our words.

## Statement #2 – what we need to do from a selection perspective

Now that we know what our end goal is, we need to think about the selection operation we need to perform. We are still working with our human mind here, so if you were going to select something to achieve the previous goal, what would you want to select?

Using the previous example where we want to clean up the spilled paint outside the image, we may say *I need to isolate the blue that is bordering the red brick outside of the image.*

This will achieve the desire from the first statement, but now we have rephrased it in terms of how we have to work it out to get there.

## Statement # 3 – what we need Affinity to do

In this statement, we take what we think we have to do to achieve the result and translate it using only color, lightness, and position.

For example, we may say *I need to isolate the broken-up light blue color that borders the red color only in the area of the lower frame.*

## How would I approach selection using this framework and Figure 5.1?

Here is what I now know to help my selection:

- I have a consistent color blue that doesn't vary much in shade
- The color I need to select is always bordered by a heavily contrasting color, so no adjustment is needed to improve the dynamics
- The paint is not continuous so I will need a tool that works in multiple areas

Based on the preceding points, my approach would be as follows:

- Select by sampled color (attribute selection)
- Keep the tolerance for the sampled color low to only get the light blue
- Clean up anything above the bricks that were selected using a shape-based selection tool in the subtract mode, leaving only the light blue at the bottom of the image

It is okay if you don't understand the tools in the preceding points, and that is why this book exists. However, after reading the chapter, I would urge you to come back and re-read the preceding workflow, and I bet you will understand why this is such a powerful way of looking at the software. We will learn about tools such as attribute selection and shape-based selection in this chapter.

In the next section, we will understand when to use selection.

# When do you use selection?

We use selection frequently in conjunction with masking in editing. It is not uncommon that you will create the mask for subjects in your edit from the selection. In the previous chapter, we dealt only with masking in isolation; the addition of selection allows us to be more precise and perform edits that simple masking would never support. Take a look at the following figure:

Figure 5.2 – Using the three statements to define the objective of an actual project

As you can see in *Figure 5.2*, this level of precision would have been impossible to achieve using traditional making techniques. This is the power of selection using the three statements I taught earlier. Here is the breakdown:

- **What we want to accomplish**: I want to turn the figure and the symbol into a teal color
- **What we need to do from a selection perspective**: I need to isolate the white from the gray
- **What we need Affinity to do**: Select only the pure white values in the area for the girl and heart

Based on this, I chose to select based on highlights (because I only wanted the brightest whites) and then I used the selection to control where I painted teal. Don't worry, we will show you how we did it later in the chapter – this is simply how the thought process works for selection.

Now that you know the principles, let's look at how Affinity will represent selection when the tools are applied.

## What does a selection actually look like?

Refer to CiA video for an example of selection in action. The typical term for selection among editors is *the marching ants*. Notice in the CiA video and *Figure 5.3* that when the selection is made, a row of moving dashed lines appears, thus the reference to marching ants. The area inside this boundary is the selection and the area outside the boundary is not included in the selection.

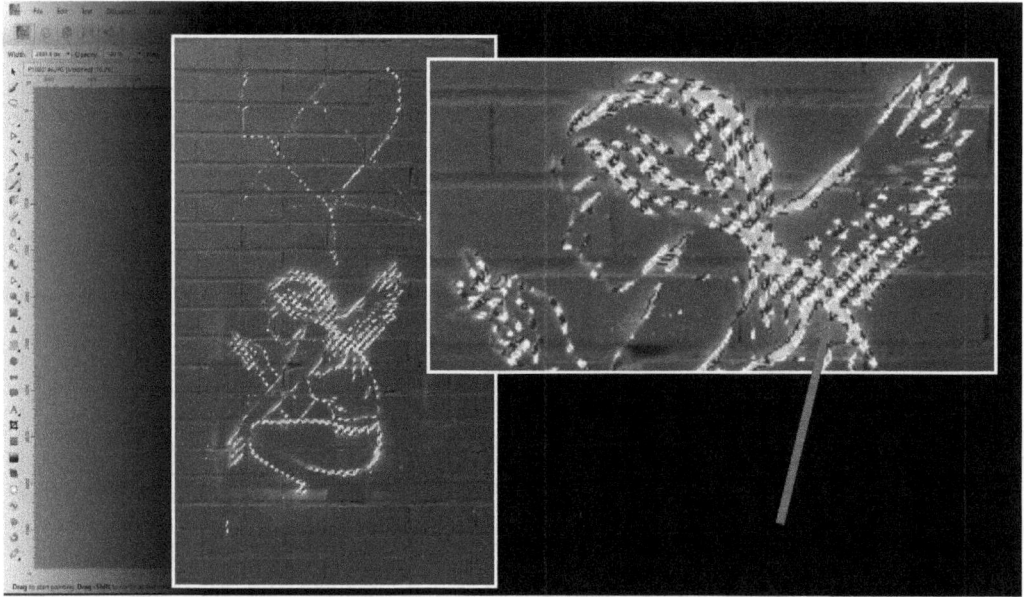

Figure 5.3 – The "marching ants" describing what is in and out of the selection

Let's see what we did in the CiA video:

1.  In the CiA video, we make a selection using the highlights in the image, and the selection appears.
2.  We then apply a **Pixel** layer above the image and paint the layer.
3.  The only part of the image that is painted is the selection, and this is why we use selection.

We will demonstrate and explain more on how to select later in the chapter, but for now, just be content with understanding what a selection accomplishes.

In the next section, we will learn about the types of selection.

## Types of selection

Typically, in a discussion on tools, I break the selection down into two types:

*   **Tool-based selection**: Uses a moving object to map the selection guided by the editor

- **Attribute-based selection**: Uses the image itself to determine where the selection occurs and then adjusts to the editor's desired area through adjustment of the % variance

These are the two types of selection we will work with. Let's understand these types in depth in the next subsections.

## Tool-based selection

All the tools used in selection are found on the left-hand side of the program in the tools area.

> **Tip**
>
> Don't forget that if you do not see the tool you want, you can customize tools by going to **View | Customize tools**, finding the tool you want, and dragging it into the toolbar.
>
> See the CiA video for a demonstration of how to do this.

The tools we will discuss for tool-based selection are shown in *Figure 5.4*:

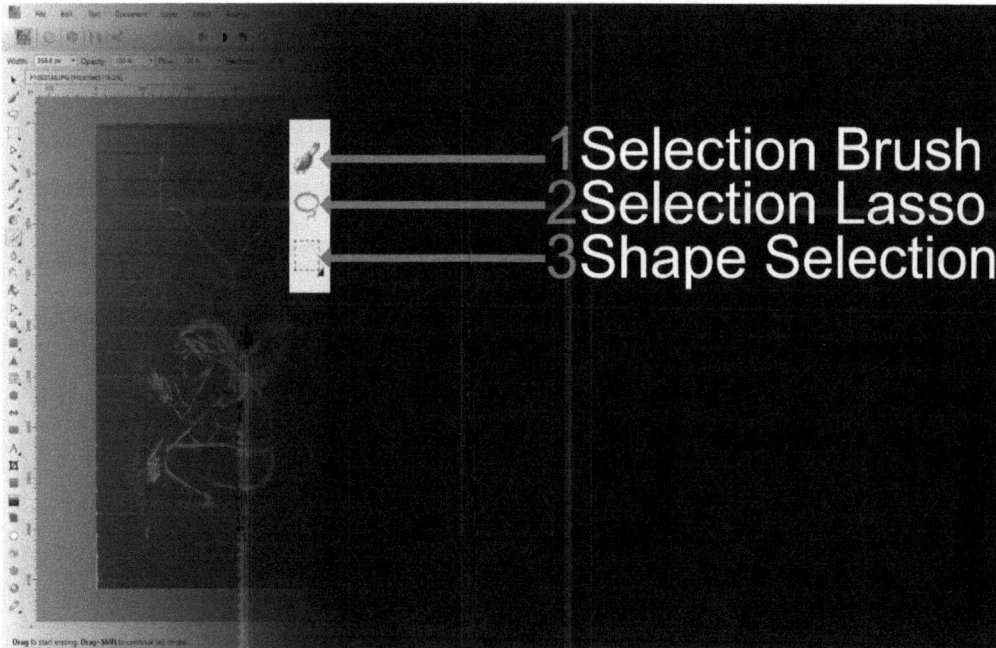

Figure 5.4 – The selection tools available in Affinity Photo

Let's go over these tools one-by-one:

- **Selection Brush**: This is my go-to selection tool 99% of the time for photo editing work. The Selection Brush looks for differences in edges and attempts to give you a selection (we will cover how to use this tool in the **Context** toolbar shortly, so hold tight for the details on how to use it).

- **Selection Lasso**: This is a selection tool that draws a simple lasso-style circle around a piece. I use this tool when I am doing illustrations and there is plenty of space around my subject.

- **Shape-Based Selection**: I rarely use this tool. For this, you need to click and drag out to select a shape from an image as the selection. A use case for this is when I replicate a section of the background sky or wall; I would grab the entire rectangle. Please note that the small black triangle on the **Shape** tools icon at the bottom-right corner implies that multiple tool shapes are available.

While these are the three most common tools, which solve 99% of problems, they all share a common approach to how to select.

All selection tools work on the idea of adding, subtracting, and refining a selection. When you click on these tools, the **Context** toolbar looks something like this:

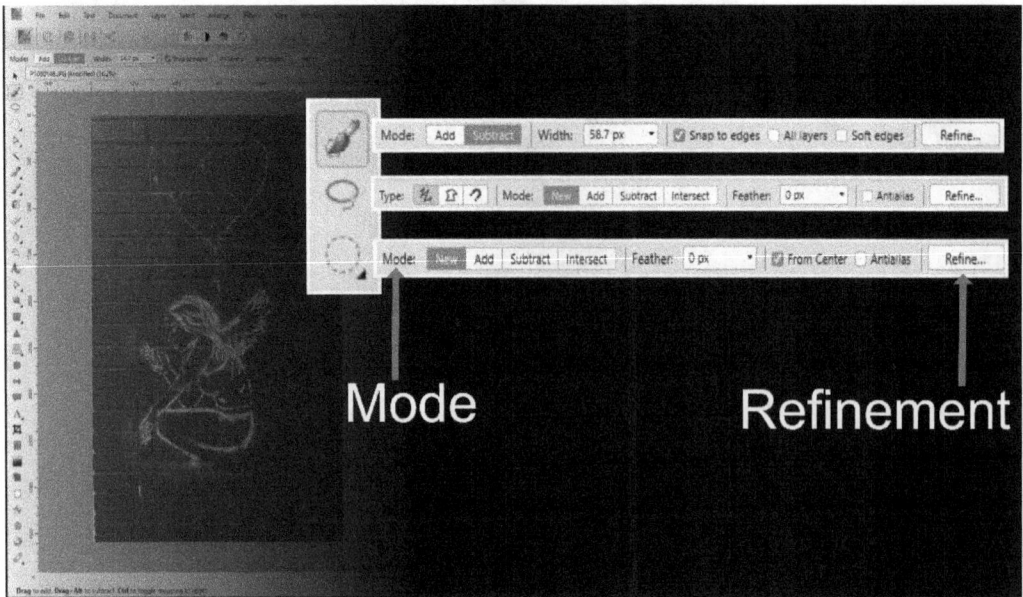

Figure 5.5 – Context toolbars for the major selection tools

Don't be overwhelmed by the choices of these tools as they all have three very similar features. Let's look for the similarities as a beginner to help you understand them:

- **Add** and **Subtract** modes: Here, you either add to the selection or subtract from the selection.

- Depending on the tool, there are ways to make the selection more "picky" or less "picky." For example, in Selection Brush, there is a **Snap to edges** option, which is only for the Selection Brush tool. This option makes the tool more intuitive.

- **Refine…**: After the initial selection is done, we can refine the selection to make it more or less precise (we cover that in the *Creating, adding to, and subtracting from a selection* section coming up soon, so don't get too far ahead of yourself).

## Attribute-based selection

While we covered the most common tools in the previous section, there is a second category of selection tools called attribute selection. Attribute-based selection uses the attributes of the image to make a selection. In general, there are three attributes that the program uses:

- Luminosity

- Color

- Transparency

Attribute-based selection options are found in the **Select** menu at the top of the program (see *Figure 5.6*).

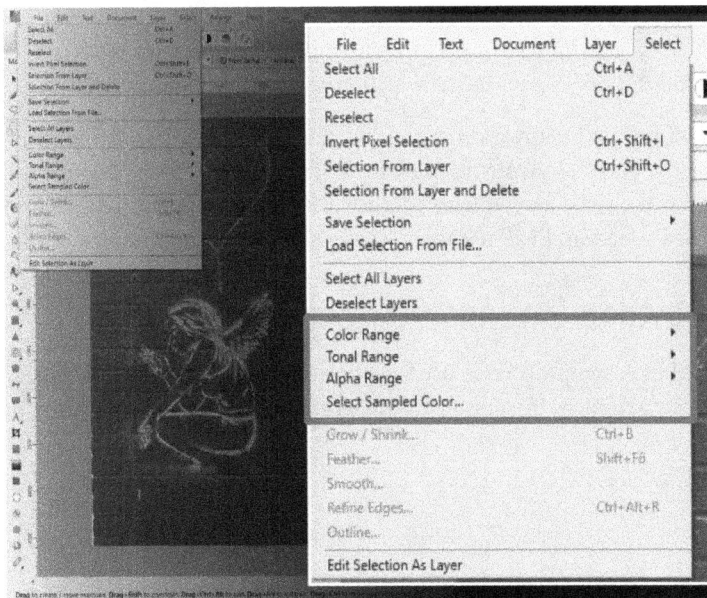

Figure 5.6 – Menu of attribute-based selection options

We will be going over all these options in the next section where we will create and add to the selection, but it is important you know here that you can select using attributes for complex applications.

So now you know about tool-based and attribute-based selection as the two dominant types, it is time to move on to actually making a selection in the next section.

# Creating, adding to, and subtracting from a selection

After learning in the previous section that you either add or subtract to a selection, and from the principles of understanding what the intent of the selection is, it is time to actually spread our wings and soar into it.

## Universal steps in creating a selection

Selection, at its core, is simply a matter of following a repeatable process over and over. Now, of course, on a case-by-case basis, there are challenges, but a simple flow will help you understand how to best meet those challenges. In my selection process, there are three steps:

- Adjusting the tool
- Creating the initial selection
- Refining the selection

The only difference in the process is the type of tool you will use to make the selection, so you just have to be familiar with how to set it up. In the previous section, we covered the **Context** toolbar for the most commonly used tools, and so in the remainder of this section, we are going to demonstrate the aforementioned steps for both tool-based selection and attribute-based selection.

I will demonstrate and explain the process for tool-based selection using Selection Brush (remember, we are only covering the most frequently used tools, so if you want to know about another tool such as **Lasso**, the process is the same; the adjustments are just slightly different in the **Context** toolbar). For the attribute-based selection, I will use tonal levels to show the application.

## Tool-based selection – using Selection Brush

In this section, we will understand when to use Selection Brush, and how to adjust it, create an initial selection, and refine the selection. To demonstrate everything, we will use the following figure:

Figure 5.7 – Selection Brush-based selection subject

You can download this image from the `Selection brush practice file` in the downloads.

## When would you use this tool?

I use this tool when I have a clear border around an object. In *Figure 5.7*, notice the pink border around the figure in the sitting position. The colors alongside it are contrasting, so it should be an easy selection for the brush.

## Adjusting the tool

Selection Brush is my go-to tool for selection; easily 90% of the tool-based selection I do is with this brush. This brush operates on three variables:

- **The mode**: The mode is what the brush is going to do when applied. There are only two modes. The brush will either add to the selection or subtract from the selection.

- **The size of the brush**: The size of the brush dictates how far Selection Brush reaches. I recommend using a smaller brush to start with, especially if you are going to use the **Snap to edges** option.

- **What the brush looks at**: Does it snap to the nearest edge? Does it look at only the layer you are on or all the layers in the edit? Does it form a soft edge or a hard edge?

*Figure 5.7* shows the brush settings I would use to select the pink figure on the wall as a starting point.

## Creating the initial selection

I have included a CiA video showing my selection process but here is a simple description:

- We cover the part of the image we want to select with the brush, watching it snap to the edge of the figure
- If we go outside the image, we simply switch to **Subtract** and then paint over the area where we strayed

Notice the following points in that video:

- I switch brush sizes frequently to get the details needed
- I switch between addition and subtraction modes

I have also captured a screenshot showing the initial selection of the figure prior to refinement. This is included because this is an initial selection – it is not perfect, and it needs to be refined.

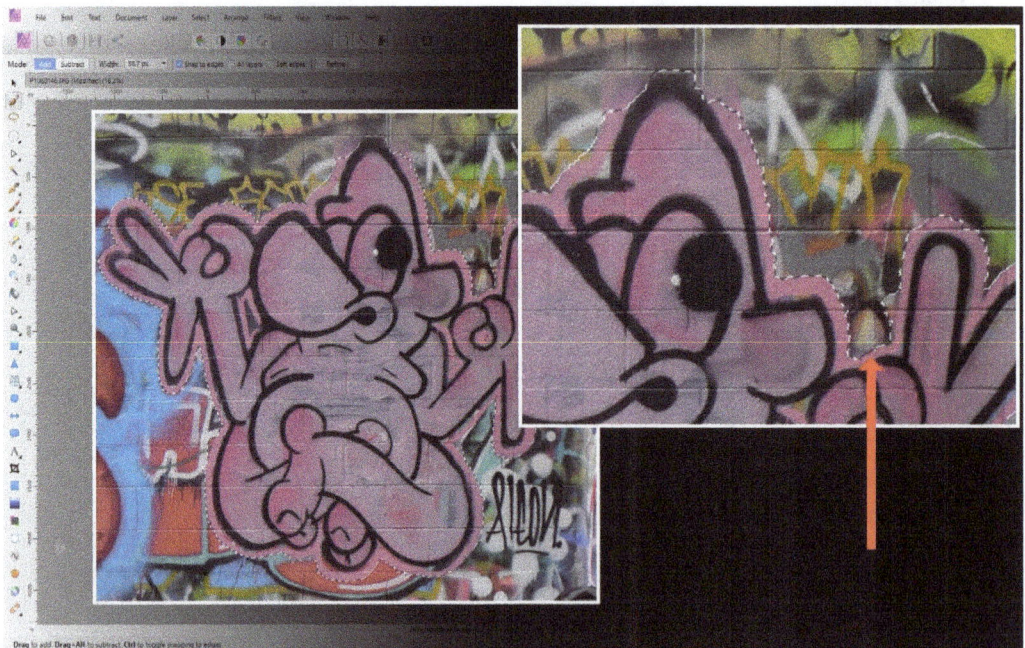

Figure 5.8 – Brush-based selection detail

## Refining the selection

Once the initial selection is created, you have the option to refine the selection by hitting the **Refine** button (refer to *Figure 5.5* to know where the **Refine** option is located in the **Context** toolbar). This is an optional step, but in effect, refining a selection allows Affinity to do a better job on the selection.

The process works by dragging a brush over the perimeter of the image, and on that perimeter, Affinity uses the settings shown in *Figure 5.9* to look for a better edge.

> **Pro tip**
>
> Smaller sections of the perimeter lead to better selections. Do not try and trace the entire image in one pass.

When you bring up the **Refine** option, the image comes up with the selection clearly shown and the rest of the image is coated in a red color. This is called the **overlay color**.

In *Figure 5.9*, Affinity shows you the selection you have made, and what is not selected is in red:

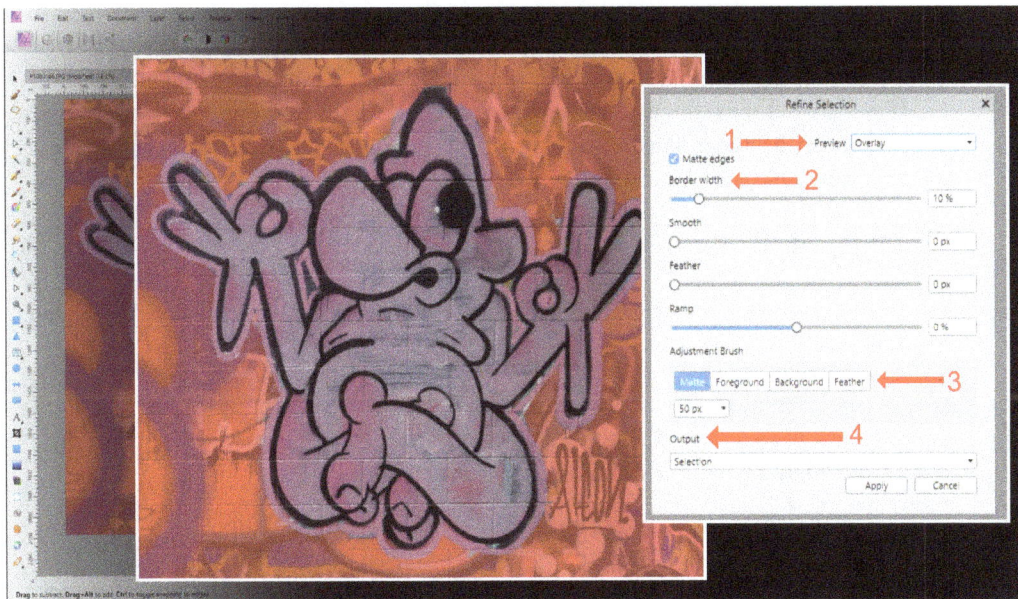

Figure 5.9 – Settings available for the refinement of the selection

Once we have initiated refinement, we need to explore the settings available to us in greater detail.

## The Refine Selection dialog box

We have broken the **Refine Selection** dialog box into four parts, and before we go into the details for each part, I want to make a simple disclaimer:

**There is no magic universal setting for any of this. Depending on the image, the complexity of the selection, and your desired output, you will have to play with it and adjust it every time to achieve the desired result. So abandon all hope of a special magical universal setting.**

At the end of each section of this project, I will tell you the levels I typically use, plus we have included a CiA video at the end of this chapter showing you how I use this in the selection flow. Let's now understand the dialog box:

- **Preview**: The different **Preview** modes give different views of what is selected and not. For 90% of my work, I stick with **Overlay**; however, sometimes when I want a hard edge and there are very fine details, I will use the **Black and white matte** option. For our work in this book, we will stick with **Overlay**.

- **Brush settings**: When you make a selection, you brush over the edges of the selection, and Affinity tries to make a better edge and a more refined selection. The brush settings will need to be adjusted on a case-by-case basis; however, I have described what each setting does and how to use them in the following points:

  - **Border width**: This contains the settings for the refinement. When you brush over the edge of the selection (see the CiA video for the demonstration), it looks for a border, and so by adjusting these settings, you are in effect altering what Affinity looks at. My best advice for this is if you are having problems getting a reined edge, raise the brush size a bit to try and get Affinity to look at it a bit more and evaluate when searching for a border.

  - **Smooth** and **Feather**: These refer to how soft the refinement is. I typically leave it at **0** and only use it if I am looking for a fuzzy selection.

  - **Ramp**: This describes how aggressively the selection goes from hard to soft. I do adjust **Ramp** a bit if the selection is tricky, but not much.

For each of these brush settings, you will have to play with them and find what works for the image. In the CiA video, you will see I do make some adjustments to the softness because the pink is not a very firm edge. I also adjust the border width because of the top area.

- **Adjustment Brush**: This has the following options:

  - **Matte**: With this mode active, you can trace an edge and Affinity will look for differences in the path of the brush to try and improve selection on the image.

  - If you choose **Background** or **Foreground**, you are telling Affinity that anything you paint over is either the background of the image (will not be included in the selection) or it is the foreground and needs to be included in the selection.

  - A practical application of this is that sometimes a selection will tag part of the image clearly inside the area as being not applicable. In this case, during refinement, I will change the mode to **Foreground**, and paint over the area. This tells Affinity it is clearly in the scope of the selection.

  - The **Feather** option creates a fuzzy border around the selection. I use this in compositions where a hard edge will look weird, but typically keep the feather super small. Remember, Affinity does not know your image, it only sees values and colors, and so you have to tell Affinity how to help you.

- **Output**: This is a very important section. There are various ways you can output this selection; in the dialog box, Affinity uses the term **Apply**:

  - If you are making an edit, I would apply it as a mask, as this mask can then be applied around multiple layers (see later in the chapter for the process for saving a selection as a mask, where we break it down).

  - If you are doing a composition and want to isolate the image (for example, isolate a flower to put the flower in another image), then output it as a separate layer giving you an isolated image.

  These are the two most frequent exports I use daily in my work.

I have included a CiA video here to show you the actual process using an edit.

Please note the following points from the video:

- I used the **Background** option to hit a spot in the lower right of the fire, telling Affinity *this spot is clearly background*

- I used the **Smooth** and **Feather** options since I wanted to blend this image into another wall, so I did not want a hard edge

- I adjusted **Ramp** until I got a good edge, and then I could just clean up a spot or two

We have included the layer structure of this image in *Figure 5.10*. Notice the output was applied to be a mask; now this image is still complete, but masked with the selection. Notice this layer is identified as a **Mask** layer; this layer can be applied to other layers as well.

Figure 5.10 – Display of the selected output as a Mask layer

Now, let's move on to the attribute-based selection tool.

## Attribute-based selection – tonal selection

For the attribute selection, there is no brush, so there is no setup or adjustment – you just click and select. For this, with the image that we will work with, we want to change the white spray paint to a light blue teal color. To choose the selection method, I follow the three principle questions we started the chapter with to make my decision. Because we have a good white color as a highlight, I am choosing a **tonal** selection.

In the example in the figure, we have a lot of gray, and while there is some good strong white, the dark blacks are not very well defined (see *Figure 5.11*). So I will select using the highlights of the image because these are the most prominent features.

I have included the file for you to follow along in the downloads for this book (see the Attribute selection practice file). The following figure shows the red arrow where I would click on the image to choose to sample the color I am asking Affinity to select the blow up image

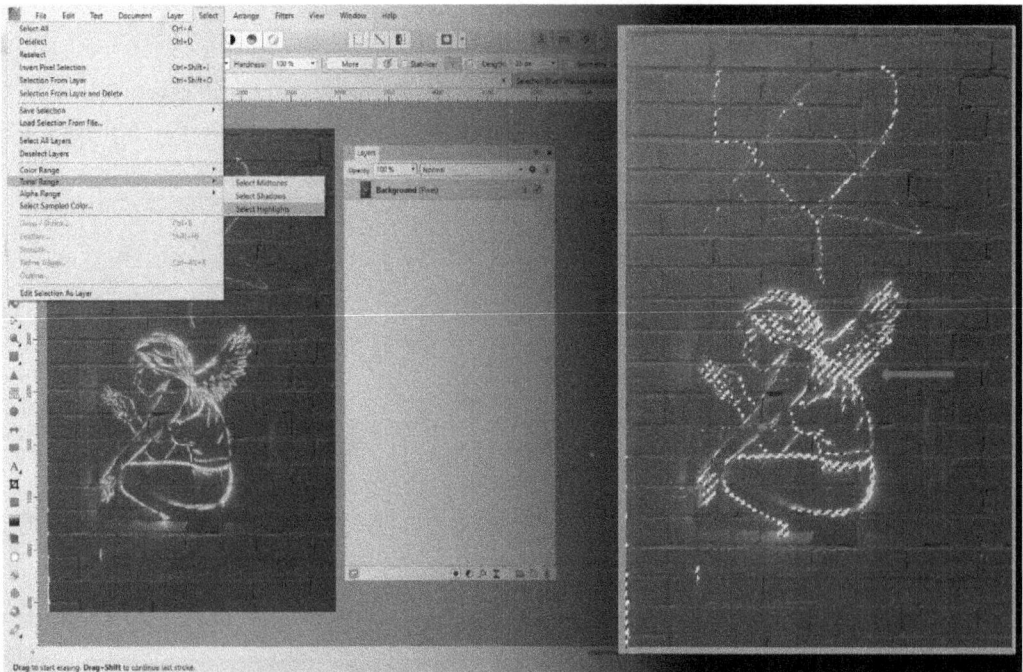

Figure 5.11 – The location of the tonal selection menu

With the highlights selected, we will add a **Pixel** layer above the image and begin to paint on the **Pixel** layer. The only place the paint will show is inside the selected area (see *Figure 5.12* and the *CiA video*):

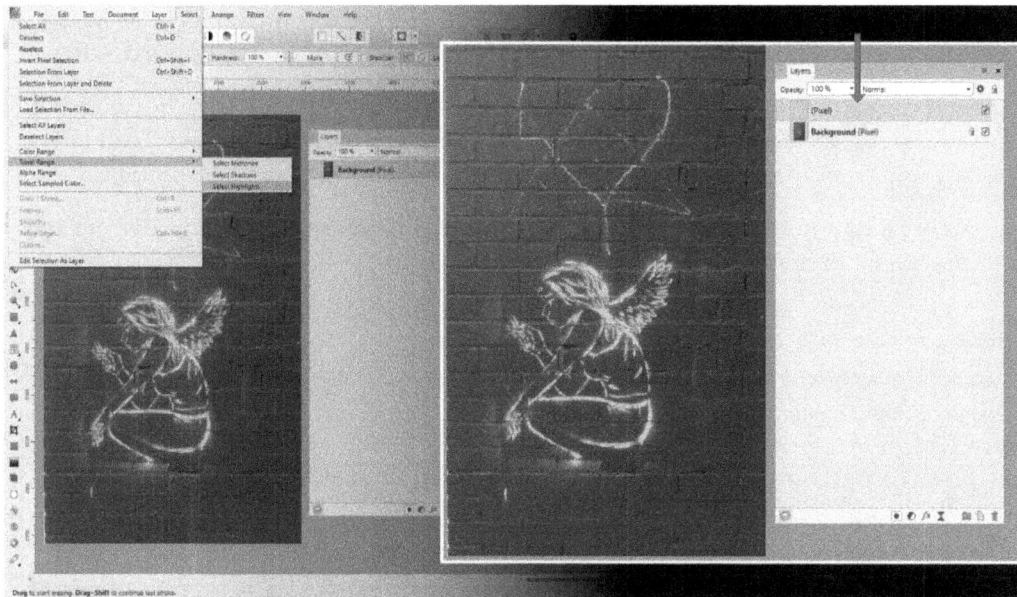

Figure 5.12 – The Pixel layer that has the color in the selected area

Once we are finished with the painting, we will return to the **Select** menu and deselect.

In reality, what we did is selected the white color in the **Pixel** layer and asked Affinity to look for other values of white that are similar. If they are similar, then include them in the selection. Lastly, once the selection was achieved, we asked Affinity to fill the selection only with teal. Notice how this language matches the way Affinity sees the world, and not the way we see it. Now, let's look at things we can do to save the selections we make so we do not have to constantly reselect.

# Saving a selection as a mask or channel

Once you have a selection, it is very strongly recommended you save the selection for the project. Even if you are really good at selection, chances of being pixel-perfect time after time are slim. As an example, if we change the angel to blue as we did in the previous example, and we later want to make her red, we will never get the exact same selection. So it is advantageous to save our selections in a project.

## What are a channel and a spare channel?

So now you know about the idea of selection. A common practice is to save a selection to use on other layers. In the attribute-based selection example, when we selected the highlights, it created a selection that contained the foreground. In this section, we are going to show you how to save that for later. Let's say you have a foreground and a background, and you only want the adjustment layers to work on the foreground, but the foreground is very detailed, and saving a mask for later use is a great idea.

In this section, I am going to give you my best method for saving selections, which can be used later in editing to mask an area through the use of **channels**, and more specifically, the creation of **spare channels**.

All images are made up of **red, green, and blue (RGB)** colors, and each color can be referred to as a channel. We can create spare channels where we can store selections we wish to reuse in edits.

Channels have their own studio tab (see *Figure 5.13*). For the **Channels** tab, I have moved it out into the studio space for the image demonstration.

> **Important note**
>
> We have a separate section on the use of channels in color (see *Chapters 10* and *13*), so for the purpose of this chapter, the color channels do not matter to you; stay focused on the selection aspects of the information.

In the following figure, I used Selection Brush to select the sky and separate it from the city:

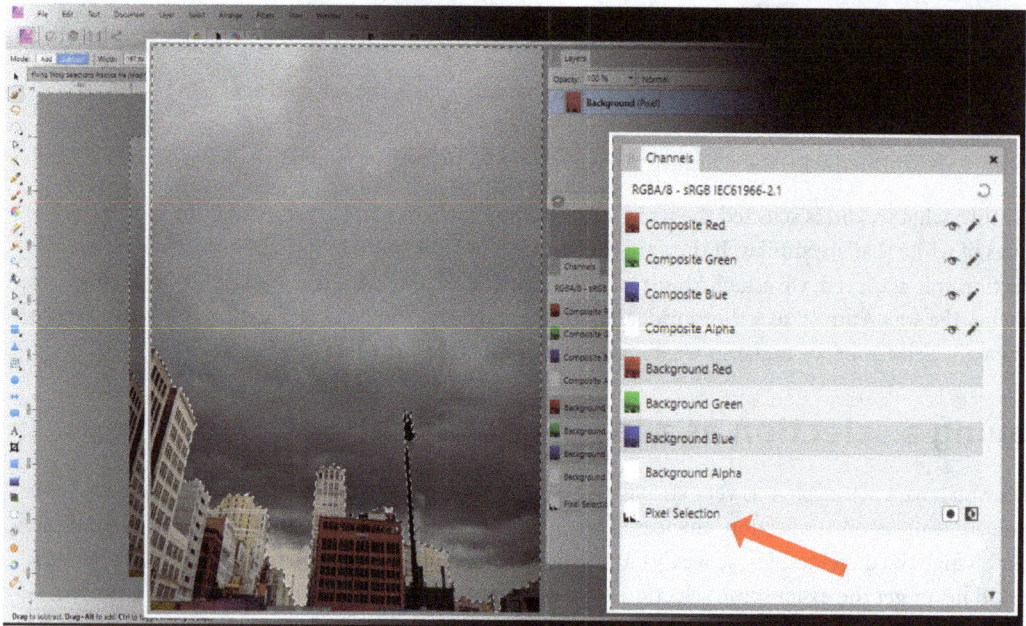

Figure 5.13 – The Channels tab showing the pixel selection of the sky

From the preceding figure, notice the following points:

- There are marching ants around the sky area (this means it is selected).

- In the **Channels** panel, there is an option named **Pixel Selection** at the bottom. This is the selection you currently have.

- Notice the black and white thumbnail by **Pixel Selection** in the **Channels** tab (what the red arrow is pointing at). Does that look like a mask? Remember, black conceals (the city is in black) and white reveals (the sky selected area is white).

What is occurring here behind the scenes is that Affinity has saved the selection in the **Channels** tab and it remembers it. What we have to do is save it somewhere as a spare channel.

So, what we need to create from **Pixel Selection** is a spare channel. A spare channel is like a selection you can keep coming back to again and again, and Affinity remembers it in the edit. Let's take a look at how to create this spare channel.

## Creating a spare channel

**Disclaimer**: We pick up this edit with the selection already made, as this section is about what to do after you have made the selection (we used Selection Brush and **Output** as a selection during refinement).

Once the selection is made and **Pixel Selection** is present in the **Channels** tab, we need to go to **Select | Save Selection | As Spare Channel**. Once you select **As Spare Channel**, the **Channels** tab will show the spare channel being created:

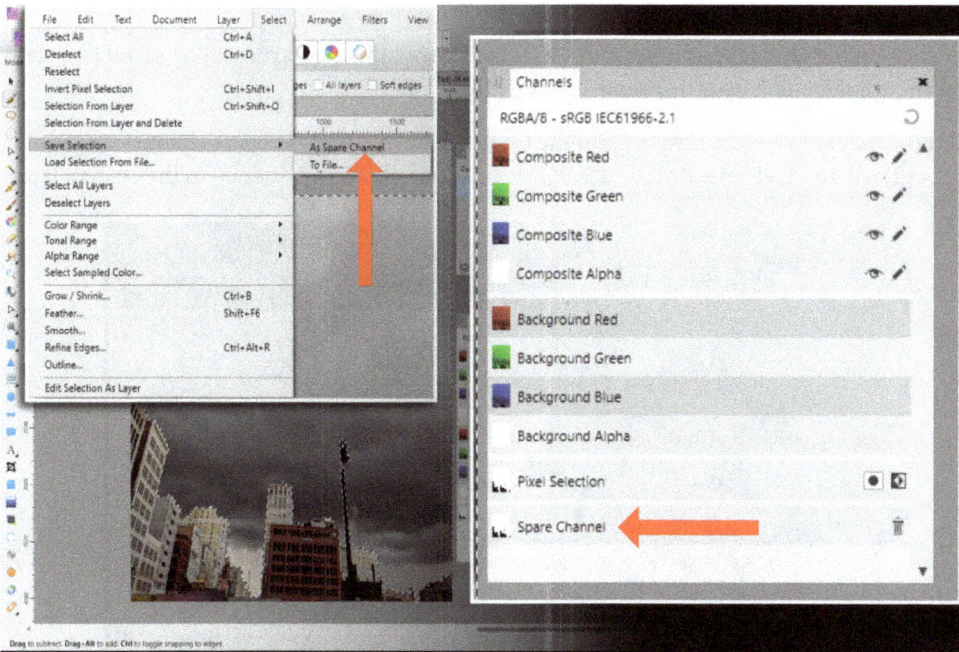

Figure 5.14 – The spare channel created in the Channels studio tab

Once created, this spare channel can be renamed – I am going to rename this one SKY for future images. Rename by right-clicking on **Spare Channel** and selecting **Rename**:

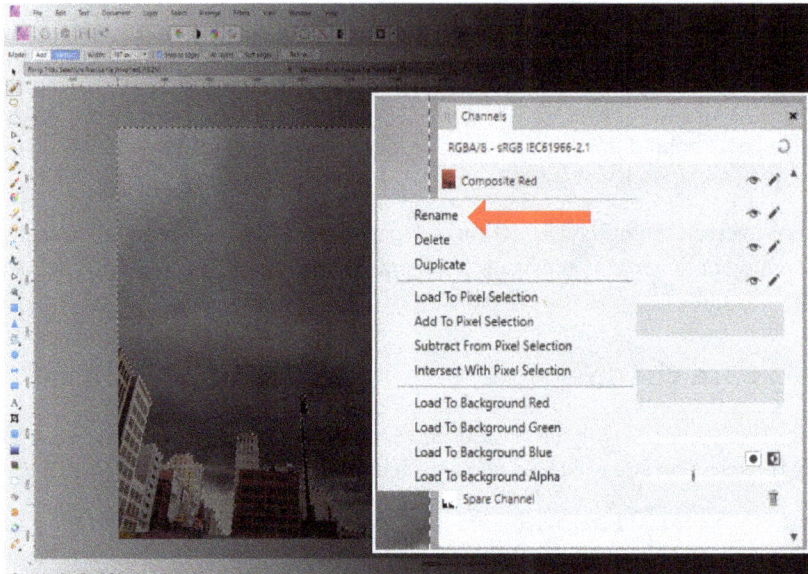

Figure 5.15 – Renaming the channel

Now, if click on the **Deselect** option from the **Select** menu, this sky layer will be saved as a spare channel and we can pull it up at any time we want during the edits.

To add the selection to a layer, simply go to the **Channels** menu, find the spare channel you want to load, select **Load To Pixel Selection**, and BOOM… the selection will be added to the current layer:

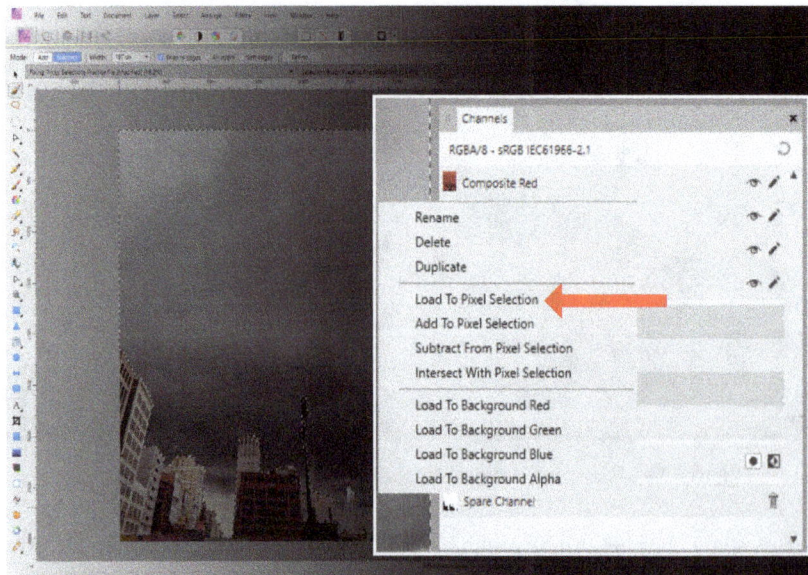

Figure 5.16 – Loading the channel to create a selection

I have also inserted a CiA video showing the process from the start with the completed selection, and the creation and loading of the spare channel to create a gradient overlay for the sky.

## When to use this technique

It is not uncommon during a complex composition edit to have several channels for isolated areas or subjects. I usually have around seven or eight, depending on the number of elements that I am juggling in a project.

Not every selection will be as clean as the examples we give here, and some are downright challenging, so I would like to offer up some tips for tricky selections.

# Tricks for refining a tricky selection

Not all images lend themselves to easy selection. As a matter of fact, for you, the first selection you try to make outside of the guidance of this book will most likely give you issues. Selection is a lot like problem-solving and so it is not uncommon that you may have to prepare images to get better selections. So in this section of the chapter, I have prepared two examples of strategies from my personal work that I use for tricky selections.

## Example #1 – adding adjustment layers to increase differences and make better-defined edges

In the example that we covered in the *Attribute-based selection* section, we chose the highlights as the attribute selection. However, what if we could have chosen the attribute selection called *select sampled color*? This selection would have been tricky because of the way the image was shot. Let's analyze the image:

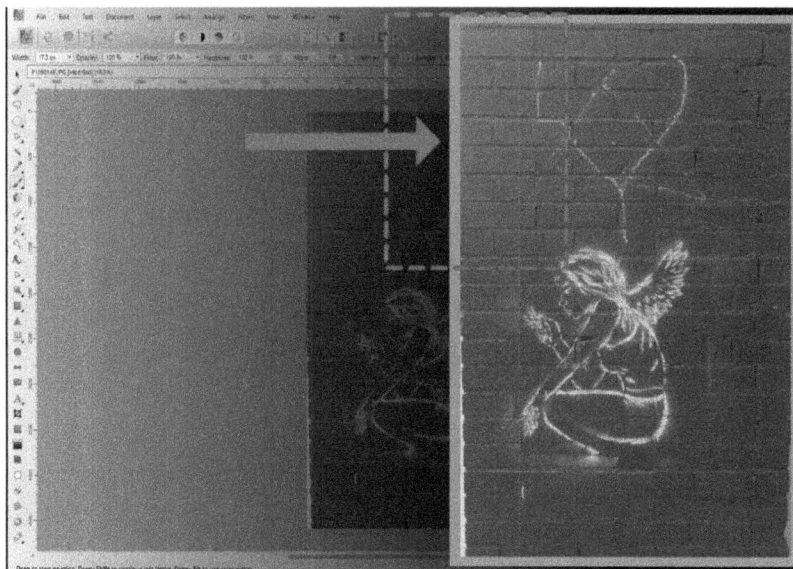

Figure 5.17 – Shows the location of "hotspots" in the image

Notice the lighter color on the upper-left side of the image coming from the natural light on the day I shot it. This will cause a lighter patch if I tried to select only the white.

So to overcome this, we can add a **Levels Adjustment** inside the image (this is a critical step – the placement of the adjustment layer inside the image is crucial because when we go to select, we are not selecting on the adjustment layer, but on the image. So make sure this is correct.)

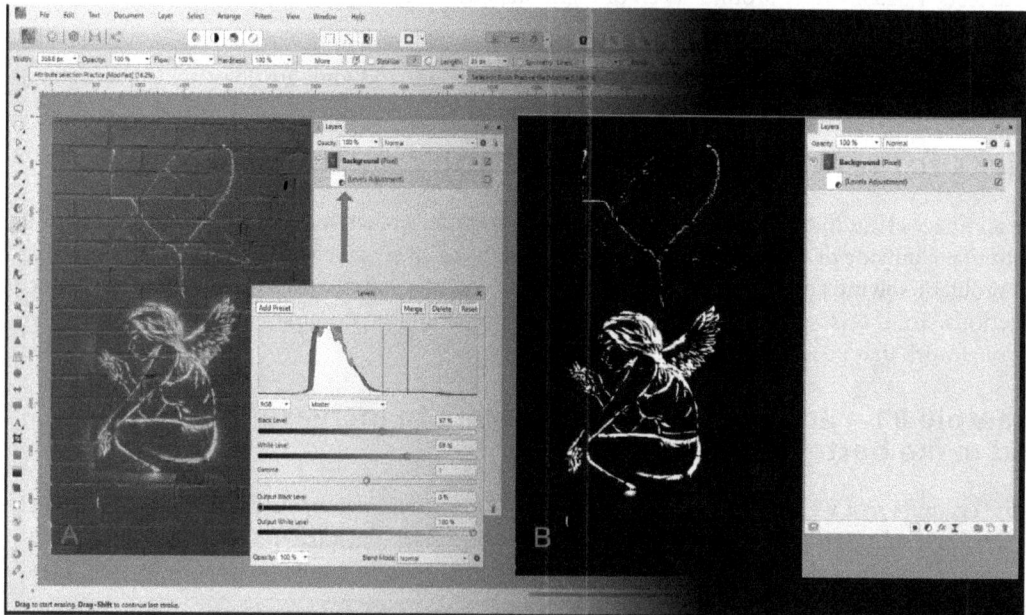

Figure 5.18 – Levels Adjustment being applied to help selection

Notice the adjustment layer properties in the image; this in effect negated the lighting variance in the image and made the selection possible. This is shown in *Figure 5.18* as the difference between A and B in the image.

Once the selection is made, we can eliminate the adjustment layer and continue with the process.

I have included a CiA video here to show you the process. You already have this image from previous activities in your downloads if you want to follow along.

## Example #2 – adjusting a selection using the menu (Grow / Shrink...)

In some areas, we may have selected well, but the small imperfections still exist no matter how many times we refine them. In these cases, a simple uniform adjustment of the selection may be appropriate (notice I said *uniform adjustment*, not *refinement*).

This is accomplished by using the **Select** menu and choosing the **Grow / Shrink** option. This option will expand or contract the selection by pixels, allowing you to get rid of those artifacts left after selection.

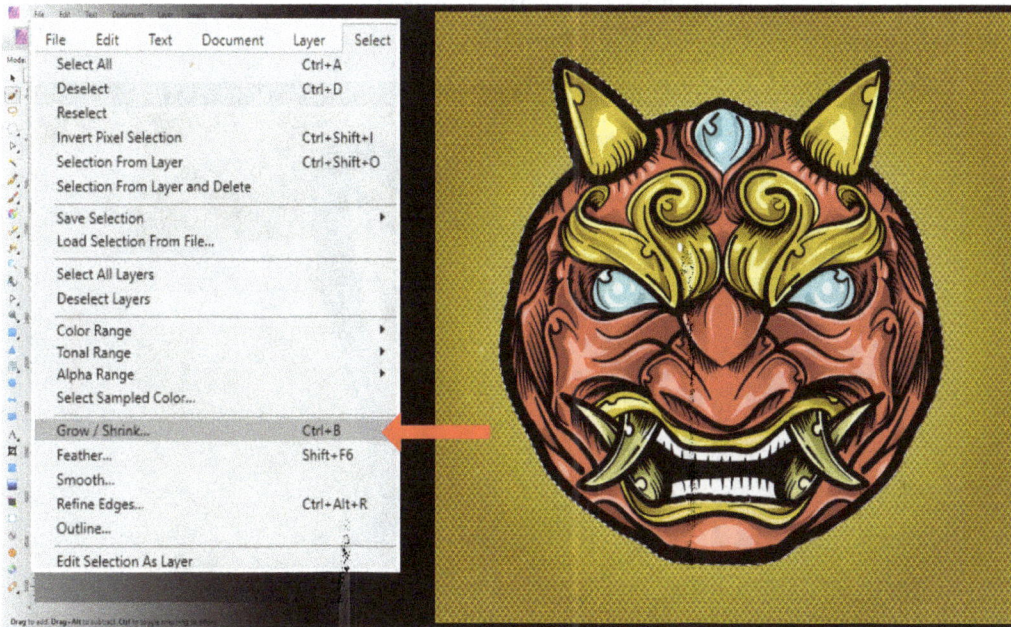

Figure 5.19 –Where to find the Grow / Shrink... options

In *Figure 5.19*, we used Selection Brush to make a selection and assumed we had a good black solid border; however, even after refinement, we still had artifacts from the halftone pattern of the background (see *Figure 5.20*). Notice the shadows in the green background to the far right.

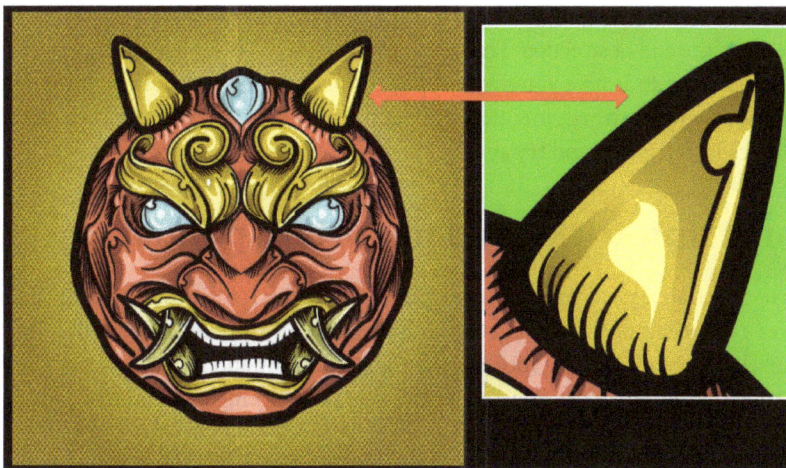

Figure 5.20 – Highlighting left over artifacts from selection

To remedy this, if we use Selection Brush and then after refinement, shrink the selection to **7.1 px**, we can move the marching ants safely inside the black border, and then the complex background becomes a non-issue for the selection (see *Figure 5.21*). We have included this file if you want to follow along (the name of the file is `Selection adjustment`).

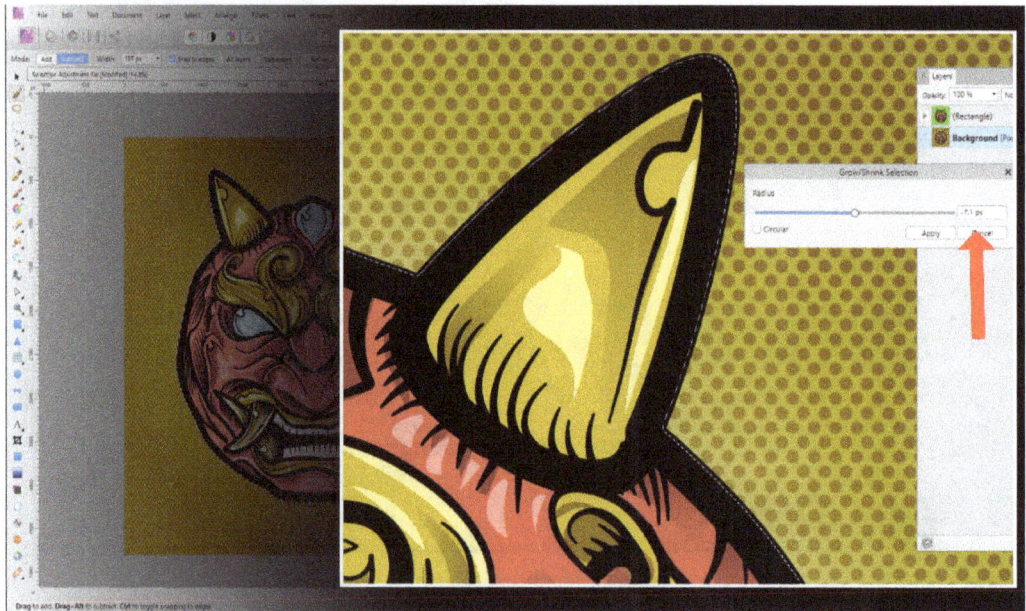

Figure 5.21 – Shows the shrinking of selection

I have also included a CiA video for you, the link to which is mentioned in the *Technical requirements* section.

Well, we have reached the end of selection. I have given you some tricks for less-than-perfect selections, and at this point, all I can do is give you my best professional pointers to get you moving on to the next step of your journey. It is worth mentioning that the only way to get better with selection after you know the basics is to try, fail, and try again.

## Professional tips, tricks, and important points

When it comes to selection, I need you to understand that you are trying to solve a problem, and each problem requires a different tool. Do not get so wrapped up in the process of selection that you end up burning yourself out. Find the right tool and the right workflow and move forward. The following are some of my most important tips and lessons after years of working as an editor:

- Do not assume the program knows the object you are selecting. While artificial intelligence has come a long way, it still cannot tell whether you intended to include or exclude that piece of leaf, so think like the program.

- Brush size is the most important variable in getting a good selection with Selection Brush. A good initial selection makes refinement easier.

- Increase the dynamics between objects, tones, or colors using adjustment layers to help selection; also, changing blend modes sometimes helps.

- Use spare channels to make sure adjustments are consistently applied.

- I use feathered softer edges when selecting things for compositions to help with the blending. In the next chapter, when we perform a composition, you will see my selection is not perfect, and is largely flawed, but it will not matter.

## Summary

This chapter concludes the strict operational fundamentals you have to know as an editor. There is no easy way to teach complex subjects such as masking, selection, and layers… so you just have to go through it and practice.

So far in this journey, we have covered the fundamentals of layers and the basics of masking, and with this chapter coming to a close, you know how to select, revise, and deselect from a variety of tools. You also know the fundamental principles of what the software sees to form a foundation to build from.

In this chapter, we did several selections if you followed along with the downloads, and so I would like to formally turn the page and introduce a chapter after this by taking everything we have learned and talking about workflow and composition – the more artistic side of editing.

In the next chapter, we will look at the role of cropping in composition, as how you crop your image tells the story for your viewer. So we can leave behind the super-technical, get our macro view hat on, and begin thinking like an artist.

# Part 2: Fundamental Concepts Used to Create a Simple Edit

In this part, we cover some fundamental concepts but begin by dipping our toes into the pool of the artistic side of editing. Topics such as composition, workflow, and the use of brushes will begin to give you the opportunity to take the technical skills you developed in *Part 1* and apply them to actual edits. We will keep the edits simple in this part, slowly building in later sections to make sure you can read and follow what the program is telling you as the layers begin to stack up.

This part comprises the following chapters:

- *Chapter 6, Cropping and Composition*
- *Chapter 7, Basics of Workflows and Balancing Dark and Light*
- *Chapter 8, Blend Mode Fundamentals*
- *Chapter 9, Basics of Stock Brushes in Affinity Photo*
- *Chapter 10, Working with Color in Affinity Photo*

# Cropping and Composition

In the previous chapters of the book, we have dealt with getting you up, running, and editing in the quickest, most logical way possible, and to accomplish that, we have been holding the assumption that the image as shot is acceptable and does not need anything removed, or it does not need to be *cropped*. In this chapter, we will remove this assumption because a key portion of a workflow (what we will be working with in the next chapter) is cropping the image to create the right *composition*, the right structure and flow to draw the reader's eye through the elements of the image you want them to see.

This chapter will be technical, covering what a crop is, how to get the desired output from your cropping activity, and how to correct it with the tools inside the cropping function of Affinity. After that, we will take a look at the composition elements in your image. We will cover things such as the rule of thirds and the concept of the golden spiral, and we will practice telling different stories simply by changing the focal points of an image.

In this chapter, we will cover the following topics:

- Cropping fundamentals
- Using cropping tools to adjust photos
- Basics of composition
- Professional tips, tricks, and important points

## Technical requirements

The CiA video for this chapter can be found at https://packt.link/Hwvzr.

You will also need the Chapter 06 project to follow along with me.

# Cropping fundamentals

In this section, we will cover the act of cropping or trimming the area that we have for the photo. While it may seem simple, cropping is actually very important in the composition of the image and plays a role beyond the simple utility of making the image fit a boundary for display. So, let's dive in and learn about how to make this work.

## What is cropping?

Cropping is a way to remove parts of the image to only allow the areas to be visible that you want your audience to see. Now, this sounds an awful lot like masking, but actually, cropping only allows you to reposition the image, meaning you cannot "crop out" the center of the image; however, you can reposition the image so that the non-value-added parts of the image are removed. This can be done for several reasons, and some of the most popular are as follows:

- Removing the areas around the image that are undesirable, such as a busy street, when the focus of your image is the building in front of it

- To call attention to an element of the image and to give it importance as the focus (i.e., placing a model's face in the center of the image clearly shows that they are the focal point of the image)

Cropping an image is an essential part of evoking the response you want from the viewer, and if you want them to read your image correctly, cropping should take them on the right journey through your photograph, and that is why you cannot separate cropping from the composition. Cropping provides part of the journey through the composition of your photo.

### Prerequisite knowledge for cropping a photo

I hope that this book was entertaining enough to read straight through, but knowing the reality, you may have skipped ahead to this section because you need to crop on a project. To understand cropping, there are a few prerequisites we will be discussing that have been covered elsewhere in the book, and we will not spend much time walking backward. So, here is a list of some of the terms that we will be using from previous discussions:

- **Dots per inch (DPI)**: This refers to how many dots there are in one square inch, a higher count means better print quality (up to a maximum of 72 for screens and a minimum of 300 from print work)

- **Aspect ratio**: The relationship between the height and width of the image

- **Unit of measure**: The native units for the document

- **Document**: The actual workspace

- **Canvas**: Additional space added that can contain several documents

Now that we know more of the technical aspects, we need to know about some of the visual aspects, such as what the overlays are talking about.

## The mental model for cropping a photo

The Crop tool is in the tool area on the left-hand side of the interface (see *Figure 6.1*) in Affinity Photo (you should know how to locate tools you do not have by now, and if you do not, refer to *Chapter 1*):

Figure 6.1 – The overlay

When you choose the Crop tool, an overlay will be placed over the image (*Figure 6.1* shows the overlay). This is a *thirds-based grid* image (the overlay is divided into three sections vertically and horizontally). The overlay visualization can be adjusted for various applications using options we will discuss later. The preceding overlay is shown as a thirds-based grid (we will discuss those soon).

When you select the Crop tool, the context toolbar will change, and it can be very confusing to proceed to ensure the image you intend to get out actually comes out the way you want when you print. In the next subsections, I have developed a simple walkthrough to help you do just that. To help with this, we will use *Figure 6.2* to explain the areas of the context toolbar for the cropping function:

Figure 6.2 – The context toolbar for the Crop tool

### Step 1 – knowing your end result

In this step, you need to know your intention—whether you plan on printing the image locally, sending it off to a printer for a high-quality or large format print, or using this image in digital applications. This matters because, once you know this, the next step is to choose a size.

### Step 2 – choosing a size

Once you know the intended output, the next step is choosing the size. Size comes in two versions: absolute and ratio. It is important to know which one you are trying to achieve.

For example, if I am trying to make a crop that will take the existing image and place it nicely on a 5" x 7" printed page, then this is an absolute size (i.e., 5"x 7"). The ratio is not important because what I am after is the print dimension that will go to the final image.

If I am working on a web tile that needs to be square, the ratio is important because it has to be 1:1 (1 unit in width must also be 1 unit tall). Most web programs can scale up or scale down an image, but the ratio *must* be maintained.

So, knowing what you are trying to accomplish is the most important part of choosing the size.

## Step 3 – choosing the mode

Once you know the size, we now choose the mode, and the good news is, once you know the desired size, this should be easy. You will notice in the context toolbar that there are various modes to choose from in the **Modes** area (see *Figure 6.2* for the location of the **Modes** option in the context toolbar). We will cover all four modes, with their benefits and limitations, so you can confidently choose your mode and get the right result.

### Unconstrained mode

This option allows you to adjust everything; there is no ratio maintained, there is no set size, and you are free to drag the boundaries wherever you think they need to go. The following are some key pointers for you:

- **When to use it**: This is best used when you know exactly the size you are looking for because you can enter the absolute size (either pixels or another unit) directly in the context toolbar

- **Benefits**: You can change the units in this mode, so if you want a different size for the document, change the units in the context toolbar, and then you are in business

- **Limitations**: You cannot change the DPI in this mode—changing the DPI requires resampling, so the DPI will not change. You can enter a new value, but when you apply the crop, nothing will change.

In the CiA video (the link to the video is provided in the *Technical requirements* section), we apply the **Unconstrained** crop to the image and then verify the changes by going to the **Document** tab and clicking on **Document Step Up**, to see whether anything changes and confirm the changes. We have included a side-by-side comparison showing the original and the new changes in *Figure 6.3*, as well as how it looks in the **Document** tab, so you can see the exact same points:

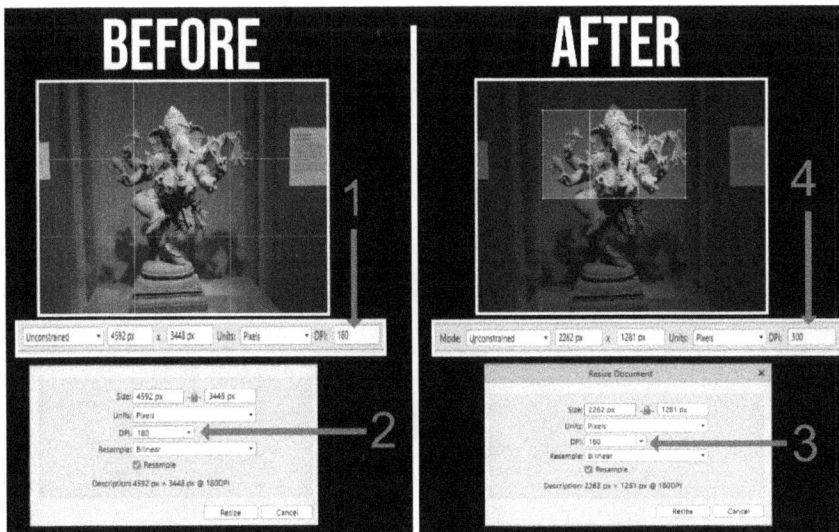

Figure 6.3 – Verification that cropping has changed the size

Please note the following points:

- Prior to the crop, the image was **4592 px** by **3448 px**, giving us a ratio of 1.33. After the crop, the image was **2262 px** by **1281 px**, giving a ratio of 1.76. This is what is meant by the **Unconstrained** crop: you can make any size.

- Prior to the crop, the **DPI** value was **180** (see items **1** and **2** in *Figure 6.3*). When we cropped it, notice we asked for **300** DPI in the context toolbar (item **3** in *Figure 6.3*). However, when the crop was applied, we still had only **180** DPI (see item **4** in *Figure 6.3*). This is because you cannot resize the DPI of the image without resampling. You must resample to change DPI; if not, the document will not gain any additional data to support.

## Original Ratio

This will hold the original ratio of the image and will not allow free movement of all the nodes. This is desirable when you have a desired size and simply want to highlight a different area of the photo. Let's look at an example.

In *Figure 6.2*, the pixel count reads **4592 px** by **3448 px**; this means that the image is 4592 wide and 3448 tall. So, the ratio is 4592/3448 = 1.33. So, for every unit of height that we add, we add 1.33 units to the length. Let's go through its usage, benefits, and limitations:

- **When to use it**: We can use this mode when the original image is already in the correct ratio as originally shot. An example of this may be a thumbnail image that was already formatted at the 16:9 ratio (such as 1920 x 1080), and you are simply looking to grab a part of the image.

- **Benefits**: If the image is already the right ratio, you can pinpoint the size you need and ensure it will fit the right ratio requirements when applied to the website.

- **Limitations**: Much like the **Unconstrained** mode, the DPI cannot be adjusted in this mode.

In the CiA video, I took the same image we have been working with and applied the **Original Ratio** crop to it. We have also included the summary of that video in *Figure 6.4*. In both cases, the image began as **4592 px** by **3448 px**, at a 1.3 ratio.

As we changed the size of the cropping box, the Pixel count changed, but not the ratio. Notice at the end of the crop, the size was **2175 px** by **1633 px**, still a 1.3 ratio (see items **1** and **4** for where these values show up in the context toolbar):

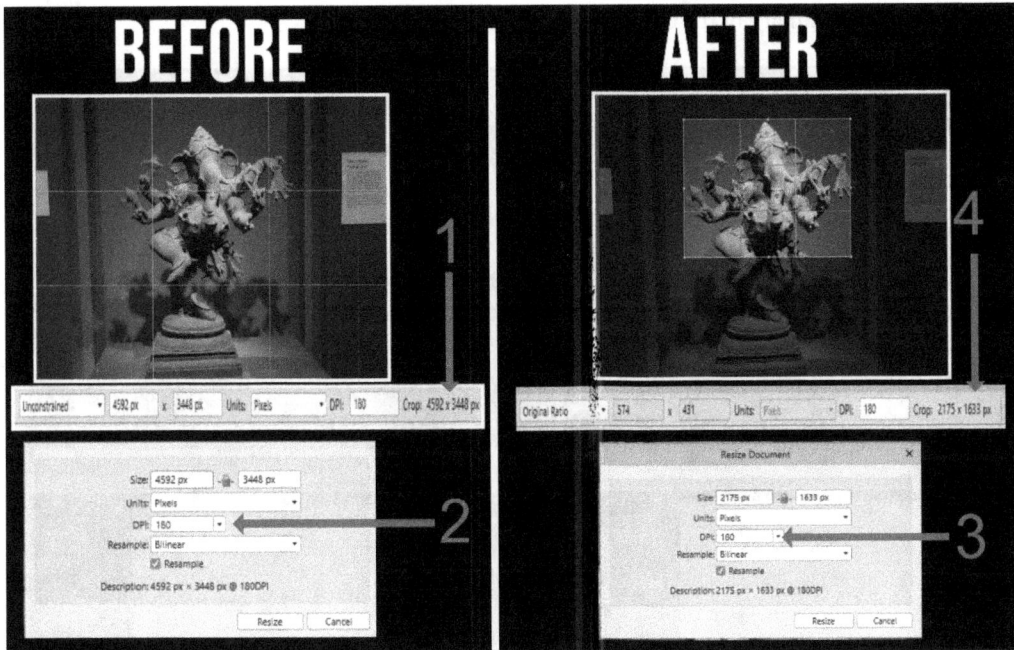

Figure 6.4 – Holding the ratio the same

Even though we changed the **DPI** value to **300** as we made the change, notice the **DPI** option remained unchanged, again because you must resample the image to change the DPI (see items **2** and **3** in *Figure 6.4,* where it shows the DPI did not change post crop).

## Custom Ratio

In a **Custom Ratio**, we adjust the units to the desired state, then define the desired numerical values in the text boxes. This created the **Custom Ratio**.

- **When to use it**: When I have taken an image on the camera, and I need it to fit a certain size for print (such as an 8" x 10"), or if I am doing digital work and I need a 4:3 ratio for a Facebook image, then this is your application
- **Benefits**: It allows you to fit the image to the ratio
- **Limitations**: If these are image elements you want to include in your crop, but they are outside the desired ratio, they will not be included

In *Figure 6.5* (see the CiA video too), I need to crop the original image to a 4:3 ratio, clearly different that the 1:33 ratio it was shot in, and I need to plan the best look I can. The size of the image is not an option because the website will take the image and compress it, but the ratio has to be right (think of this as Instagram).

I have included the summary of the context toolbar settings in *Figure 6.5*:

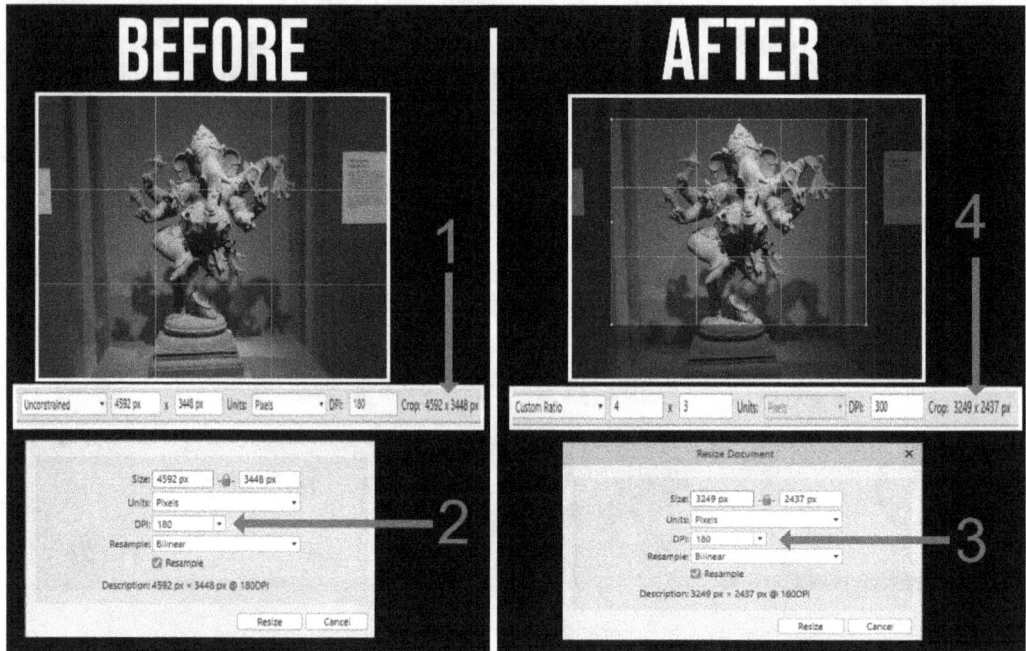

Figure 6.5 – Changing the ratio

In the preceding screenshot, notice that the **DPI** value didn't change. For Instagram and digital websites, this is not really an issue, as **180** DPI is more than adequate to cover the resolution, and the program will compress the file size to meet the platform's requirements.

## Resample

The **Resample** mode is when you convert or resample the picture to change the **DPI** value. This is a critical step if you want an image to change the **DPI** value, and you need to adjust the size to a smaller or larger ratio.

Resampling asks Affinity to review the image and either add or subtract pixels based on the desired outcome. So, Affinity reviews the image (the example we have been working with is **180**), and let's say we want it to have a value of **300** DPI in the area we cropped for a printed application. We will need to resample, and Affinity will attempt to create new pixels and sample from the pixels it has to make this happen. This is why it is called resampling.

We will not go through the benefits or limitations of this mode, as it is self-explanatory if you walked through the preceding modes. This is the one I use all the time for print work if the image I shot is inadequate in resolution.

Let's take a look at one example of resampling.

In the image we have been following, the **DPI** value shot for the image is **180**, and the size is given in terms of pixels. I want to place this into a 5" by 7" picture, so it needs to be in inches instead of pixels in ratio—this one goes in absolute values.

The following order is crucial in this process, Affinity is picky about the order of operations in this area so the right process is as follows:

1. Change the mode to **Resample**.

2. Change the **Units** option to **Inches**.

3. Enter the width value (for me, it was 7 inches).

4. Enter the height value (for me, it was 5 inches).

5. Reposition the crop box where you want it.

6. Lastly, enter the **DPI** value change you want to see.

7. Hit the *Enter* key to apply the crop.

I have also included a CiA video showing the process, the link to which can be found in the *Technical requirements* section. In the CiA video, the post-crop image, when checked in the **Adjust Document** tab, is **2100 x 1500**.

If we see the **DPI** value is **300**, then we can divide the previous pixel count by 300 and arrive at a 5" x 7" crop.

Also notice that the **DPI** value, when checked on the **Document** tab, is **300**. We have included the summary of this shift in *Figure 6.6*:

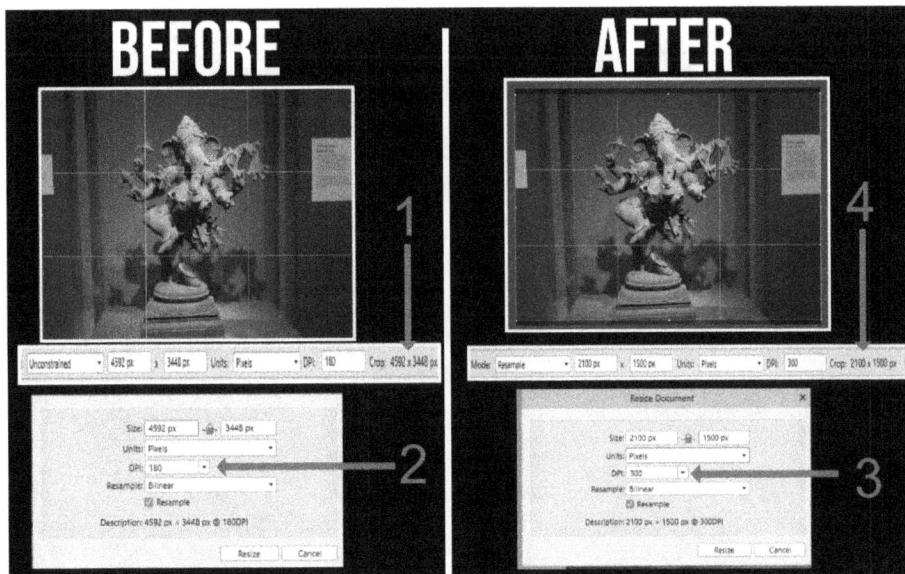

Figure 6.6 – Conformation the DPI has changed post resample

> **Important note**
>
> Even though the **DPI** value has changed, realize that Affinity added some pixels and geometry to fill in the gaps, and so there will be an almost imperceptible loss of quality because this data has to be generated, but it is nothing the naked eye will see.

After all of that, you may want to save the preset as you most likely do similar work time after time. For this, we can save cropping presets, so let's look into how exactly we do that.

## Using presets for cropping

While this book wants to give you the *why* and *how* behind the different modes, Affinity has made it simpler to crop through the use of presets. As the name implies, presets are already pre-dispositioned to get you to the right result if you just select the right one.

> **Note**
>
> The Crop tool presets are located at top of the document's left side, just below the Develop Persona icon. You need to select the Crop tool first to see the preset icon, which looks like a small gear shape (see *Figure 6.7*).

## Reading the presets

As you can see in *Figure 6.7*, the process of reading the preloaded presets in Affinity Photo is aligned with the model we had developed to go through the modes—this just augments and solidifies the concept:

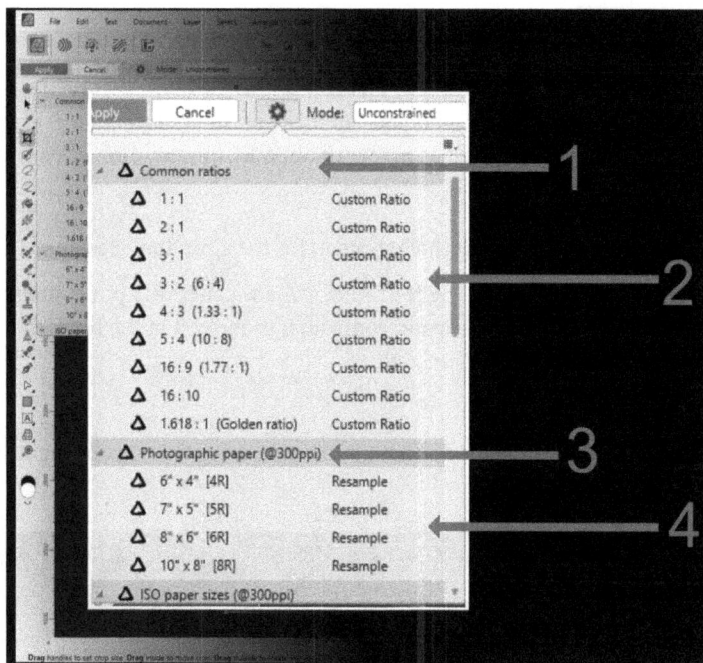

Figure 6.7 – Preset location and graphic

Notice the following details in the preceding figure:

- There are groups that are based on the ratio
- Each preset shows the mode they are used in—in this case, it's the **Custom Ratio** mode
- Photo-based presets have **300** DPI called out
- Resampling is used for the **300** DPI presets to make sure there is adequate resolution

So, this covers the preloaded presets in Affinity Photo, but what if you want to make your own? What if the work you do is so unique that the ones that come loaded by default are inadequate? Well, we can make our own, so now, let's cover that topic.

## Creating your own presets

Sometimes, when you are working on a project, you need to crop more than one image to fit a certain need; in this case, creating your own presets will be helpful. In my case, I frequently need to create Facebook group headers. A quick search reveals that the size needed for these headers is 1640 x 856 for a ratio of 1.91:1, so let's make one.

## Creating a category

The first step to creating your own presets is to create a home for them by creating a category. A category contains all the presets you make, and you can make as many as you need. I have presets for Facebook, Instagram, and YouTube, as well as my work on WIX for common sizes and formats. To create a category, follow these steps:

1.  Click on the gear icon in the context toolbar menu for the Crop tool (see item **1** in *Figure 6.8*).

2.  In the drop-down menu, click on the four lines (this is called the panel menu; we Affinity users call it the *hamburger* menu. To avoid confusion, in the rest of the book, we will call it the hamburger menu). See item **2** in *Figure 6.8*.

3.  Select **Manage Presets....**

4.  Choose **Create Category** and name it. That's it, the new category will appear at the bottom of the list:

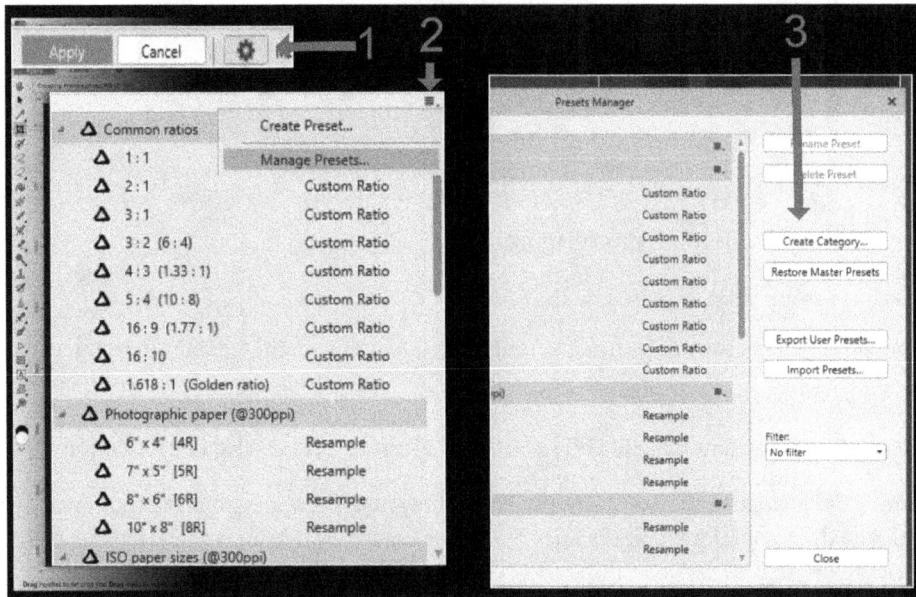

Figure 6.8 – Creating categories

In the next section, let's understand how to create an individual present.

## Creating an individual preset

Follow these steps to create your own present:

1.  Enter the values to the crop that you want in the preset (see item **1** in *Figure 6.9*).

2.  Go to the gear icon and then click on the hamburger menu.

3.  Select **Create Preset** (see item **2** in *Figure 6.9*).

4.  Choose a name and a category if you do not have a category and click on **OK** (see item **3** in *Figure 6.9*):

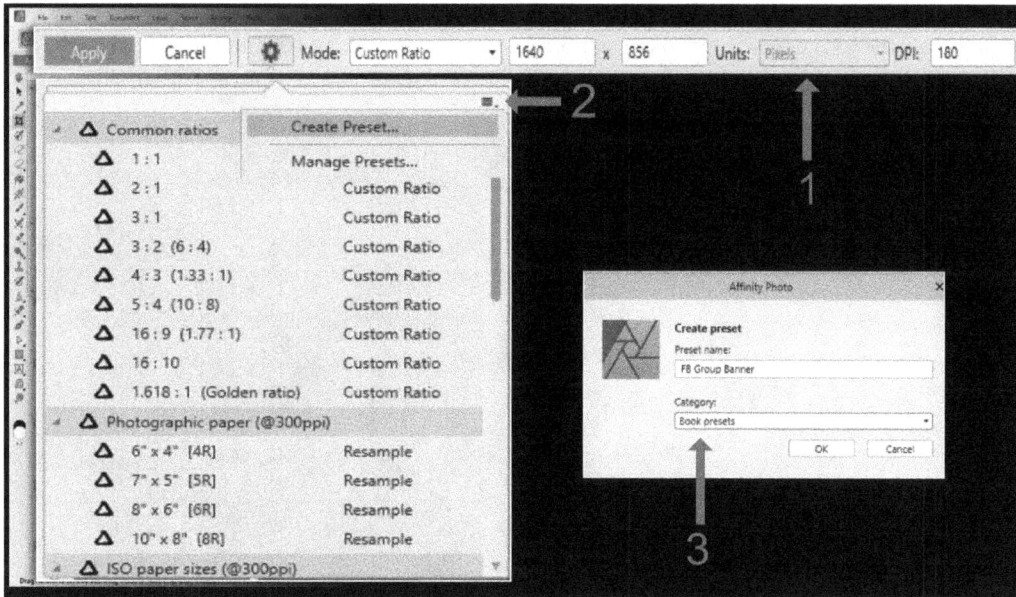

Figure 6.9 – Creating a preset, part 2

I am not even going to pretend that you will get the crop right each time, and as you edit, you may want to reposition, or in the case of a piece that you may show in a variety of formats, you may have to re-crop from time to time. When this is the case, we may have to *release* the crop and take another attempt at it or reposition it. So, let's take a look at how we do that.

## Releasing or resizing a crop

Now that we know how to crop, we have to realize that not every crop you do is going to be good and that you may have to do what is called *releasing the crop*. In short, releasing the crop allows you to re-position or just eliminate it entirely. Because Affinity Photo uses non-destructive methods to crop, your crop is still available to change. In the following subsections, we will see the two most common ways this is done.

## *Releasing the crop*

When you crop an image, it is important to realize that what you have done is *pin* the document to the canvas (you may remember these terms from the earlier chapters, this is why it was important to get those out of the way first prior to discussing the cropping function). So, the way to think about *removing the crop* is to *release the crop* from the canvas, thus allowing the entire image to be seen again.

In *Figure 6.10*, the image was cropped to a ratio, and we want to bring it back to its original size. To do this, follow these steps:

1. Go to the **Document** tab in the menu section.

2. Choose **Unclip Canvas**:

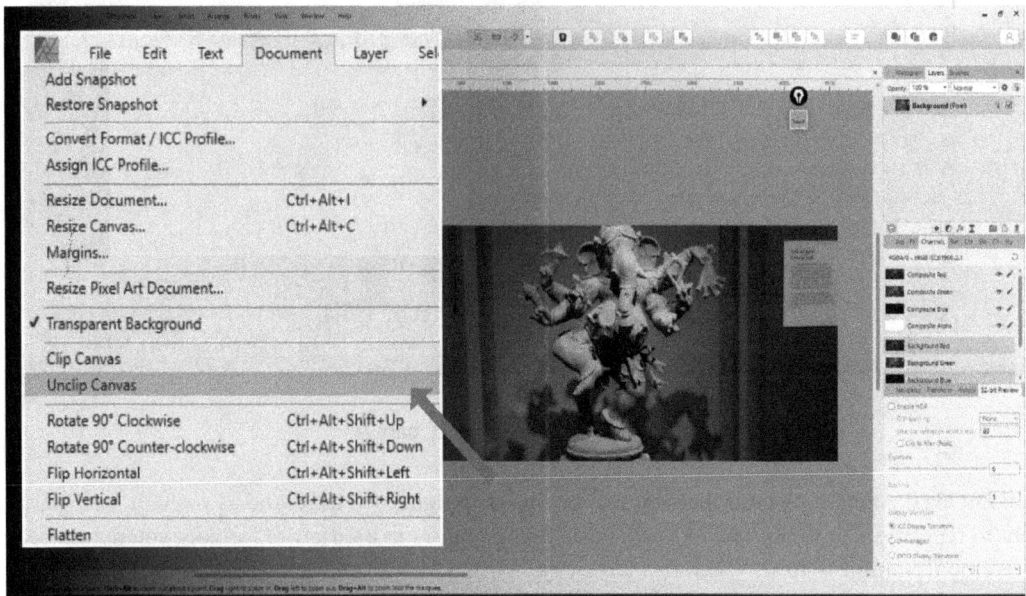

Figure 6.10 – Unclipping the canvas to release the crop

There are other methods, such as using the context toolbar to reveal the full image, so let's show you how to do that.

## *Hitting reveal to resize*

In the context bar, the Crop tool has an option called **Reveal**. This option will allow you to see the image borders that are cropped out, and then you can reposition the crop where you want it.

To bring it back to original ratio, simply drag the overlay back to the edges. This **Reveal** option is shown in *Figure 6.11*:

Figure 6.11 – Revealing the original ratio

In this section, we covered all of the mechanics you need in order to be successful, including the following:

- What cropping is
- How to set absolute and ratio crops
- How to resize and resample an image, including adjustment of the DPI
- How to release and reapply the crop
- Creating presets for your crops

The rest of the chapter is a lot more art and fun than tech and settings. We are moving Into some advanced overlays, and in the last part of the chapter, we will revisit the more theoretical aspects.

## Using cropping tools to adjust photos

Moving beyond the basics of cropping, there are various other tools inside the cropping menu and some more advanced overlays that can assist you as the editor in telling your story. In this section, we move past the mere mechanics into the optional tools to help you as an editor.

## Overlays

There are various overlays available in the cropping function to assist in getting the right crop, and you have to know the theory behind them to figure out which and when to use one. In the following subsections, we will cover the basic types and theory for the three overlay options.

> **Important note**
>
> Now, these are just rules and theory, so do not take them as musts. If these rules do not work for you, of course, do your own thing.

The **Overlay** options are in the context toolbar, as shown in *Figure 6.12*:

Figure 6.12 – The Overlay options available in cropping

Let's go through these options one by one.

### *Thirds Grid*

There is a rule in photographic composition called the *rule of thirds*, and the rule implies that the photographer should do their best to position the point of interest in the intersecting points of a grid if you were to divide the grid into thirds.

Affinity provides a thirds grid to assist in your composition to allow you to position the subject you want the viewer to see in the right places.

This thirds grid is shown in *Figure 6.13*, where we position the subject of interest in the thirds area and not the center. I wanted the reflection and the light source to be focal:

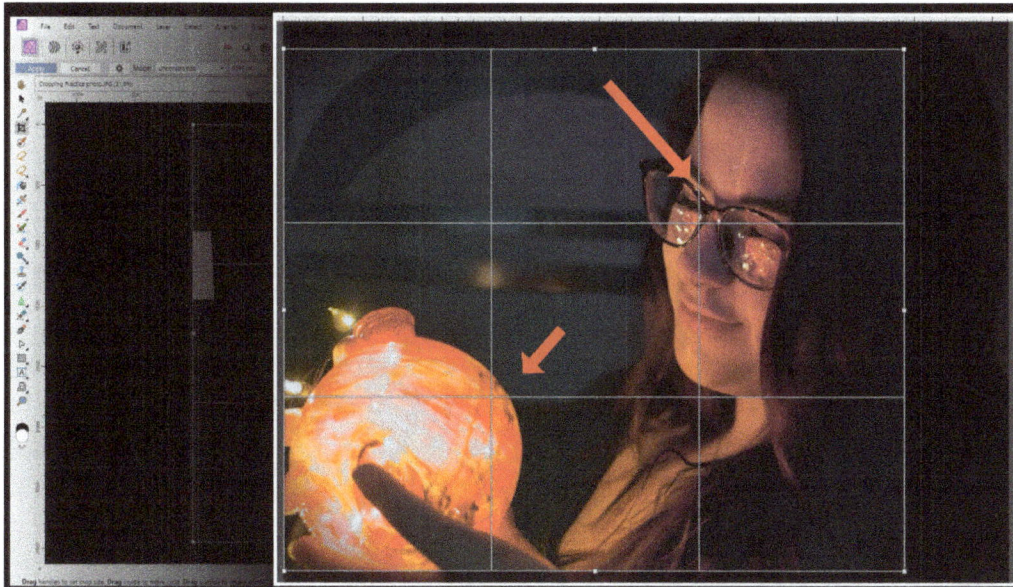

Figure 6.13 – Positioning the composition in the thirds grid

## Golden Spiral

There is a ratio in mathematics that is essentially repeated throughout the natural world, referred to commonly as the golden spiral. This ratio is created by identifying the pattern in math, as explained here:

- 1 + 2 = 3
- 3 + 2 = 5
- 5 + 3 = 8
- 8 + 5 = 13, and so on

It eventually stabilizes at around 1.61, and if we trace this out, it forms a spiral. This spiral pattern can be used to describe how the human eye works through an image, and good composition can lead the eye by taking advantage of this concept.

In *Figure 6.14*, we have switched **Overlay** to **Golden Spiral**, and you can see the path that the composition will lead the viewer on. Many studies have been conducted on how humans process images, and the spiral keeps coming up as a constant.

When editing, I use the techniques we cover later in the chapter (see the later part on composition technique) on composition to help lead the viewer on this journey:

Figure 6.14 – The Golden Spiral overlay

## Diagonal

You will want to use this when you ideally have a symmetrical design with strong lines. It really doesn't work well with organic shapes (I would never use this overlay for a bird in a tree). In *Figure 6.15*, I have added this overlay to an image that I shot in a market. Notice the strong structural elements that are symmetrical in nature. This is the best use of the **Intersections** overlay.

Figure 6.15 – Diagonal overlay

Notice in the crop that I have hit the intersection points and placed the image directly in the centerline, centering it inside the frame.

Now, let's move on to our next cropping tool: the straightening tool.

## The straightening tool

If you are anything like me, sometimes the excitement of taking the photo and being in the moment shows up during editing in the form of images that are not straight and need correction. The option to straighten an image is found in the Crop tool context menu (see *Figure 6.16*). I have included *Figure 6.16* in downloads for you to follow along:

Figure 6.16 – Pre and post straightening

In the preceding image, we want to straighten the horizon line level as it is a focal point of our image, and to do this, we will follow these steps:

1.  Select **Straighten**, and then click where we want to begin pulling a line.

2.  We then hold the click as we pull it out along the horizon line—this tells Affinity that *this is supposed to be straight*. Remember Affinity does not know that this is a horizon; it does not know this is the ocean; it simply sees it as a collection of pixels.

3.  At the end of the line, release the click. Affinity will adjust the photo to the straight line.

See the before and after screenshots in *Figure 6.16*. I have also included a CiA video for you to review the process (the link is mentioned in the *Technical requirements* section).

We can finish the chapter with a brief discussion on composition related to cropping, as there are aspects of theoretical composition that are directly impacted by your crop. So, let's review the more artistic side of cropping.

## The basics of composition

While laying out this book, I thought it was prudent at this point to introduce some of the artistic aspects of photo editing, as cropping is a major theme in composition. When we discuss composition, we are referring to, in this case, how the visual elements of the image are laid out (not about taking multiple images and combining them, which we will get to in the next chapter).

To get more ideas on cropping images and creating great compositions, you should start studying photographs of master photographers, studying the old master painters (illustrators), and just start studying the fundamentals of design compositions to improve image editing and creating.

How you crop and where you put the focal points in your composition drastically affects how your viewer travels through your image. Do not overlook the fact that they are taking a journey when evaluating the image. We have covered a few of the universally accepted principles for composition as we went through the cropping lessons, such as the following:

- **The rule of thirds**: The focal points of your image should be in the intersection points of your image
- **The golden spiral**: The viewer's eye travels outward at a known rate of expansion to absorb your image

I want to introduce you at this point to the hierarchical ranking of the three elements of an image. In any image, there are primary, secondary, and tertiary elements of the composition that are largely created by the cropping of the image and the level of importance that a subject has in an image.

In the hierarchical scale of importance, the following points should be taken into account:

- The primary shape or subject is the thing you want your audience to notice first; it is the dominant feature of the composition and the story
- The secondary feature is a smaller detail that usually tells a supporting story and provides context to the main subject
- The tertiary subject pulls the story together and fills in the holes in the image so that the viewer can see what it is

To illustrate this concept, let's explore an image. I performed two crops on the same raw image (as shown in *Figure 6.17*). Notice they are telling the story of two very different subjects and journeys:

Figure 6.17 – Before and after crop

What is the first thing you notice in each photo? These are the primary subjects:

- The image on the left: the path
- The image on the right: the trees

Simply by cropping, we have told the reader the dominant subjects in the composition of our image.

These are the secondary objects:

- In the left image, the rail along the left side of the image assists the viewer in getting to the focal point down the path
- In the right image, the large bush on the left-hand side of the image keeps the viewer in the image and implies *there is nothing past this point; stay in the scene*

Lastly, on the tertiary elements:

- In the left image: The arch of green bushes tells the viewer there is depth, there is intrigue, and there is more; the creation of the primary shape of the path assists this.
- On the right side: The background, I believe, plays a bigger role because the dominant tree is clearly screaming obstacle, foreground, and overshadowing anything else. I would most likely elect to accentuate the "behind" using lighting as a focal point.

Now for the last section of this chapter, we will cover some tips and tricks and some important points.

## Professional tips, tricks, and important points

For professional tips and things when it comes to cropping, Affinity Photo has taken care of most of them, from the overlays to the non-destructive nature of the crop function (releasing and revealing the image). If forced to give you my professional opinion over the years, the most common things we see people struggle with in their growth are the following:

- Not knowing the result they are after regarding print versus digital. Remember the DPI value is different if you are printing the image, and that image may look great on screen, and then the print is less impressive than expected. Be intentional in the desired output.
- Secondly, people fail to strategically plan their image with regard to their focal point and subject hierarchy, and then the image has less than stellar appeal or confusion on behalf of the viewer.

# Summary

In this chapter, we covered the first fundamental skill of composition and also gave you the first look at some rules around composition that will help you grow as an editor. In this second portion of the book (*Chapter 6* and beyond), you have all the core tools you need to perform really great edits (fundamentals such as masking, layer knowledge, and so on), and these middle chapters will be just as much art and theory as technical information.

Think of these middle chapters as being tools that will help *augment* the fundamentals and get your composition skills straight so that your composition is strong in the more advanced chapters, such as color and brush techniques.

In my years as a tattoo artist, it was drilled into me that the bones of a good tattoo are the outline, and while there were fundamental skills to tattooing, a poorly structured outline will never provide for long-standing work, even if the technical skill is good, the structure is weak, and these composition chapters and tools are your structure.

In the next lesson, we will look at the idea of workflows and explore actual compositions using the tools you already know. We will tackle things such as double exposures and placing objects in scenes and making them look seamless, so we are moving right down the road on the strong structural concepts in the second portion of this book.

# Basics of Workflows and Balancing Dark and Light

Welcome to the first workflow-based chapter in this book. Up until this point, we have been working on fundamental techniques that you needed to understand to get here (things such as masking and selection). Now, we will turn our attention to the idea of workflows – the quickest way to teach a technique is in the context of workflows. So, in this and other chapters dedicated to workflows, we have divided them into three topics:

- **Concepts related to workflows**: In this chapter, we will cover the three pillars of atmospheric perspective

- **Doing an edit using a workflow**: In this chapter, we will be doing a flower edit

- **Expanding on tools you could use in the edit**: In this chapter, we will cover multiple ways to adjust the lights and darks of your image

By structuring in this manner, you will learn repeatable proven workflows that get you moving quickly while increasing your tool knowledge by having an actual context through which to deploy it.

I will tell you that "the quickest way to kill motivation is to teach all the tools before letting the editors work on their projects." We will continue learning technical aspects later in this book, but right now, appreciate how much you have learned and how much you can accomplish. We have a saying in the office: "*You do not have to learn everything to make something.*"

By the end of this chapter, you will walk away with a repeatable workflow and a firm awareness of various tools we can use to adjust darks and lights inside of the workflow. Let's get started.

## Technical requirements

The CiA video for this chapter can be found at `https://packt.link/yZT7G`.

You will also need the `Chapter 07` project to follow along.

# What is a workflow?

I like to think about a workflow as a culinary activity. There may have been a certain dish your mother or father made and if they made it many times, hopefully, it tasted similar and consistent each time. However, down the street, your neighbor might make a similar dish and achieve an acceptable result but utilize a completely different recipe. This is how you need to think of workflows. A workflow is simply a pre-defined series of steps that you take to accomplish an output or an image.

Now, as I mentioned earlier in this book, the question of *what is the best workflow?* always comes up, and there is no right answer because there are endless ways of working in Affinity Photo. I guarantee you that five editors have developed five different ways to do the same sort of edit… remember, it is the result that counts.

As an example, when I do a simple edit that does not require multiple subjects, I create an overview of the workflow. This is a great one for beginners because whether it is a wedding photo, a travel photo, or a family photo, this workflow requires no selection and no cutting and pasting, and simply allows you to crop and mask out to get the right composition.

## Cropping the image to get the focal points right!

I usually figure out the hierarchy of the image using the priority method that we learned in the cropping chapter (*Chapter 6*). I get the primary, secondary, and tertiary elements to the right places in the intersection points and then step back.

> **Pro tip**
> Zoom out when you think you have it right and see how it looks from across the room and ask yourself questions such as *What catches your eye? Is what you wanted to be the focus actually the focus?*

## Setting the levels (using curves, levels, or Exposure Adjustment layers) for the entire image

If I am looking for a consistent level of difference and dynamics in my image, I will ask myself these questions: *Are the highlights different than the shadows, Are the "darks" dark enough, Are the "brights" bright enough?* As mentioned previously, I will zoom out and ask myself, *Can I make out the forms in the image with this level of dynamics?*

Now, it is worth noting that my work is known for being very bold – I do high-contrast edits and sometimes push too far. If that is not your style, the workflow would still apply; just make sure you tone it down a bit.

## Applying a sharpen and blur adjustment layer to the image and masking out the areas that you do not want it to apply to

I practice non-destructive workflows whenever possible so that I can change my mind whenever I want, and while there is a Blur and Sharpen brush, why would you destroy the image and not be able to walk back on my decision? The more appropriate and professional way to handle this is to apply an Adjustment layer, and then mask out the areas you do not want it to apply to.

Remember, masks can always be walked back and undone using a white color, so I choose not to use a brush, but rather always an Adjustment layer or Live Filter layer that I can readjust and turn off and on.

## Apply the color grading you want to your image

I work on my color grading after all of the structural elements are in place, such as the focal points, the levels, and the saturations. We will discuss colors in *Chapter 13*, so I just want to introduce it here as a part of my workflow, and not cover it in depth here.

## Balance and finishing

Lastly, I balance the piece to make sure that, at the end of the day, the subject I want you to see is shown, and that the image conveys the intent I had when I set out to complete it. Later in this book, in more advanced chapters, we will discuss finishing, but I wanted to mention it in my workflow here to give you an idea of where it fits.

Now that you have a workflow you can use, let's take a look at the theoretical concepts we can deploy within that workflow. These are known as the three pillars of atmospheric perspective.

# Intermediate concepts in composition – the three pillars of atmospheric perspective

In the cropping chapter (*Chapter 6*), we covered the concept of composition principles as it relates to placing elements in the image. In this chapter, I will give you another set of universally accepted principles for composition that are utilized when editing after the crop. We refer to this concept as the three pillars of atmospheric perspective, and in short, it is a list of things that help draw our viewer's eye and make logical sense of your image in the viewer's eye.

## Pillar 1 – objects in the foreground are brighter

This means that if you have a composition, the object closer to your viewer usually has a higher exposure or is brighter. Now, do not mistake this for where the light comes from. The light will still strike in natural places, but during edits, we want the background to not be as bright as the foreground overall.

## Pillar 2 – objects in the foreground are more heavily saturated

Saturation, in this case, refers to the colors closer to the viewer being more saturated – a little bit *bolder*. Toward the background of an image, the saturation is *less* saturated. Notice that we said less saturated, not black and white – there will still be color, it will just not be as *poppy* as the foreground.

## Pillar 3 – objects in the foreground have more details

Detail means texture, and those pieces closer to the viewer have more detail. Think of a person 100 yards away. At that distance, can you see the texture on the cording of their jacket from that distance? Of course, not – but if they approached you, you could see it 3 feet away. This is the concept.

A solid edit uses these pillars to make a good flow in the composition of the photograph and adds to the discussion from the cropping chapters, where we discussed cropping for composition. A good image is planned and then carried out strategically through the idea of the workflow.

In the next section, we will be applying these three pillars to our workflow. Let's get started.

# Applying the pillars

As a practical example, let's look at *Figure 7.1* and see if we can use the three pillars mentioned in the previous section.

This untouched image is available as a download for this chapter so that you can follow along (the name of the image is `Atmospheric Perspective image.JPG`; go ahead and download it to get started with the steps). We will only use four steps in this edit to illustrate the concepts, but notice how we use the concept of masking using a gradient from previous chapters.

The flow is simple. *Figure 7.1* shows the completed image side by side. (Can you spot the three pillars in the application?) Notice the increased sharpness of the petals closest to the foreground:

Figure 7.1 – The three pillars applied to an actual edit

Let's get started with the steps, where I will tell you how I achieved the result shown in *Figure 7.1*.

## A practical edit walkthrough for the workflow

*Figure 7.2* shows my settings and the **Layer** panel for the completed image. A CiA video for the process has also been included (see the *Technical requirements* section for the link). Let's look at the detailed steps:

- Step 1 – working out the brightness:

  I.    Add an **Exposure Adjustment** layer to the image.

  II.   Reduce the **Exposure** property, making the image darker.

  III.  Use a gradient to mask out the area to the left; you do not want that to be darker.

- Step 2 – adjusting the saturation:

  I.    Add an **HSL Shift Adjustment** layer for saturation.

  II.   Reduce the saturation of the image.

  III.  Use a gradient to mask out the area to the left; you do not want that to be darker.

- Step 3 – increasing the details in the foreground:

  I.    Add a **Live Filter** layer for the **Unsharp** mask to the image.

  II.   Add an **Unsharp** mask to your liking.

  III.  Mask out the area on the right this time. The area furthest from the camera should not be as sharp.

- Step 4 – adjusting the masks

  You may want to use a soft basic brush and bring back certain leaves that cross over the gradients to finish the edit. So, the last step is to go over the petals that you want to either change or highlight to make them fade back or stand out.

  This can be done by painting white on the **Adjustment** layer to remove the mask you added:

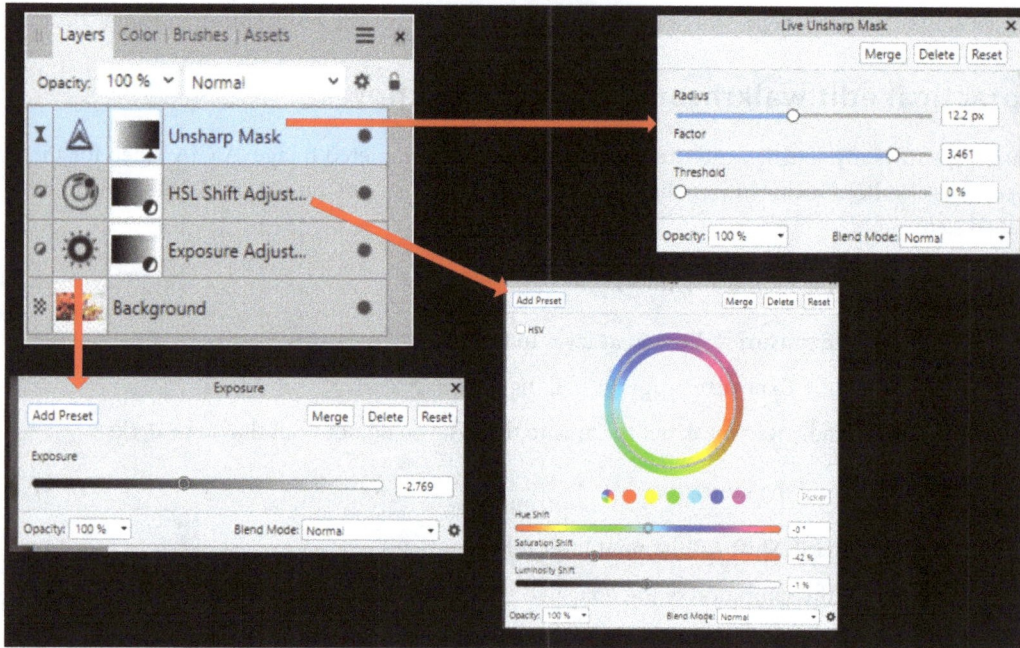

Figure 7.2 – Adjustments for the three pillars

For the remainder of this lesson, I would like to tell you about my favorite tools for some of the preceding steps. Again, this book will not teach you *all* the tools, but if you have a few for each task, you will be well-equipped to get up and running in the quickest time possible. So, let's get started and look at adjusting the lights and darks of the image.

## Tools to set exposure (levels of light and dark)

Brightness and darkness (what I am referring to when I say exposure) are part of all my workflows. In the workflow that I presented earlier in this chapter, adjusting these values was the second step. We do this early since the brightness of certain areas of the image will affect the colors, focus, and so on. While there are an infinite number of ways to adjust this, I typically use three tools over and over again – they are all **Adjustment** layers, and they are as follows:

- **Exposure Adjustment**
- **Levels Adjustment**
- **Curves**

These are ranked in terms of complexity, with the top one (**Exposure Adjustment**) being the easiest, and **Curves** being the hardest. So, if you are new to editing, please start with the easiest to keep the momentum going.

To minimize variation, we will be working with the same file for all the different adjustment types (the cat on the wall that's present in *Figure 7.3*). this way, you can see the same image with various treatments. Note that the adjustment layer's effect is not influenced by variations in the underlying image.

For your use, I have included the working file for each method, complete with the adjustments we've already made, so that you can toggle them on and off to see their effects (see the adjustment files in `Chapter 7` project folder).

## Exposure Adjustment

In my opinion, the simplest, most direct tool to adjust the brightness of your image is **Exposure Adjustment**, which is located in the adjustment portion of the **Layers** panel.

When you open the **Exposure** panel, you will be greeted with a slider that allows you to adjust the level of brightness:

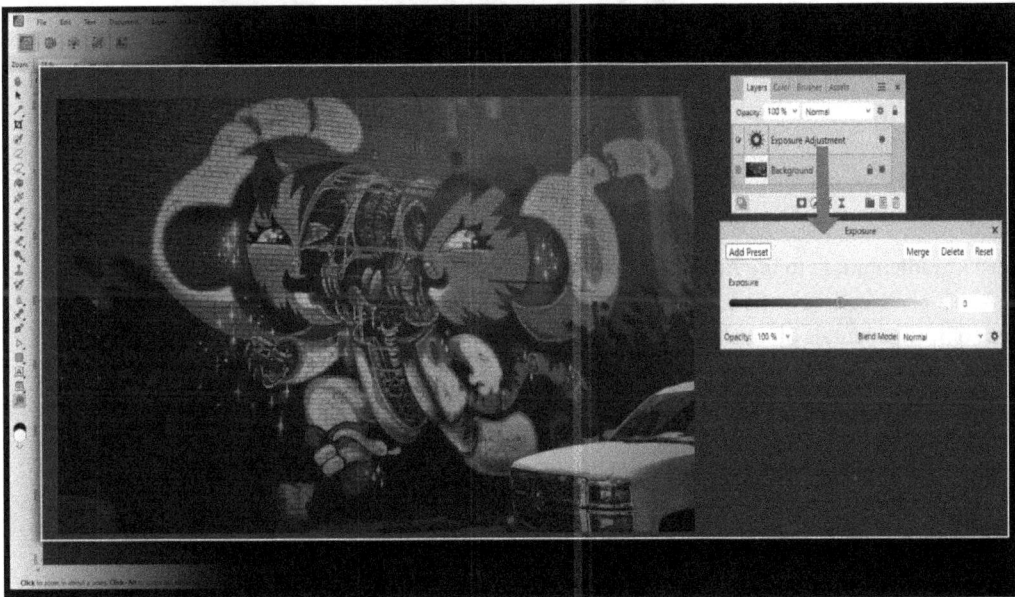

Figure 7.3 – Exposure Adjustment applied to the image

Now, after adjusting the image to the brighter setting (in this case, 1 . 4 or so), notice the dark portion of the *before* image in *Figure 7.4*. The dark portion is greatly reduced and more detail is brought out after using the **Exposure Adjustment** layer:

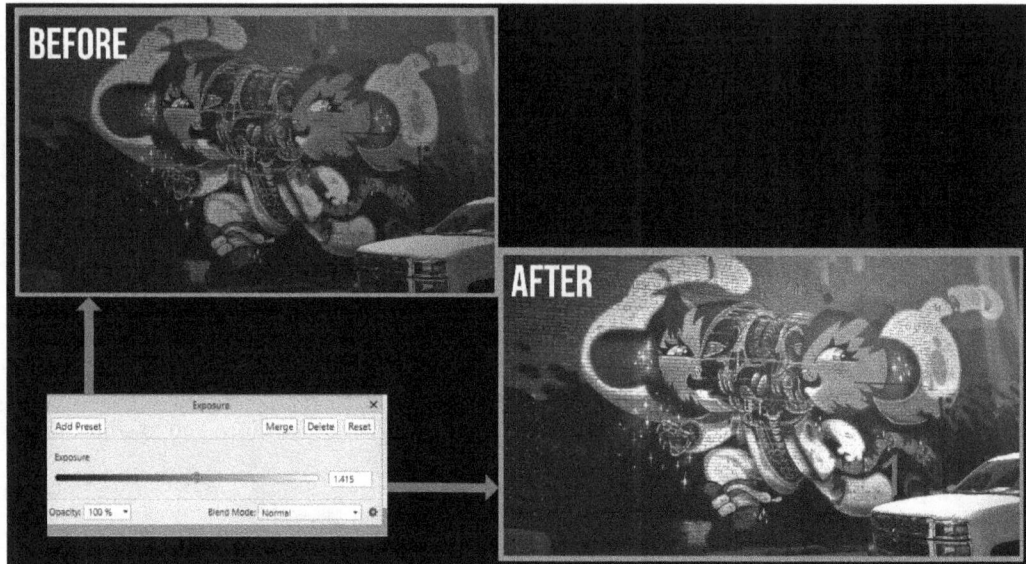

Figure 7.4 – Before and after Exposure Adjustment

Take care not to blow out the whites when adjusting the darks – this is a limitation of the exposure method for adjusting brightness. The lights and darks are pushed uniformly and cannot be adjusted separately.

I added a gradient mask to the Adjustment layer to balance it out. This way, the **Exposure Adjustment** layer does not affect the right-hand side of the image, and only adjusts the darkness on the left. (Notice the use of fundamentals of masking here, which we learned about in the previous chapters.)

## Levels Adjustment

The second form of Exposure Adjustment is the adjustment of levels. Using **Levels Adjustment** allows you to adjust the lights and darks separately and adjust the overall gamma of the image (gamma is a complicated term that simply means the brightness of the image and is a macro concept).

The **Levels Adjustment** layer is located in the **Layers** panel, and when you click on that layer, you are greeted with several options:

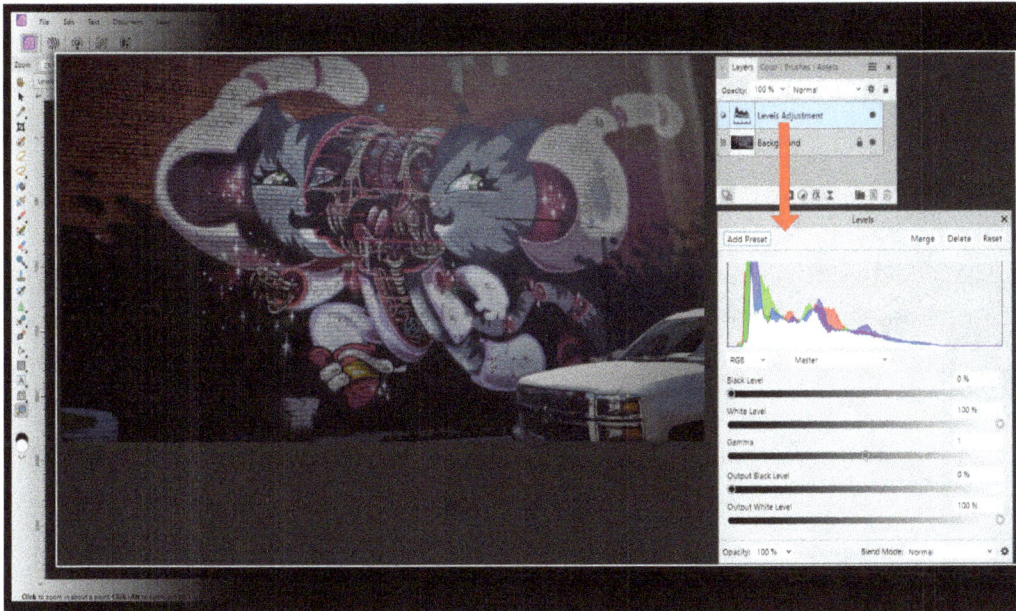

Figure 7.5 – Levels Adjustment applied to the image

Let's go over these options:

- **Black Level** and **White Level**: Here, you can increase the amount of black or white by pushing the sliders inward. This will add some more contrast to the image and help clear up areas that are over or under-exposed.

- **Gamma**: This refers to the overall brightness or darkness of the image.

- **The Output sliders**: These will affect how much overall output for light or dark there is in the blacks and whites. So, as you move inward, the amount reduces. The *before* and *after* image adjustment is shown in *Figure 7.6*. We have included this working file in the downloads for this chapter (see `Levels adjustment image`):

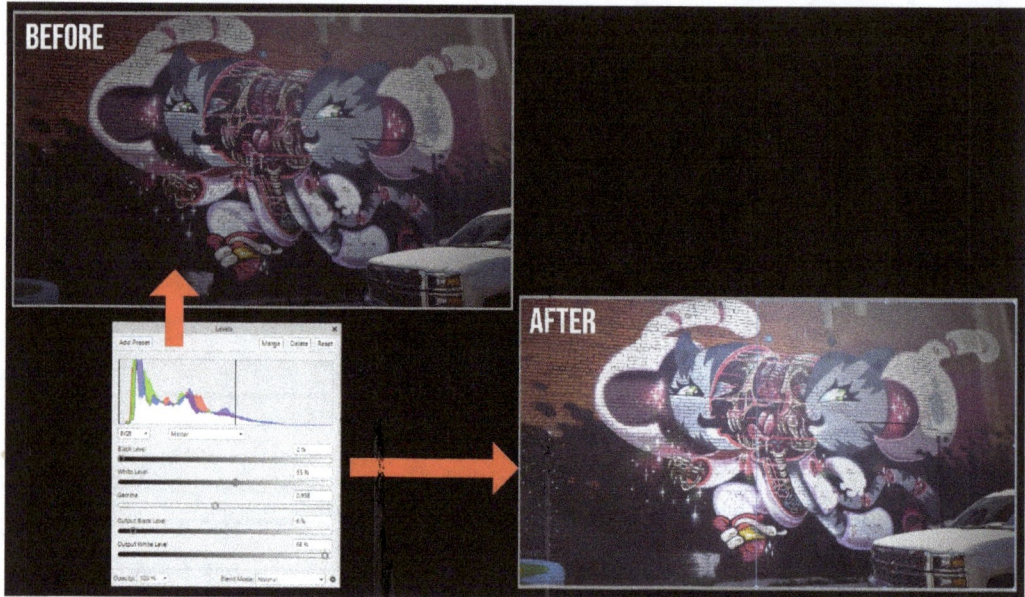

Figure 7.6 – Before and after Levels Adjustment

In the working file and image, I applied a gradient mask to the **Levels Adjustment** layer to even out the image.

The idea of **Levels Adjustment** is what I call *the lazy man's curve adjustment*. While there are more adjustment options, they certainly do not allow for the level of control present in a curve adjustment, which we'll look at next.

## Curves

In terms of light and dark adjustment, probably the most complicated but highly customizable and targetable is the idea of the curve adjustment. Curve adjustment allows you to pick out any range of dark and light and then add contrast and adjustment to only those targeted areas.

For this example, we will use *Figure 7.7*, where I have labeled the adjustment layer for curves and highlighted the following areas.

The far left of the *X*-axis shows the 100% black values, while the far right shows the 100% white values. Notice that there are not any pure bright 100% white values, but there are some *very* dark black values in the lower left corner:

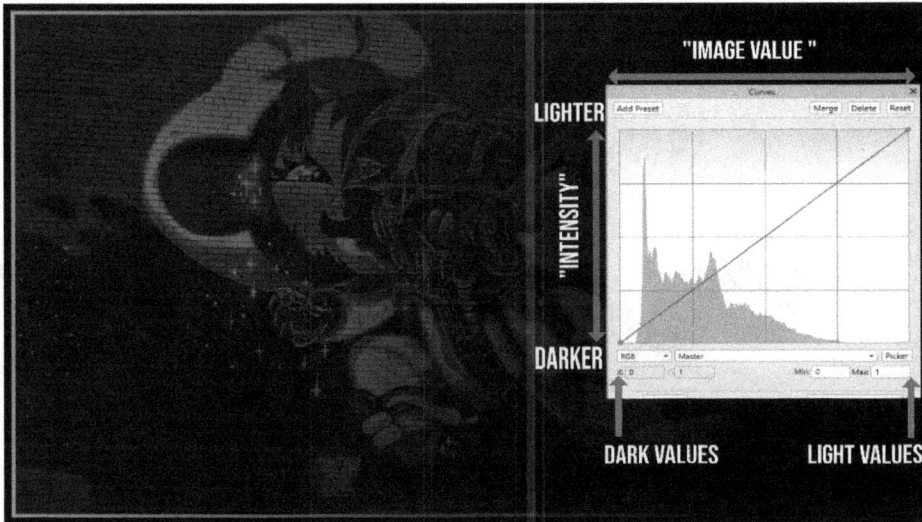

Figure 7.7 – Curves adjustment applied to the image

The way you need to look at this adjustment is best done through the following phrase: *I want to adjust the ___ values to be (brighter) or (darker)*. An example of this may be: *I want to adjust the really dark values up a bit and raise the midtone so that it's brighter.* While saying this, we know we need to raise the areas of the curve for the dark blacks, and then we need to add a node in the center of the image that only affects the highlights. See *Figure 7.8* to see the adjustment for this edit:

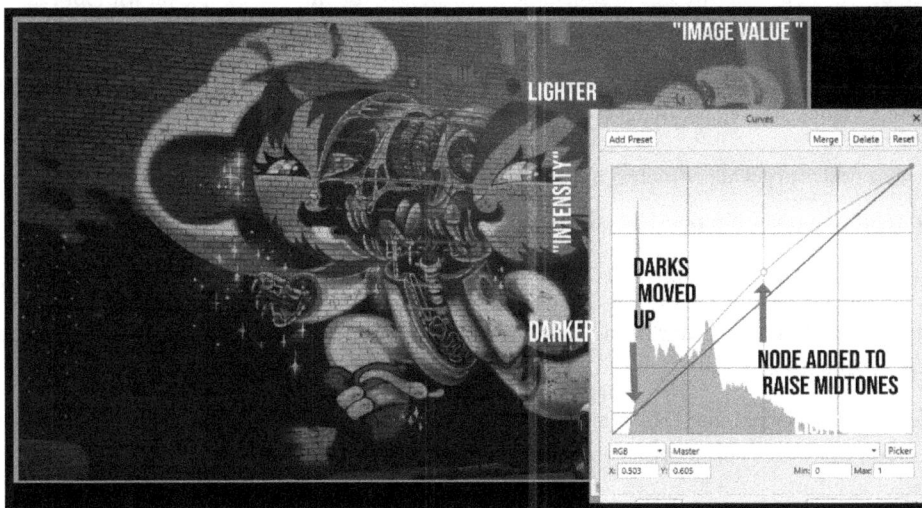

Figure 7.8 – Adding a node to a curve

Here is the result of the adjustments:

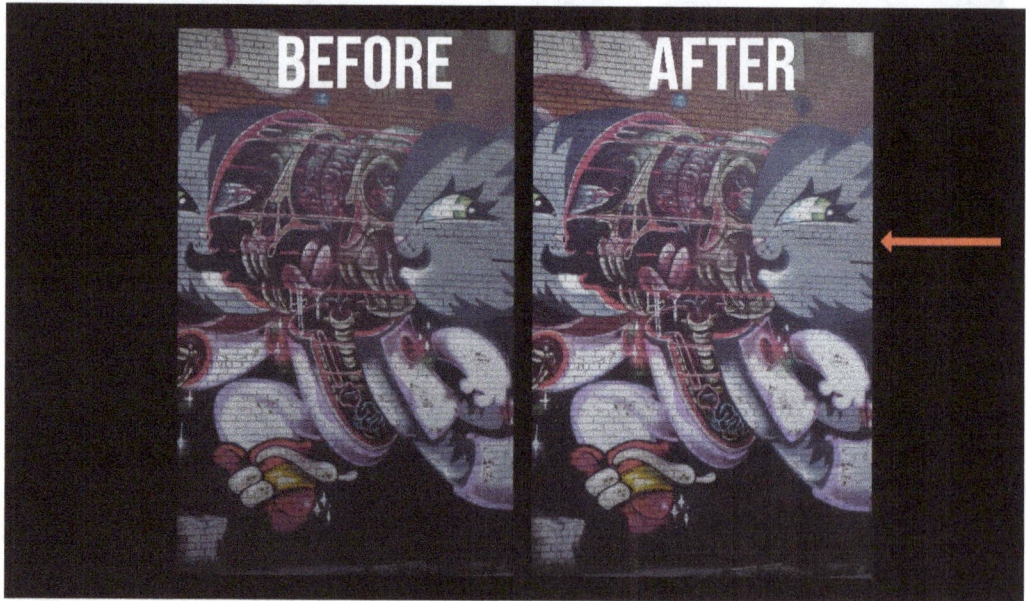

Figure 7.9 – Before and after the curve adjustment

In *Figure 7.9*, the midpoint adjustment can be seen most effectively on the gray of the cat's face. This is a subtle adjustment but it allowed us to adjust the brightness of the mid-tones up a bit without making the entire image brighter and blowing out the whites on the right-hand side.

To add or subtract nodes during curves adjustment, and target certain exposure values (shown in *Figure 7.8*), we will follow these steps:

1. To add nodes to get adjustments that are targeted, simply click on the line; a node will be created.

2. To remove a node, simply select the node and hit *Delete* on your keyboard.

I have included a CiA video to show this approach (see the *Technical requirements* section for the link).

In the video, we adjusted the darks and lights, added nodes, and then subtracted them using the preceding techniques.

Now, let's look at a very popular tool to aid in setting exposure. While it is true that you should always trust your eye, there are tools to assist you along the way.

# Histogram – an aid in setting exposure

The last helper I want to share in this section is not an Adjustment layer, but rather a helper to assist you in setting a correct exposure. This can be found in the **Studio** panel and it is called the histogram. It is used to assist the editor in setting better exposure:

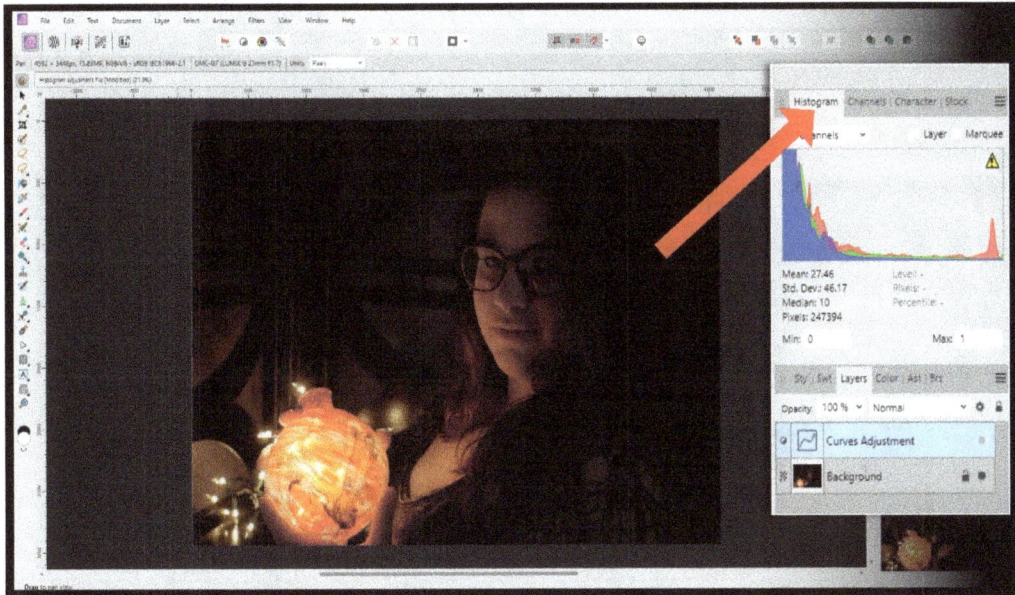

Figure 7.10 – Histogram example

A histogram is a graphical representation of the darks and lights of an image and gives you an indicator of how dark your image may be and how balanced or unbalanced your image may be in terms of light and dark. It also shows the individual color channels, so when we get to colors, this topic will come up again for color adjustment.

> **Important note**
> Never go off the histogram alone, go off your eye. The histogram is there only to help you adjust exposure. It is up to you as the editor to have the final say.

## Reading the histogram

I have included an image (*Figure 7.11*) that is significantly under-exposed (meaning it is too dark). Using the workflow I introduced in this chapter, the first thing we have to do is crop it, immediately followed by adjusting the lights and darks.

To read the histogram, the horizontal axis runs from the blackest blacks at the far left of the histogram to the whitest whites at the far right. The distributions of the blue, red, and green channels are shown in the histogram. Take a look at the following figure:

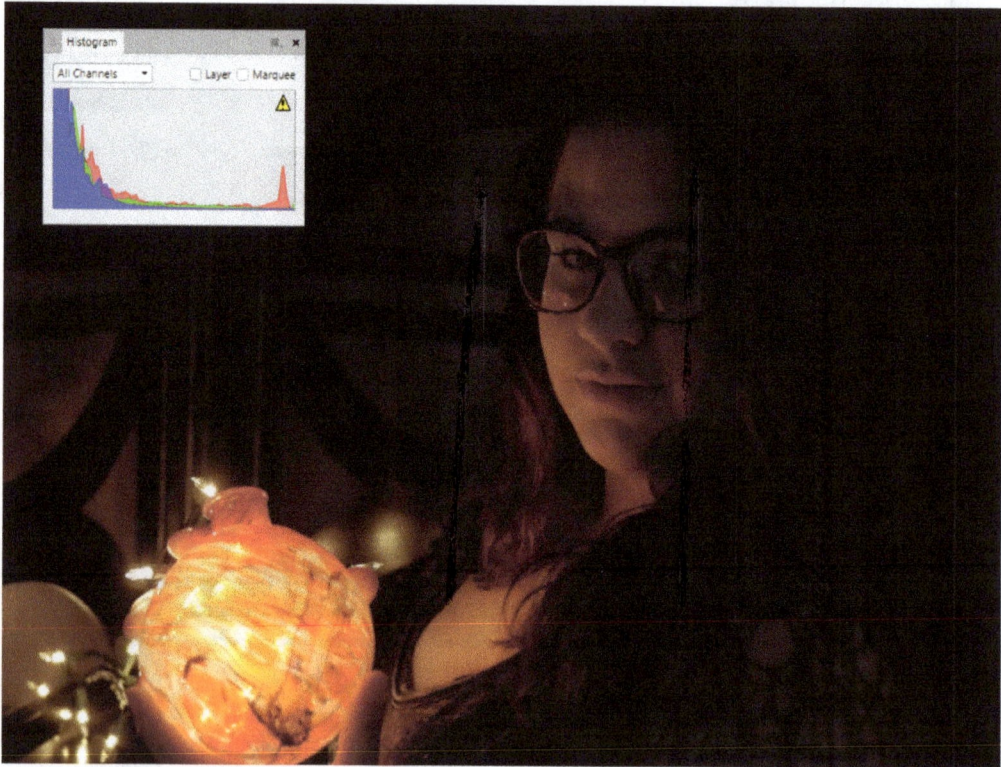

Figure 7.11 – Close up Histogram

Notice the following things in the preceding figure:

- All the colors are skewed to the far left of the histogram, which means that, overall, the image is dark

- The red channel is the only color on the lighter side of the image (this makes sense as the light source has red in it)

Now, let's edit this image using a histogram.

## Editing an image using a histogram

In *Figure 7.12*, I added a **Curves** adjustment and included the *before* and *after* views of the histogram. Notice the following things:

- The dot in the lower right of the **Curves** Adjustment layer was pushed to the right, adding more darks to some of the lighter darks.

- The mid-tones were made lighter by adding the mid-point in the **Curves** Adjustment layer. This made more of the reds lighter.

- Notice the peak in the histogram in the far-right corner as more of the reds got lighter.

In short, we smoothed out the areas from dark to light and created more dynamics in the image because there are fewer values overall:

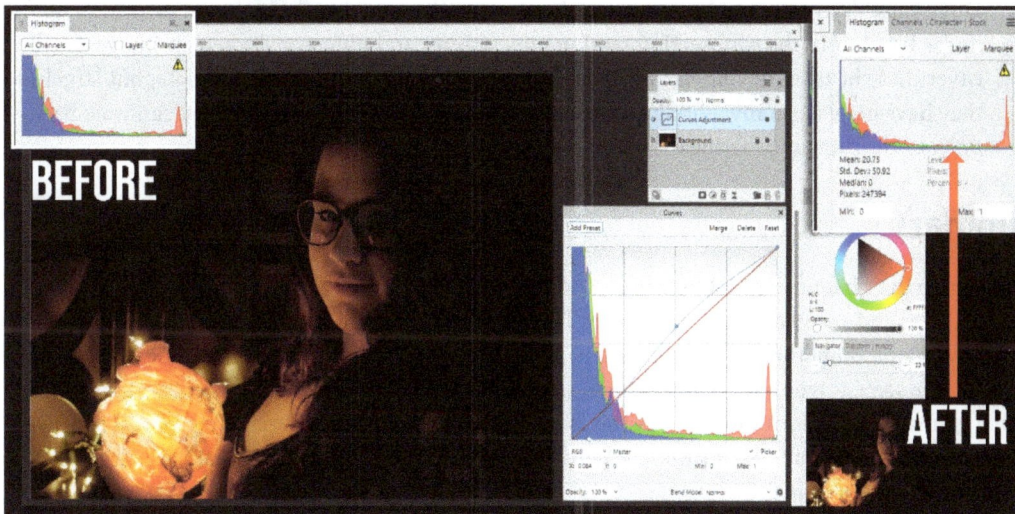

Figure 7.12 – Histogram before and after

I have included this file in the downloads for this chapter (see `Histogram adjustment file`). It is recommended that you take this file and make some curve adjustments to see how it affects the histogram.

## Professional tips, tricks, and important points

As a working professional, I wanted to provide some tips and tricks to help you with the concept of Exposure Adjustment and workflows as we wrap up. I've done this because so many new people struggle with this baseline technique and then their edit looks less than they want it to be:

- Do not blow out the whites. Over-exposure is one of the most common things I see when an editor does not apply exposure correctly. When you blow out the whites, the program has no data in those areas to go off, and you will be forced to heal the image before you can continue with exposure. If the image is unevenly lit, then use the Adjustment layer and mask out the areas where the adjustment will cause the image to become over-exposed.

- Try and use curves for light and dark whenever possible. Sometimes, you can get away with a simple Exposure Adjustment, but most of the time, it is beneficial to use curves.

- Do not use destructive techniques on a simple edit. There are brushes referred to as *burn and dodge* brushes in the tool section, and in effect, they can lighten or darken an image. I will cover these brushes in *Chapter 14, Destructive Filters and Tools in Affinity Photo*, but largely, they have no place in my workflow. Always work non-destructively so that you can walk back on your decision and readjust.

## Summary

Wrapping up this chapter, we first looked at workflows. As you mature as an editor, you will begin to develop a style and a way of working. Coming back to the analogy of cooking, not every recipe you try will work out. Sometimes, you will ruin it, and sometimes, you will add something that didn't work out. The same is true with photo editing. I will often try a new way of adjustment and the effect will not go my way. So, don't lose heart… this is why we work non-destructively. To give you one last piece of advice I would say "*Fail forward fast*," by which I mean to try it and see what happens… there is always the undo option.

In this chapter, you created your first workflow, and we covered the three pillars of atmospheric perspective: exposure (how light or dark an image is), saturation (how aggressively the colors are displayed in the image), and focus (varying amounts of detail closer to the viewer).

We also covered the three most common ways to set exposure from a non-destructive perspective.

In the next chapter, we will be tackling the idea of blend modes, so we will be going back to the deeply technical chapters. I hope you enjoyed the break – the next topic is a complicated one.

# 8
# Blend Mode Fundamentals

In this chapter, we will cover one of the most confusing and misunderstood concepts in digital art, the idea of blend modes. Blend modes are present in every type of digital art: photo editing, vector illustration, and even programs such as video production and 3D creation. While this topic can be daunting, it really does not have to be if you understand what the blend mode is trying to tell you and you let go of some of your expectations. To explain this concept, I want to frame the conversation in the right context by giving you a few fundamental rules to make sure you approach this topic correctly.

But for those that want to effectively and quickly edit, I offer up the following three principles:

- You will never know every blend mode, and you will develop a few that you know and love for your work. No one cares whether you know what each one does; they care what you make with them.

- There is no perfect blend mode for every application. When we add texture to an image, we can use overlay, add, or darken…there is no perfect one for every application. What I will give you are some of my favorites for very common applications.

- Every time you use blend modes, it is beneficial to play *blend mode roulette* (which means scrolling through them all just to make sure you do not miss out and stumble on something unexpected), because while you know what some of the blend modes do, subtle differences in every piece can make a known mode run very differently, and you might even miss out on some really cool effects and deprive yourself of growth.

In this chapter, we will cover the following topics:

- What are blend modes and where are they used?
- Using blend modes in practical editing
- Blend modes in Adjustment and Live Filter layers
- Blend modes versus blend ranges
- Professional tips, tricks, and important points

# Technical requirements

The CiA video for this chapter can be found at `https://packt.link/X384d`.

You will also need the `Chapter 08` project to follow along with me.

# Blend modes and their uses

The idea of blend modes comes after your understanding that there are layers in digital art (information covered in the first fundamental chapters). Blend modes are meant to be used to experiment in an artistic, fun, creative way. Relax and explore the various color effects. Most users only use about five common blend modes (**Multiply**, **Overlay**, and **Screen**, to name a few), so once you realize that layers are like sheets of paper, we can begin to understand what a blend mode does, and this is why the book is structured the way it is. To explore blend modes, you have to understand the following:

- Layers
- Luminance

I find it beneficial to understand what a blend mode is trying to do by using the following phrase:

*The blend mode of a layer tells the layer how to interact with the layer below it.*

When you set the blend mode, you are telling the layer what to show and what not to show underneath the layer. For 99% of blend modes, this is done using the **luminance** or **lightness** of the image (there are weird ones such as **Hue**, which use a color value).

To illustrate this, we will be using a very simple example that takes away the complexity, and then we will slowly build it back in using photos later. We have included this working file in the downloads for this section so you can follow along, and it is highly recommended you follow along as it takes a lot of practice to wrap your mind around the concept of blend modes (see the `Blend mode basic` file). Take a look at the following figure:

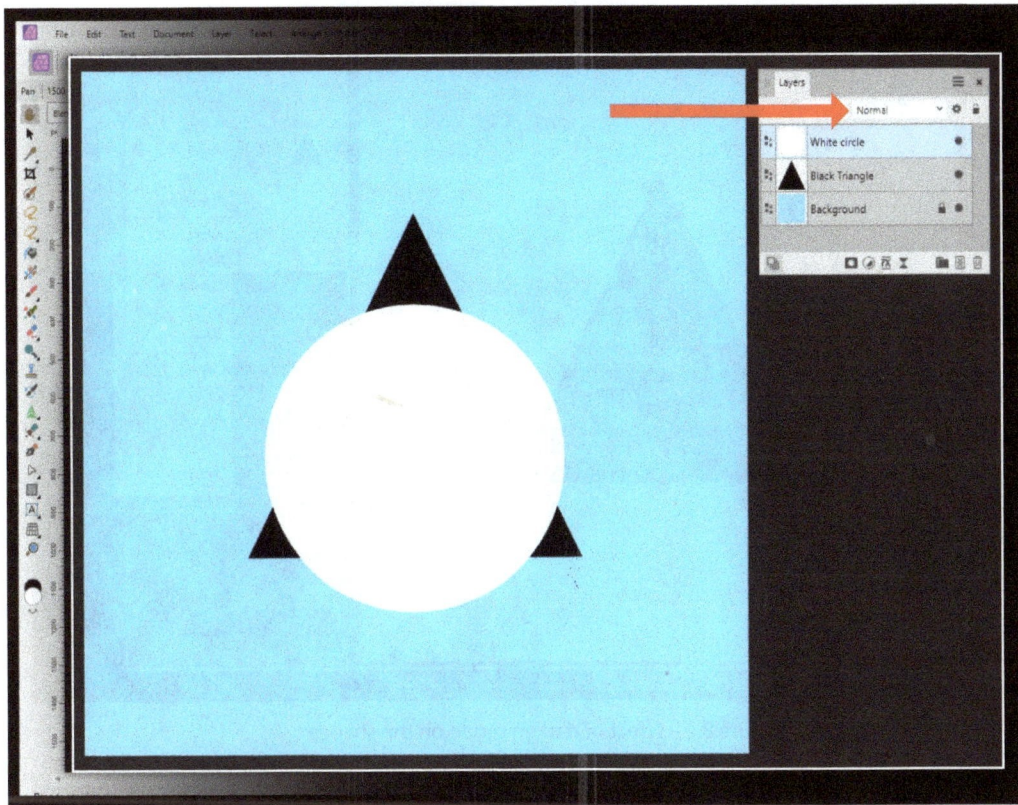

Figure 8.1 – Location and name of blend mode for the layer

In the preceding figure, we have two basic shapes: a black triangle and a white circle. Notice they are on an annoying blue background. This is to illustrate what is showing through and what is not. You will notice the following things in *Figure 8.1*:

- Notice the blend mode is set to **Normal** and it is set in the **Layers** panel

- Notice the white circle is on top of the black triangle (order is important when discussing blend modes)

- Using the definition of *blend mode* that we gave at the beginning of the chapter and by inserting the shapes, we can say that the blend mode of the white circle will tell the white circle how to interact with the layer below it

Now, if we click on the **Normal** option, we will see a drop-down list of options (and as previously mentioned, you will never know every blend mode). Let's click on **Darken**. This officially changes the blend mode of the layer:

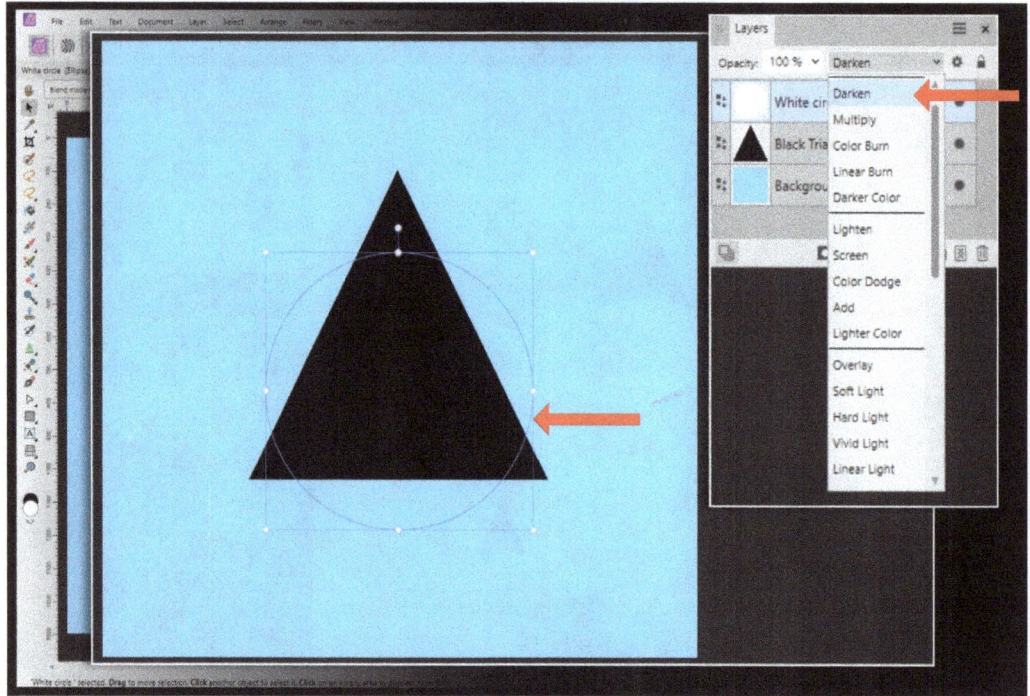

Figure 8.2 – Effect of darken mode on the shapes

Notice that when the blend mode was changed to **Darken**, the entire circle disappeared…why? When the **Darken** mode is selected, the blend mode says that *if the layers under the white circle are darker, let them show through*. So, was the black triangle darker? Yes. Was the light blue square darker? The answer is also yes.

Now, if we do the opposite and select **Lighten**, let's see what happens:

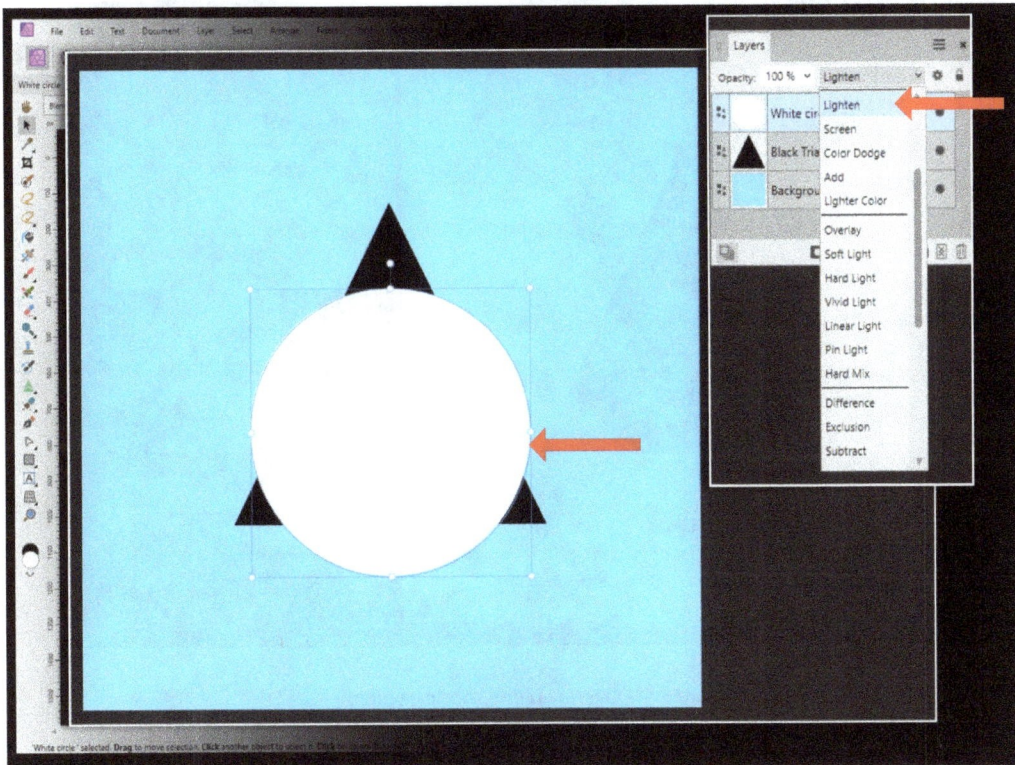

Figure 8.3 – Lighten blend mode applied

Notice how the entire circle is there, so what happened? By selecting the **Lighten** blend mode, we told the white circle layer to *show the circle layer if the luminance of the circle is lighter than the layers below it*, and in this case, the circle was brighter than the layers underneath, so the circle stays showing.

Now, this seems simple and not all that useful, so let's look at a very common blend mode called **Overlay**. In photo editing, a common application of this blend mode is applying textures, which we will do later in this chapter, so let's use the example and select **Overlay**:

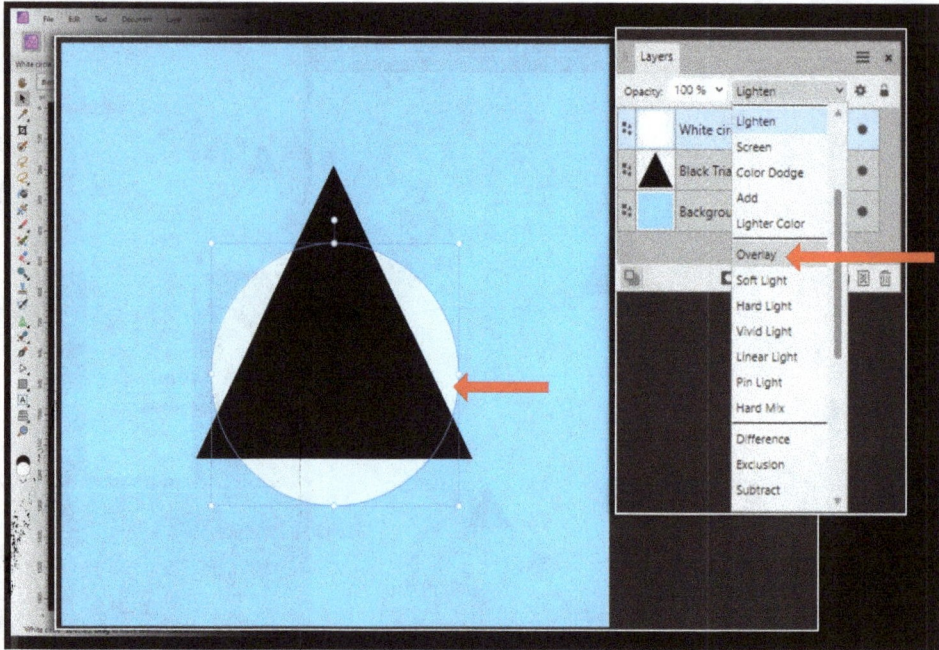

Figure 8.4 – Overlay blend mode applied

So, what happened? Well, the magic is in the math.

In the **Screen** example from *Figure 8.3*, the white circle looked at the luminance values and compared them to the black triangle, and made a decision on what to show. In *Figure 8.4*, we used **Overlay** and compared the values for the white against another value, and made a decision on what to show and how to show the layer.

The good news is we can treat them as artistic expressions and you do not need to know how the match works under the hood. After all, you do not need to know how your phone circuitry works to make a call, so do not overcomplicate this either. The magic is in the conversation of the phone call, and when it comes to blend modes, the magic is in what you make out of them.

## The role of colors in blend modes

We have described blend modes in terms of lightness and darkness so far, but the question of what role color plays will come up, and as a general rule, it does not play a role in blend modes, but the luminance of each color does.

Remember from the earlier chapters, we said color has three variables:

- Hue
- Saturation
- Lightness

If we open up the **Color** studio tab, we can explore this concept. We have included the working file for you to follow along (see the `Blend mode color example` file in the downloads for this chapter). Let's get started with the steps:

1.  Open the file and find the **Color** studio panel:

    •   Go to the **Window** menu and click on the **Color** panel if it is not already in the studio.

    •   Using the three lines (hamburger menu), choose **Wheel**.

2.  Click on the green circle using the Move tool.

    Notice that there are three values in the lower-left corner of the panel (**H**, **S**, and **L**) and they each have a value. We will be focusing on the luminance value (**L**). This is the lightness on a scale of 0 (being completely dark) to 100 (being completely light), so this green color is precisely at the 50% luminance mark:

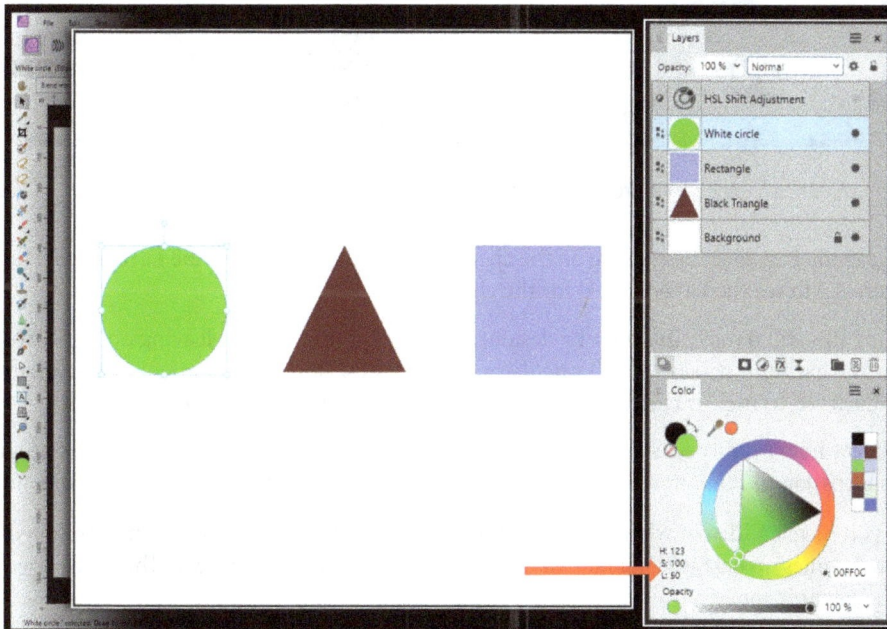

Figure 8.5 – Setting the colors for blend mode examples

3.  Repeat the same thing by clicking on the dark red triangle and the light blue square.

    Notice each of these has an **L** value. The dark red is **30** and the light blue is **77**. What you will want to take away from this is that each color has a luminance value with limited exceptions (the **Hue** blend mode, for example). The blend modes simply use the **L** value to determine which value is darker and which is lighter…just like it did in the black and white example earlier.

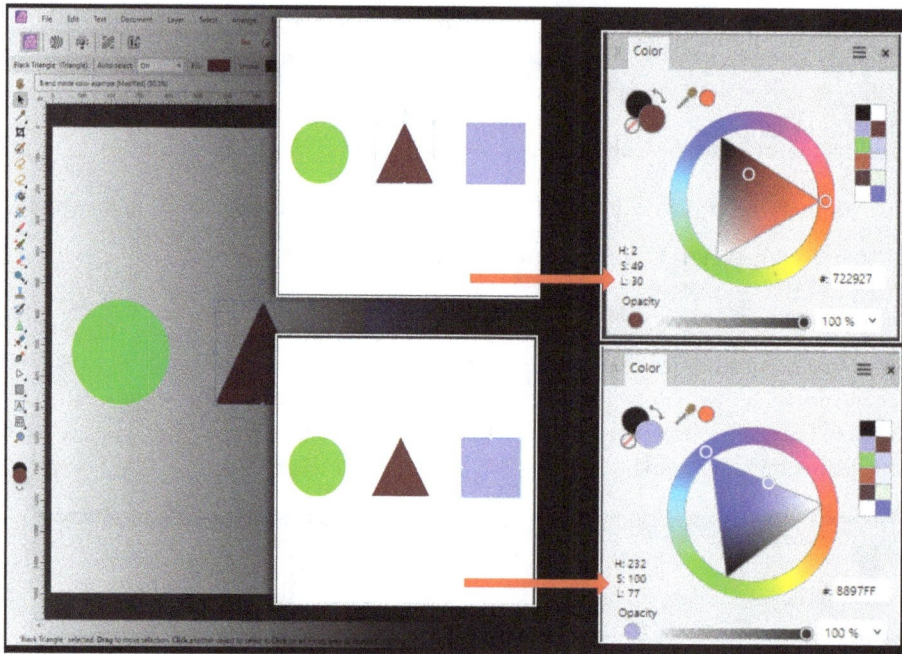

Figure 8.6 – Setting the remaining colors

4.  Engage the **HSL** layer by clicking on the checkmark next to the layer in the **Layers** panel (check *Figure 8.2* to see the **Layers** panel for the checkbox).

    Doing this will remove the color by desaturating the lower layers and leaving only gray values. This is what Affinity sees when it calculates blend modes. Notice the blue square is the lightest, the dark red triangle is closest to black, and the green circle is at the mid-gray mark.

Now you know how Affinity sees color for the purposes of blend modes. Knowing how the program sees things helps you, as the editor, anticipate the behavior of the program when assigning blend modes.

Now that you have an idea of what blend modes do and how to read and set them, the inevitable question that comes up next is *What do I do with them in an actual practical edit?* This is what we will cover in the next section.

## Using blend modes in practical editing

Now that you know the theory, it is time to do two edits using the blend modes as our primary focus for the edit. There are an unlimited variety of ways you can use blend modes, and even if I could write them all down, you would get really tired of reading them. So, in this section, we will be using blend modes for two of the most frequent applications I work with in photo editing:

- Blend modes and the application of textures
- Blend modes for the addition of elements in compositing

# Blend modes and the application of textures

During the workflow I laid out in the previous chapter, the last step was balancing and finishing, and part of my finishing process is applying textures to the top layer of an image to give it that last little bit of interest. This method requires two things:

- An image that you are happy with and ready to finish
- A texture that you can apply over the top

A texture is an image that has dynamics that give the image a *feel*. Examples may be canvas, cement, or pretty much any image that has good dynamics that you want to transfer to the image underneath.

We have included two images in the download files for this chapter (see `Practical Blend mode Edit #1`). These are an image I did for a commission and a piece of cement that I shot on the sidewalk outside of my apartment. Let's get started with the steps:

**Step 1**:

    I.    Open the base image in Affinity Photo.

    II.    Place the cement image on top of the base image.

    III.    Drag the cement image to fit the base image.

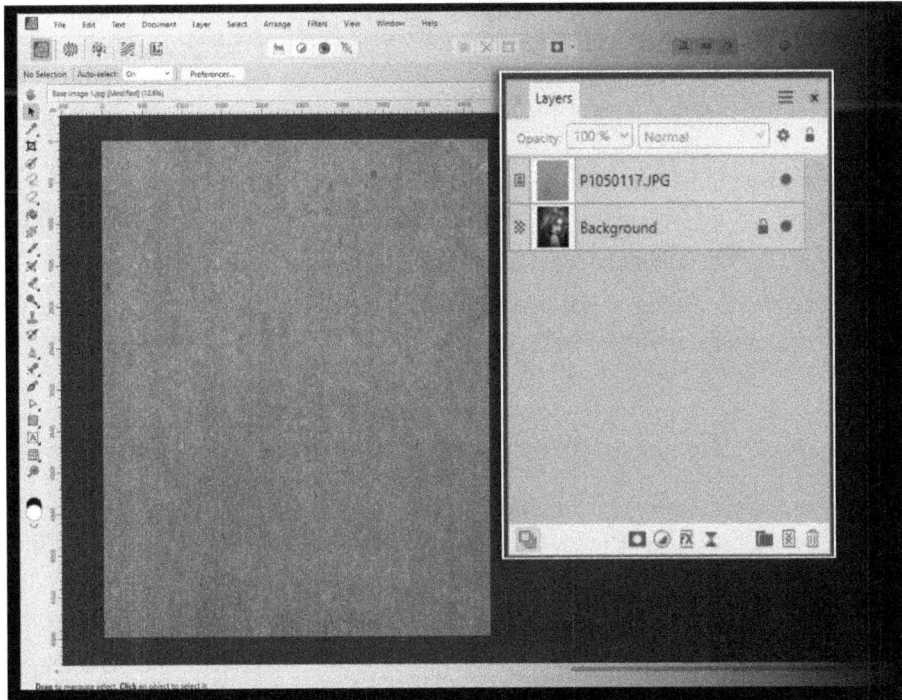

Figure 8.7 – Texture layer applied over an image with no blend mode

**Step 2:**

    I.      Change the blend mode of the cement layer to **Overlay**.

    II.     Adjust the **Opacity** setting of the layer to taste.

Figure 8.8 – Application of blend mode to texture overlay (before and after)

If you want to use another blend mode, feel free to make the piece your own. I like **Overlay**, but **Screen** is a common one for working with textures. There is no 100% right method; it depends on the image and what is going on underneath.

Notice the difference in visual interest between the before and after images. These subtle touches are the difference between a basic edit and one by a master of the craft who understands it. It is the little touches that make the viewer look at it for just a little bit longer.

## Blend modes for the addition of elements in compositing

The second frequent application of blend modes in an application is the addition of elements in compositing. We will be using this technique in a more advanced way later in the book, but I wanted to introduce it to you here.

When we shoot stock elements (elements we reuse in various composites), it is not uncommon for us to shoot them against a solid color background so that we can easily select them and mask them out. However, if the situation is right, we will often use a blend mode to remove the elements.

We have included the images for this activity in the downloads for this chapter (see `Practical Blend mode edit #2`). In this edit, we will be taking the lighting and applying it over the graffiti wall to create an atmosphere for a composite. Let's get into it:

## Step 1

1. Open the wall image in Affinity Photo.

2. Place the lighting image above the wall.

3. Drag the cement image to fit the base image.

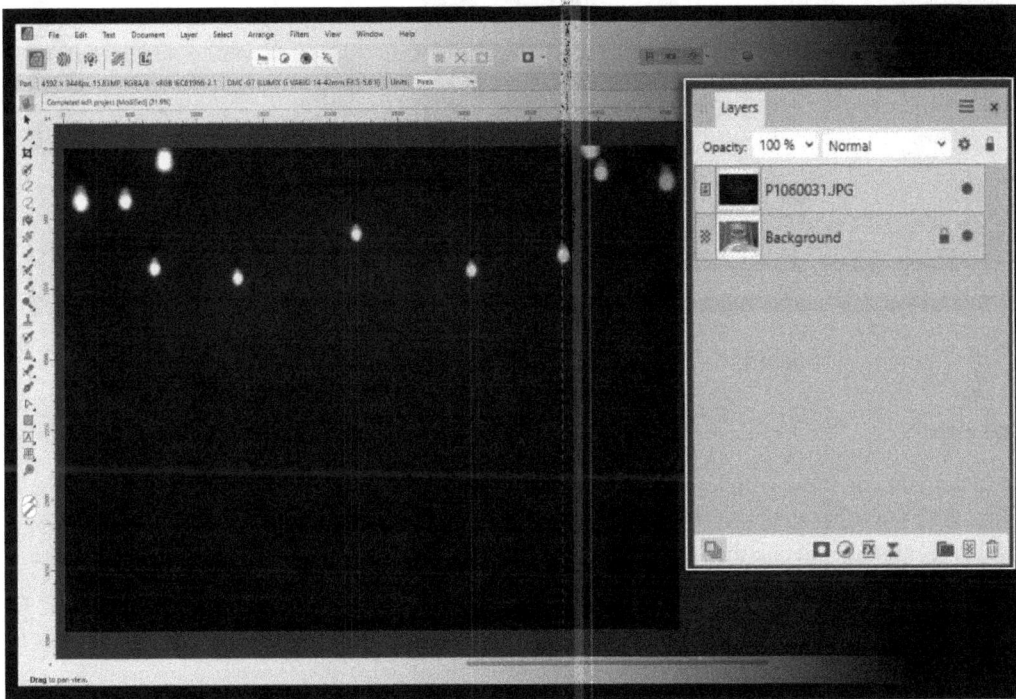

Figure 8.9 – Staking of layer order

## Step 2

1. Change the blend mode of the light layer to **Glow**.

2. Adjust the **Opacity** setting of the layer to taste (**89%**).

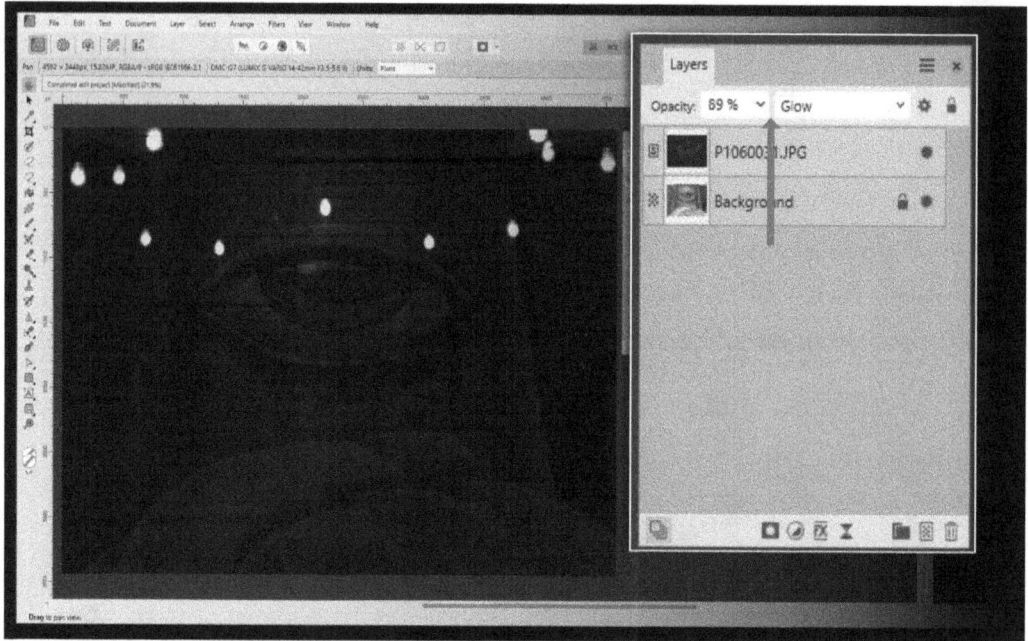

Figure 8.10 – Blend mode adjustment for the lighting layer

**Bonus steps**

To help get you one more edit, let's finish this one up and show you how this can be kicked up by introducing a few other steps. We will introduce each of these adjustments here, but in later chapters, we will dive into more detail; the important thing right now is that you see how they can be used to make an incredible environmental edit. So, we would like to add the following:

- HSL Shift Adjustment layer
- Lens Filter Adjustment layer
- Clarity Live filter

**Step 3**

Add an HSL layer above the background and adjust the luminance.

This will brighten the background without affecting how much of the light came from the lights above.

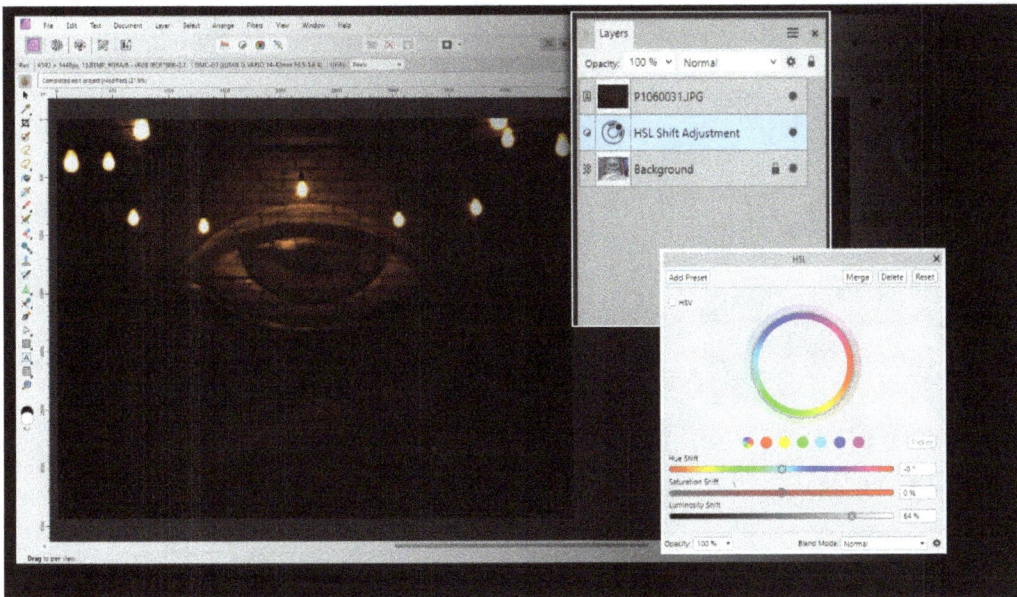

Figure 8.11 – HSL Shift Adjustment layer

**Step 4**:

1. Add in a **Lens Filter Adjustment** layer at the top of the stack. This will add a color hue to the image.

2. Select a green color from the small color square on the screen.

3. Change **Optical Density** to **74%**

4. Change **Blend Mode** to **Screen**.

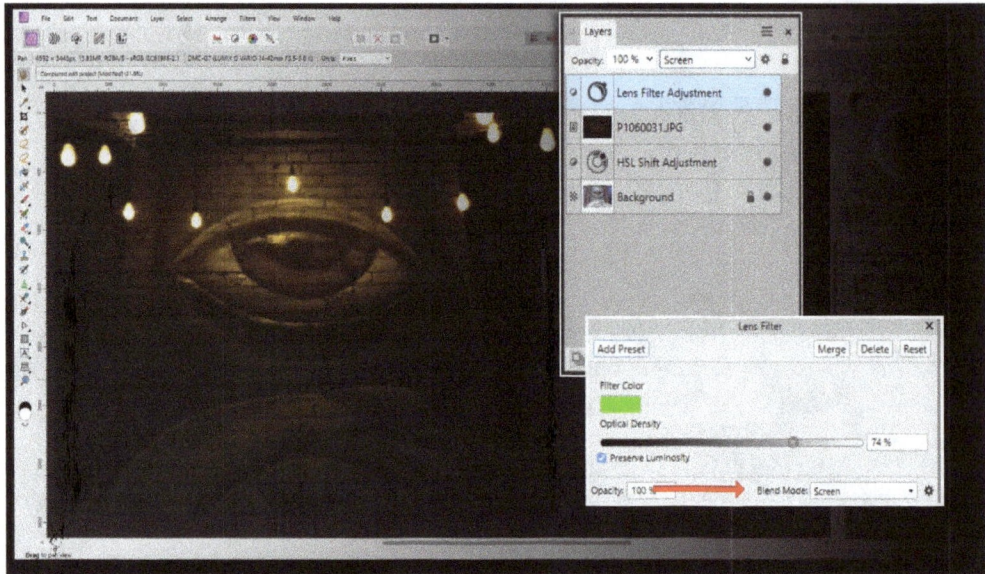

Figure 8.12 – Lens Filter Adjustment

## Step 5

1. Add a **Clarity** filter to bring out the brick.

2. Set **Strength** to **90%**.

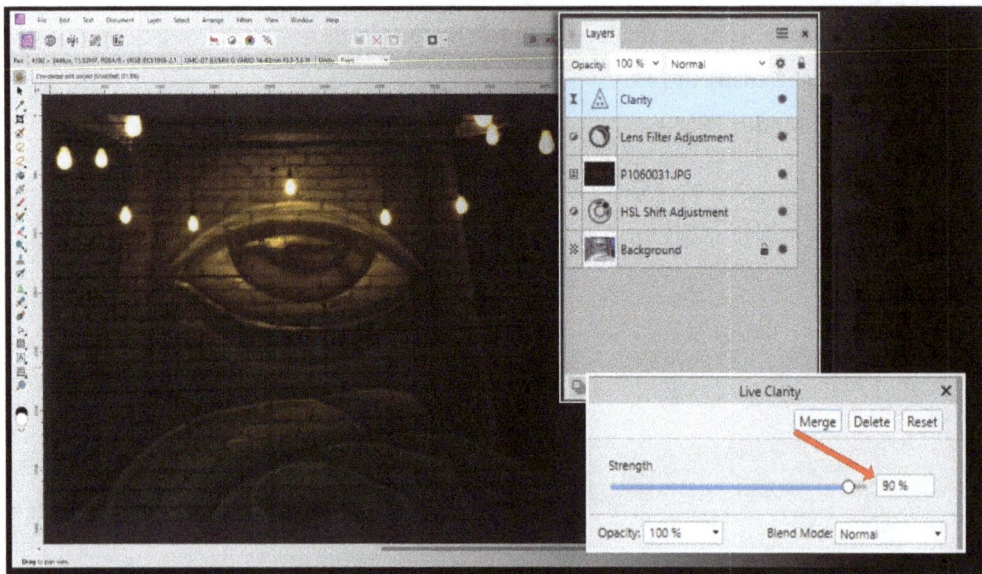

Figure 8.13 – Strength adjustment

We have included the working file for this in the downloads as well so you can follow along and see my settings and edits (see `Completed edit file` inside the `Practical blend mode edit #2 folder`).

You saw how knowing how to set a few blend modes allows you to create a very atmospheric edit with just a few buttons. Now that you know how simple this can be, let's combine the concept of blend modes with the opacity adjustment present in the layers.

## Using blend modes and opacity together

A common union in the setting of blend modes is to adjust the opacity of the layer when adjusting the blend mode. Sometimes, the blend modes are so powerful that the effect needs to be turned down. We saw this when we added the texture layer to the image in the practical edit previously.

In the *Figure 8.14* edit, notice the use of blend modes with opacity adjustments in the following areas:

- Engraving of the geometry into the wall was accomplished by a blend mode change with an **Opacity** adjustment (this is shown in *Figure 8.14*)

- The lens flare and the resulting stars are simply a **Screen** blend mode

- The texture layer is applied and then masked out with a gradient mask, and blend mode and **Opacity** adjustments are made.

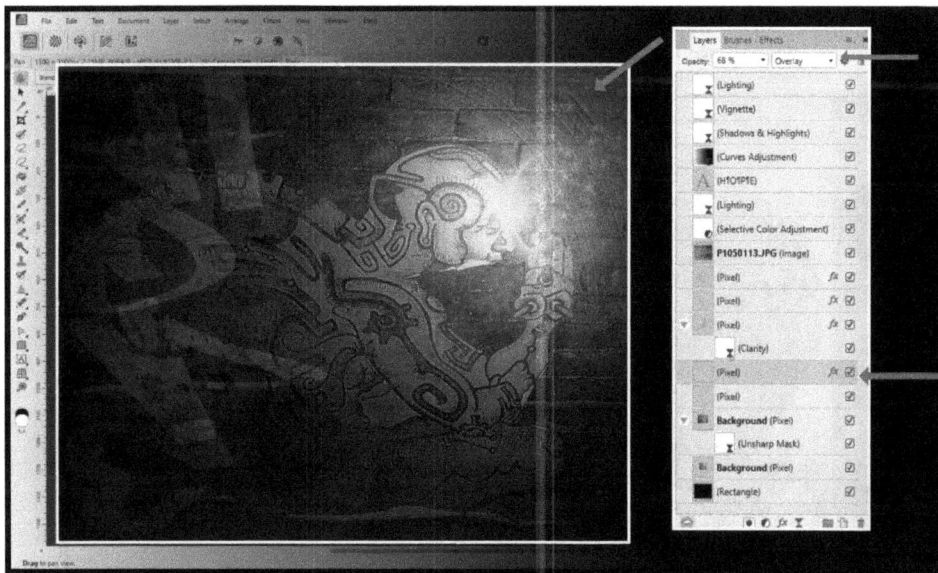

Figure 8.14 – Modification of Opacity in the Pixel layer

The bottom line here is that blend modes are augmented and made more powerful by adjusting the opacity.

Now, blend modes are not only for **Pixel** layers and **Shape** layers but can also be combined when using **Adjustment** layers and **Live Filter** layers to bring about some really great effects in your work.

# Blend modes in Adjustment and Live Filter layers

In previous chapters, we covered the different types of layers, including the following:

- **Pixel**
- **Adjustment**
- **Live Filter**

Up to this point, we have only really talked about blend modes using a **Pixel** layer, as in the images we placed over the others in the practical edit. However, both **Adjustment** and **Live Filter** layers have the option for blend mode adjustment as well, and blend mode adjustment on these types of layers can lead to some unexpected and beautiful effects in your work

## Practical example

In *Figure 8.15*, we have a picture of a hot air balloon, typically a colorful object, but I have applied a **Black & White Adjustment** layer to it, turning it into shades of gray. We have included this working file in your downloads for this section if you want to try this for yourself (see the `Blend mode for adjustment and filter example` file). This example will only look at an **Adjustment** layer, but as the process is the same for **Live Filter** layers (and the adjustments are in the exact same place in the dialog boxes), we will not be doing a separate one for those.

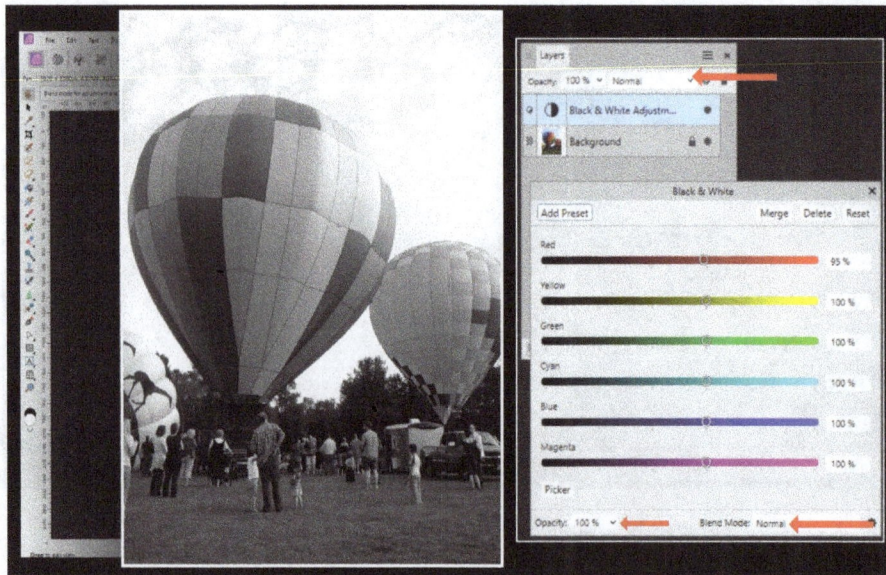

Figure 8.15 – Application of the Black & White Adjustment layer on the image

Notice that the blend mode is set to **Normal**, which means that it is fully present and covering the layer below.

However, if we switch the blend mode to **Multiply**, we get the color image back (see *Figure 8.16*). This is because the **Multiply** blend mode darkens the image but retains the hue, so it is a popular way to darken an image without jeopardizing the color values.

Figure 8.16 – Targeting the individual colors using blend modes on the adjustment layer

### Why would you want to use this? What is the advantage?

I use this technique because I have a color-specific way to adjust only certain color-specific portions of the image. In this image, if I wanted to adjust the darkness of only the reds, I can. If I to only affect the yellows, I can. This is an advantage over other methods, where you must apply an **Exposure** adjustment and then mask the areas you didn't want to be affected.

In the CiA video (the link is mentioned in the *Technical requirements* section), I have shown how I use this in the actual editing process. It does a nice job of only influencing the colors I want and gives me more control than some other methods. Notice that it depends also on the image; without the bold colors, I may have chosen another method to adjust the exposure.

Now, as you progress as an editor, you may hear the term *blend ranges*, or you may want to target only certain values (for example, only the darks) to be blended. This is where blend ranges come into play. This is what we will be covering in the next section, but it was essential to build the previous understanding before going to the more complex topics.

# Blend modes versus blend ranges

So far in this chapter, we have been talking about blend modes; however, there are such things as **blend ranges**. A blend range differs from a blend mode in the following ways:

- You can target certain dark and light values of an image to blend only if they are part of that value range (say you only want the black parts to work with the **Darken** adjustment)

- You can create specialty curves to target these values, whereas blend modes apply to the entire layer without exception

So, in this section, we will be covering where to locate them, how to read the graphs they produce, and where we would actually use them in the world of editing.

## Where do blend ranges show up?

Blend ranges show up in the **Layers** panel, toward the top:

- In *Figure 8.17*, we see a cog/gear icon next to the blend mode in the **Layers** panel, and this is where the blend ranges exist

- In *Figure 8.17*, we also see graphs that come up for blend ranges

> **Author note**
> You will never use most of the settings—I have covered those that I actually use frequently.

Figure 8.17 – Location and visualization of blend ranges

# How to read the blend ranges

The way you want to think about blend ranges is *I only want to apply the blend mode to the (dark, mid, or light) values.* This way of thinking allows you to plan which portions of the image you want to see blended. The graph is broken into two parts:

- The layer you want to set the blend range on
- How you want it to blend

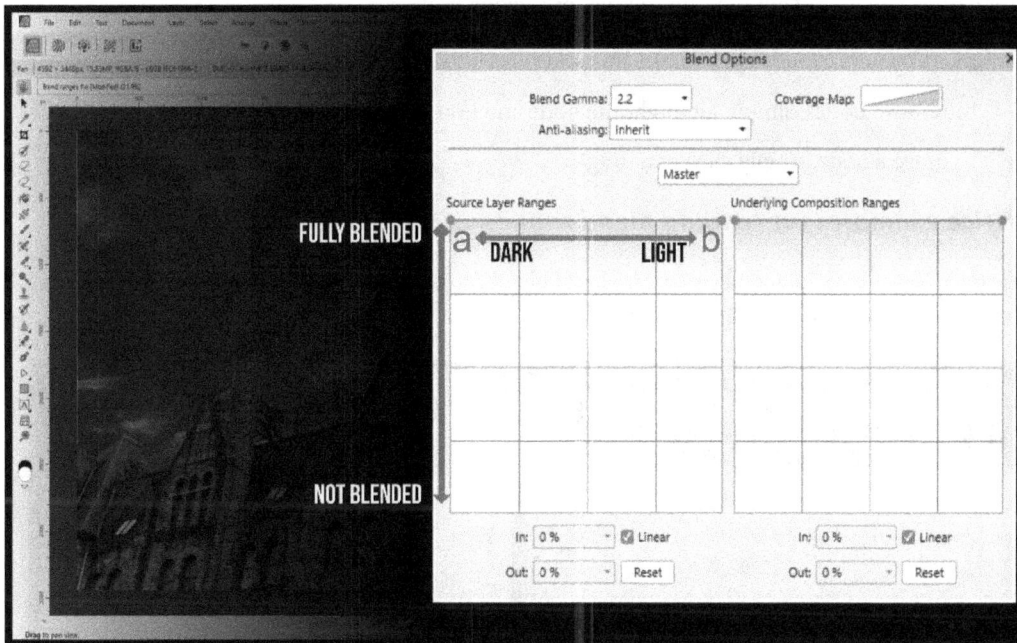

Figure 8.18 – How to read the blend ranges

## *Setting the layer*

There are two options for which layers the blend range uses in its calculation:

- You can set it to blend using the value of the layer you are on
- You can set it to blend with the values of the layers underneath

I personally use only the layer I am on; however, if you have a lot of darks in the underlying image, and you only want to blend in that area, you may choose the underlying composition if it makes sense – feel free to experiment, as experimentation is the only way to really try out the technique.

## Setting opacity and managing nodes

The blend range graph shows the darks to the lights running from left to right (see points **a** and **b** in *Figure 8.18*). The opacity of the blend is shown from top to bottom, so moving a point closer to the bottom reduces the blend.

### Modifying points for advanced controls

To create a more customized look, you can add points along the line to target very specific values. This is simply a matter of adding, subtracting, or moving points:

- To add points, simply click on the line and this will add a new node
- To remove nodes, simply highlight the node and hit *Delete*
- To move a node, simply click and drag

## Practice examples for reading blend ranges

In *Figure 8.19*, we have included two examples of blend range graphs, and above each graph, we have added the way you should see them as an editor:

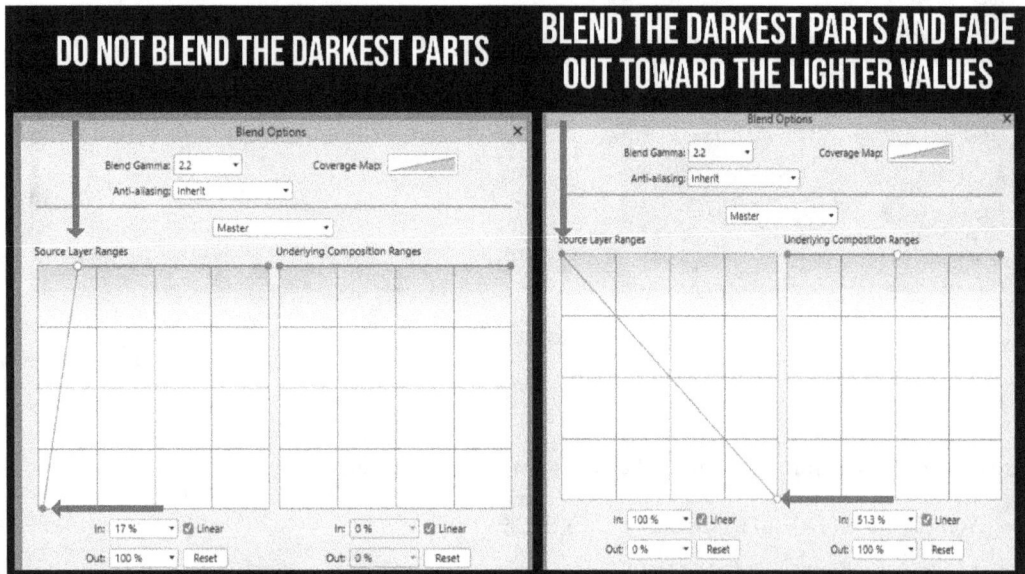

Figure 8.19 – How to read the graphs for ranges (two different curves)

Let's understand the graphs in the preceding figure:

- **Left-hand graph**:

  - The darkest parts of the image will not be blended because the node on the left-hand side of the graph is fully dropped to the floor

  - A node was added and raised up, meaning that apart from the deepest, darkest parts of the layer, everything else will be blended fully

  - The lights are up at the top of the graph, so the lights will be fully blended

- **Right-hand graph**:

  - The darks will be fully blended at the deepest, darkest level, but everything else will be left unblended

  - The blend opacity will get weaker and weaker until we get to the lighter areas of the layer

  - The brightest parts of the layer will not be blended

## Practical application of blend ranges

We have included a working file (see `Blend ranges file` in the chapter downloads). In this file, there is a layer structure as shown in *Figure 8.20* and described here:

- There is a church in the foreground with a mask attached (see the isolated church)

- There is a sky image, which we will be using to show the blend range

- The complete image is in the lowest layer

The blue sky is a bit overpowering, and so we need to only blend some areas and not the entire thing. This is the perfect application for blend ranges. We will be using blend ranges on the sky layer to blend only parts of the sky and make it match the image of the church.

I have included a CiA video (link mentioned in the *Technical requirements* section) for you to see how it was done in each step.

### Edit steps

Complete the following steps in the working file:

1. Change the blend mode of the sky layer to **Multiply**.

2. Open blend ranges through the cog/gear icon for the sky layer in the **Layers** tab.

3. Add a node in the middle of **Source Layer Ranges**.

4. Pull the node down until you are happy.

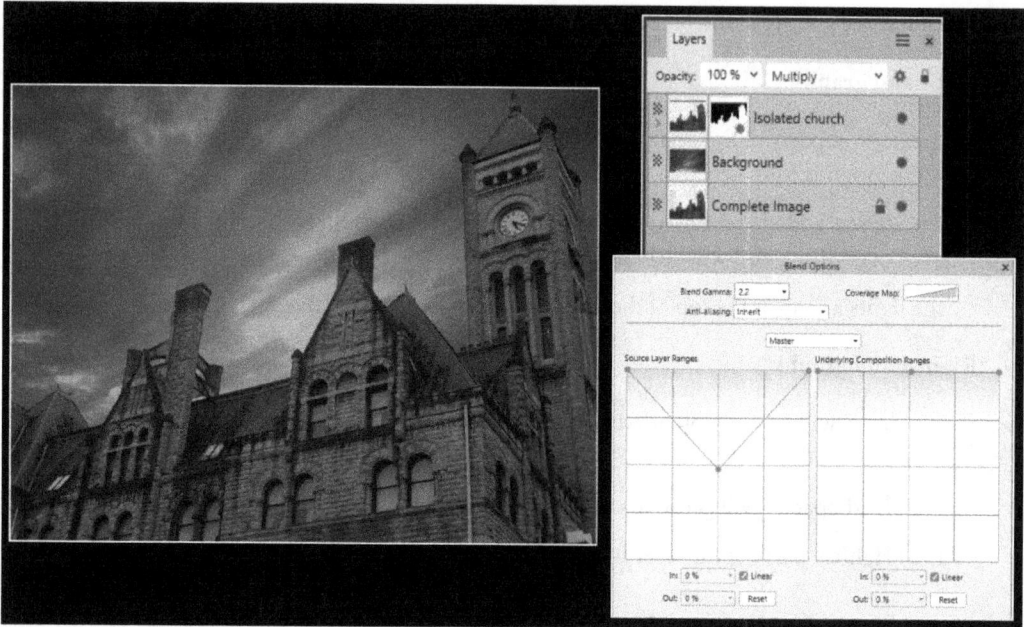

Figure 8.20 – Sky replacement using only blend ranges

What just happened? We wanted a more dynamic sky but not necessarily that overall darkness (the full darkness made the blue far too dark). Thus, we reduced the power of the blend mode on the mid-tones by creating a node in the middle of the graph and pulling it down. This is the power of blend ranges as opposed to blend modes.

## Professional tips, tricks, and important points

As we close, I want to give you some guidance on common blend modes for common tasks. I do want to note, however, that there is no perfect rule or perfect blend mode for every task. There have been many times in my edits where I shaded using **Darken** instead of **Multiply**, and many more subtle blend modes can be achieved by using a more aggressive mode and then reducing the opacity. So, the following list is a starting point, but not the end. These are common blend modes for common tasks:

- **Shading and darkening: Multiply**
- **Adding overlay textures: Overlay**
- **Heavy lightening: Screen**
- **Dynamic darkening: Linear burn**
- **Dynamic lightening: Color dodge**

## Tips to make your work more dynamic

Lastly, as a professional, some of my most dynamic edits have been when I combined a layer blend mode with an **Adjustment** layer blend mode; that interaction adds a ton of interest and value and defines your personal style. The following are a few things to try as you grow as an editor:

- Combine blend modes for layers; with the **Adjustment** and **Live Filter** layer blend modes, this makes for some awesome interaction.

- Use blend ranges where possible. They blend only parts of an image and then stack them with the **Adjustment** and **Live Filter** layer modes. This creates some wild effects.

- Use **Opacity** with the blend modes as well, even if the blend mode is too dynamic; tone it down until it looks right in the composition.

# Summary

We have completed one of the most fundamental and confusing concepts in digital art and came out on the other side. Earlier in this chapter, I used the term *blend mode roulette*. This statement is as true as when I began editing, because you will never know everything about blend modes, and you will never be able to fully anticipate the effect, given the endless variety of the images they apply to.

We need to treat the blend modes as our spices in our cooking analogy—a bit of **Multiply** and a dash of **Lighten**—in order to make our edits stand out to be unique and really be ours. So, play around, try some stuff, and if it doesn't work, just change the blend mode back.

Lastly, the application of blend ranges allows us to be very specific in the application of the technique but requires a firm understanding of how to arrive at the right blend mode as a prerequisite. We learned that the application of blend ranges in complex images (such as the sky) allows us with just a few clicks to do what would previously have been tedious.

With this topic behind us, we can continue to the most fun topic when it comes to fundamentals, the thing that makes digital art what it is: the topic of brushes. We had to understand blend modes first because brushes have blend mode options that we will set. So, let's get into the art side of Affinity Photo in the next chapter, on brushes.

# 9
# Basics of Stock Brushes in Affinity Photo

In this chapter, we will explore the feature we all got into digital art to play with – that is, the idea of brushes. Brushes are infinite in their possibilities, configuration, and uses. Similar to blend modes, an artist's brushes can make the edit easily identifiable as belonging to them. In later chapters (toward the end of this book), we will cover the details of making highly artistic brushes for custom applications, such as particle brushes, fog brushes, or flares. However, in this chapter, we will cover common terms and provide a high-level overview relating to stock brushes – those that come native in Affinity, or that you can buy pre-made from third-party marketplaces.

With this scope in mind, we will begin with the brush studio panel so that you know how to read the brush previews and manage your brushes. This will get us up and running in the shortest possible time. From there, we will dig into the details of the brush adjustments present in the brush itself so that you can make these stock brushes work for you, followed by general adjustments using the context toolbar. Lastly, we will cover how to make a brush from scratch using all of the techniques we learned earlier in this book.

By the end of the chapter, you will have made your first custom brush (that we will use in the next few chapters to edit), so you will be well on your way to being a creative editor.

We will cover the following topics in this chapter:

- Introduction to the brush studio panel
- Brush management
- Exploring the individual parts of a brush
- Making your first brush
- Adjusting brush settings by utilizing the context toolbar
- Applying the basics of brushes to an edit
- Professional tips, tricks, and important points

# Technical requirements

The CiA video for this chapter can be found at `https://packt.link/UQgSM`.

You will also need the `Chapter 09` project to follow along.

# Introduction to the brush studio panel

Fundamentally, our exploration of brushes should begin with the studio panel that houses the brushes, so we will start by exploring the **Brushes** panel. The **Brushes** panel is a **Studio** panel that can be accessed by going to **View | Studio**.

Make sure the check is next to **Brushes** – if it is not checked, check it.

This is the **Brushes** panel, and it is your gateway to the various brushes that you have in Affinity Photo that you can use to add visual interest to your art.

Now that we have the **Brushes** panel present, let's learn how to read the panel and dive into exactly what it is telling us.

## Reading the Brushes panel

The **Brushes** panel is divided into a few areas. These three areas can be seen in *Figure 9.1* and are coded by number:

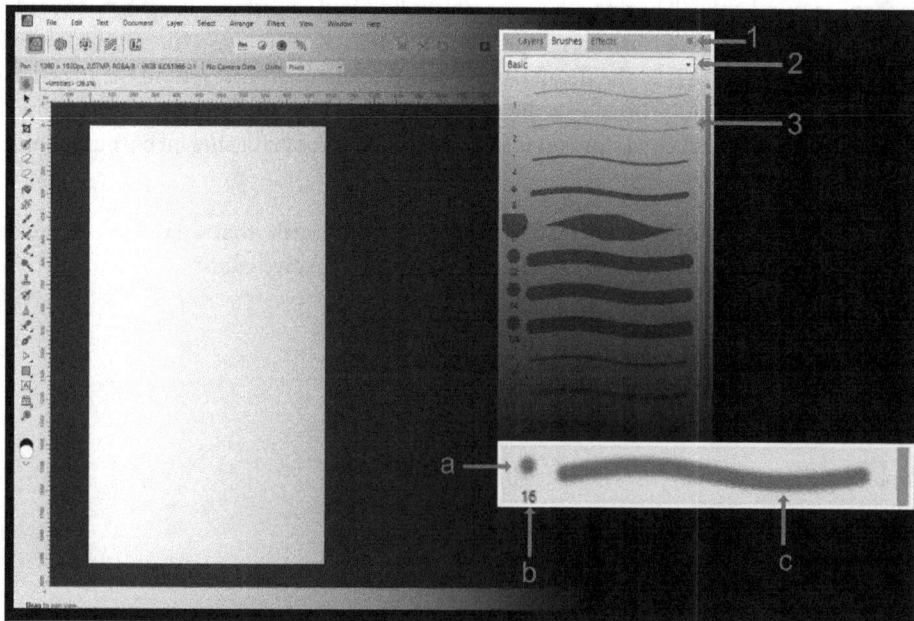

Figure 9.1 – The Brushes panel

Let's explore this menu in more detail:

- (1): The menu for the **Brushes** panel:

  This is where you will be able to add categories, remove categories, duplicate, as well as import and export (these functionalities are covered in a separate section named *Brush maanagement (installing, deleting, and so on)*).

- (2): The categories of brushes:

  Scrolling through the categories of brushes that come stock loaded with Affinity will give you a good idea of what is available. Each category has different individual brushes.

  Personally, for the default brushes, I typically use three categories for a majority of my work:

  - **Basic**

  - **Texture**

  - **Sprays and Spatters**

- (3): The brush preview:

  The brush preview shows you things about the brush such as its size, shape, and more. As you scroll down along the right slider of the panel, you can choose the brush that works for you.

  At the bottom of *Figure 9.1* is the detailed view of the brush preview, denoted by the letters **a**, **b**, and **c**. It includes the following:

  - The shape of the brush (**a**):

    Later in this chapter, we will discuss brush shapes, but in *Figure 9.1*, the brush is round – simply a circle. However, it is a soft brush, as shown by the fuzzy borders on the circle, which tells us as the viewer what sort of shape it uses in the stroke.

    If you scroll up through the **Basic** category, you will see that some circles have hard edges. This means there will be a firm border on the areas the brush affects.

  - The size of the brush (**b**):

    This is given in pixels. **16** in *Figure 9.1* means that by default, the brush is 16 px wide. This is important because if you are dealing with a 2,500 px document, this is going to be like a pin size and you will have to adjust it to make it noticeable. The brush size is adjustable; we will cover that later in this chapter.

  - The behavior of the brush (**c**):

    A brush has a shape and a size, and when you drag it across a surface, it also has a stroke. This part of the preview shows you what the behavior of the brush is going to be. In *Figure 9.1*, the stroke will be continuous and linear so that there is consistency. These behaviors can be set in the brush menu, which we will explore later in this chapter.

Each brush will have a preview, and aside from artistic editing, I use the soft round brushes 90% of the time with the default settings. So, before you go adjusting your brushes, realize that for basic editing, masking, and so on, the default brushes will work for most of your needs.

Now that we know how to communicate with the **Brushes** panel and what it tells us, we need to learn how to get brushes in and out of the panel.

# Brush management (installing, deleting, and so on)

When it comes to brushes, you can manage them (including their import, export, modification, and so on) by clicking on the four lines icon, which can be found in the top right-hand corner of the **Brushes** panel (previously shown in *Figure 9.1*). In the following sub-sections, I have covered everything you need to know to get started with brushes in Affinity Photo. So, let's get started.

## Importing and exporting

Importing and exporting brushes refers to introducing or distributing brush categories from outside sources. Let's say you bought a new set of brushes from the Affinity website; they will come in a ZIP folder, and there will be an Affinity brush file that you can import.

We have included a sample brush file in the downloads for this section that you can use to follow along. This is the brush we will make at the end of this chapter so that you can see the beginning-to-end workflow (see the 7SS-Bokeh brush file).

I have also included a CiA video showing this process using the sample pack provided. You can find the link to this video in the *Technical requirements* section.

In *Figure 9.2*, the images at the bottom show the options for either importing or exporting; the top images show the result:

- If you import the brush file, you will see the **7SS-Bokeh** brush in the pane (see the left-hand side)
- If you export, an **Affinity Brush Files** format will be created that you can share:

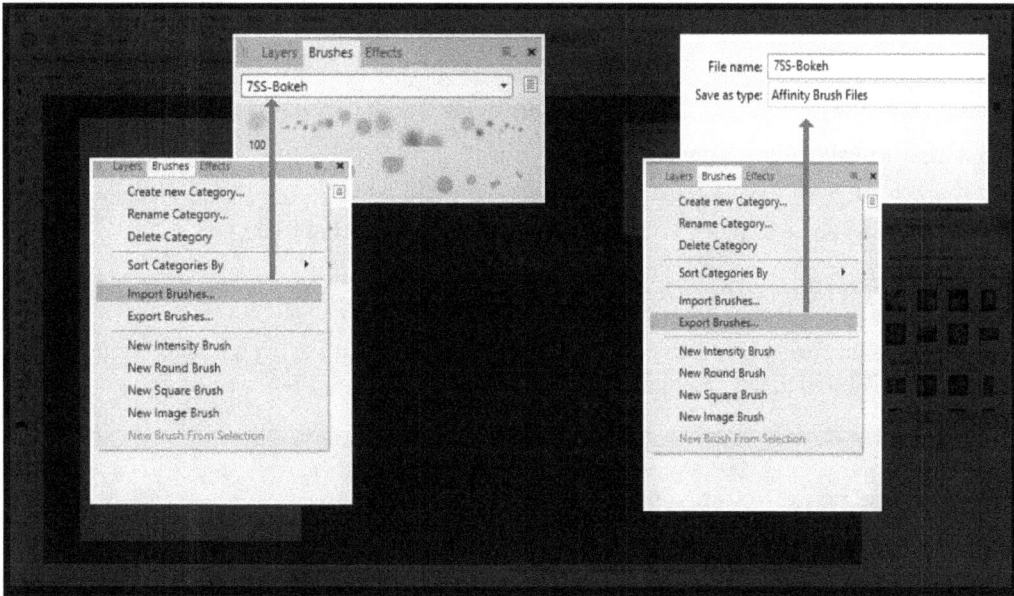

Figure 9.2 – The import/export options and the resulting file

In the next section, we'll learn how to import a brush.

### Importing a brush

To import a brush, simply go to the menu in the top right-hand corner of the studio tab and select the **Import Brushes…** option (see *Figure 9.2*). A dialog box will open, asking where you stored your brush file; then, you are good to go.

Note that only raster (pixel-based) brushes can be imported into Affinity Photo. Affinity Photo does not accept vector-based brushes as it is not a vector program. Remember back in the introductory chapters where vector art was used in programs such as Illustrator and Affinity Designer? This is because it is based on matches, not pixels. Affinity Photo exclusively works with brushes that are pixel-based.

On successful import, you will see your brushes present in the **Brushes** category (see **7SS-Bokeh** in *Figure 9.2*).

Next, we'll learn how to export a brush.

## Exporting a brush

Let's say you have a brush set that you would like to share. For this, you will need to select the **Export Brushes…** option, which will create an Affinity brush file (see *Figure 9.2*). This is the file you will send or post to your desired location.

Notice *Figure 9.2* and the file type that is created. Save it wherever you want and you will be all set. Raster exports can also be installed in Affinity Designer and Affinity Publisher.

## Adding and deleting categories

If you want to modify existing brushes (for example, change their settings), it is best to create a new category. This is covered in the first three steps of *Figure 9.3*:

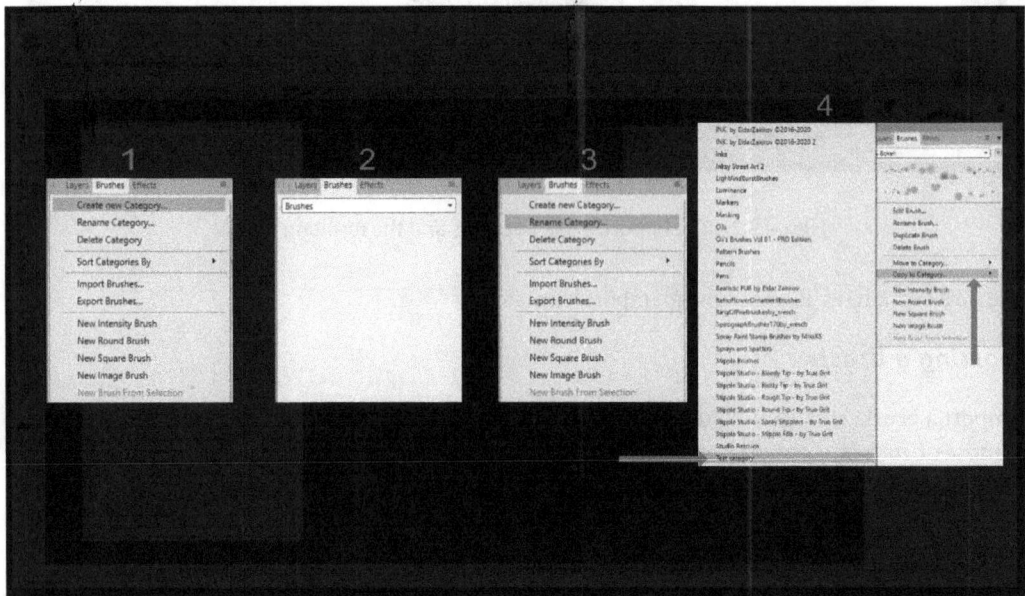

Figure 9.3 – The proper way to duplicate a brush to adjust it

I have included categories in the CiA video for you to follow along.

To create a new category, follow these steps:

1.  Select **Create new Category…**. This will create a new category in the **Brushes** menu.
2.  Select **Rename Category…**. We will call this Test category.
3.  Then, duplicate your brushes into this category (shown in the next section) and modify from there.

Next, we'll learn how to duplicate our brushes so that we can modify them.

*Duplicating brushes to modify them*

When modifying brushes, it is best to copy them into the new category you created (this is shown in *step 4* of *Figure 9.3*).

We have also included duplicating into the category as part of CiA 9-2 for you to follow along. Let's get started with the steps:

1.  Right-click on a brush you want to duplicate in the brush preview.

2.  Click on **Copy to Category...** and select the category you want to put the brush in.

    This will place the brush in the category of our choice. In this case, I moved a Bokeh brush to the **Test category** area.

Now that we know all about brush management, let's dive in and explore the brushes individually in more detail, including how they are made up and how they function.

# Exploring the individual parts of a brush

Now that you know how to add, import, and export brush categories, we need to look at the individual portions of the brush. To accomplish this, we will be working through a basic 16 px soft round brush (this means that it will be in the **Basic** category that comes with Affinity Photo). If Affinity removes this brush in later updates, rest assured any soft round brush will do, so the following instructions still apply.

If you click on the soft brush, the individual adjustments for the brush will appear. There are four adjustment tabs:

*   **General**
*   **Dynamics**
*   **Texture**
*   **Sub Brushes**

As this is a basic chapter, we will only cover the **General** and **Dynamic** tabs. We will save the **Texture** and **Sub Brushes** tabs for the advanced brush-building chapter later in this book (see *Chapter 15*).

# The General tab

This is what the **General** tab looks like:

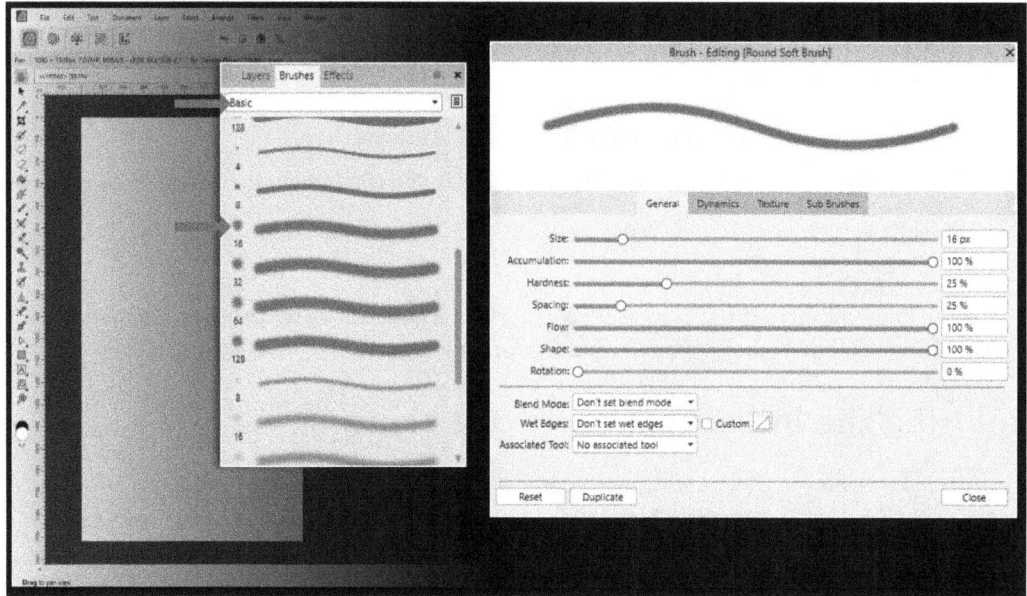

Figure 9.4 – Overview of the General tab

Let's go over the options under the **General** tab:

- **Size**:

    This dictates the size of the brush. In *Figure 9.4*, the size is set to **16 px** at the far right of the screen. Adjustments to size are constant, so if you modify the size, it will not go back to its default.

- **Accumulation**:

    Accumulation specifies how opaque the stroke is. To show this, play with the slider, turn it down, and notice in the preview window how it gets a lot less noticeable.

- **Hardness**:

    Hardness refers to how crisp the edges are. If the hardness is turned down, then the brush is more on the soft side; if it's brought to zero, it is at its softest.

- **Spacing**:

    The round brush is made up of circle-shaped layers that are placed over one another, so adjusting the spacing makes this more visible. Play with this by adjusting the spacing; notice that the circles begin to break apart.

- **Flow**:

  Flow is how much paint is applied with each stroke of the brush. This is different from **Opacity**. With low flow, you can layer the effect by applying the stroke multiple times. A high flow means the effect (paint application) is applied at full strength each time.

- **Shape**:

  In a round brush, the shape means how round it is. To illustrate this, move the slider to the left and notice that the brush becomes a series of lines. This is because the shape is not being fully realized.

- **Rotation**:

  In brushes that have a non-round shape (remember that all brushes have a shape that is shown in the preview of the brush on the **Brushes** panel; some that are more oval can be rotated). Rotation turns the brush – for example, calligraphy brushes typically have a slanted oval shape. For the round brush in this example, roundness does not play a significant role.

- **Blend mode**:

  If you want to set a blend mode for the brush by default, this is where you set it. I do not set blend modes for normal brushes since I set them in the layer.

## The Dynamics tab

The **Dynamics** tab refers to what the brush does when you apply a stroke. So, while the **General** tab applies to the stagnant (non-moving state) state, the **Dynamics** tab says to the user, "*When we use this brush, and it is in motion, this is how it behaves.*" So, when you pull a line along the canvas, this is the behavior of the brush.

The following figure illustrates the **Dynamics** tab:

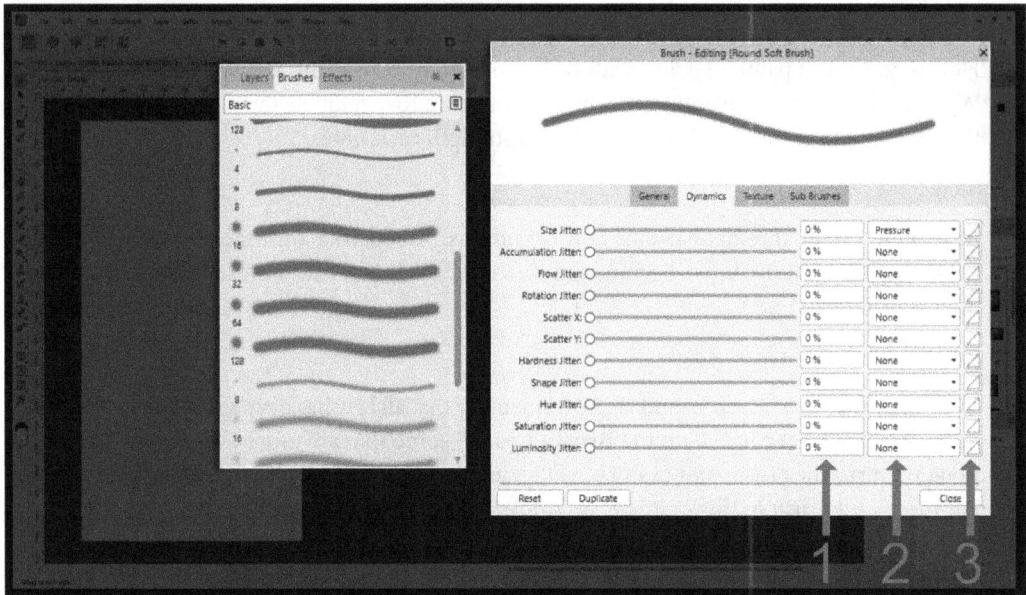

Figure 9.5 – Overview of the Dynamics tab

There are three universal settings for every dynamic adjustment (covered under the **General** settings section). Then, each of the adjustments individually uses the term **Jitter**. Jitter refers to movement, so if it says **Size Jitter**, think of it as how much the size fluctuates while you are drawing.

## General settings

Learning how to read these settings is important. % tells Affinity how to drastically vary the adjustment, the controller tells Affinity how to vary the adjustment (for example, the pressure speed), and the control curve tells it how to linearly adjust over that frame. Let's go over this in more detail:

- %: This shows how aggressive the dynamic is. As an example, a size of **100%** means that it will go from the default size – in this case, **16 px** – to **0 px**. As another example, if the saturation is set to **100%**, it will go from being in color to a desaturated gray.

- **Control method**: There are various ways to control the adjustments, and almost all of these require a graphics tablet. As an example, unless you have a pressure-sensitive device, you cannot control it with pressure unless you have a pen, and tilt doesn't work for a mouse. However, certain controllers, such as **Direction**, **Random**, and **Speed,** can be used with a mouse. In my opinion, aside from the **Pressure** controller, I have never set another controller other than **Random**.

- **Controller curve**: This represents the linear nature of the adjustment. For example, if the size has been set and we look at the graph, the size will decrease in a predictable linear manner. Again, as a working artist, I have never set a curve other than the straight linear curve.

Now that we know what the **Dynamics** panel is trying to do and how it works, we can explore each of the options in a bit more detail. If you are a brush builder, there is no substitute for trial and error… I have spent hours playing with perfect settings for how I work, all for only one brush.

## Specific settings

The following jitter settings simply reflect how much the brush will deviate from the initial settings during strokes. Because the following terms were defined in the **General** tab, we will not repeat them here:

- **Size Jitter**
- **Accumulation Jitter**
- **Flow Jitter**
- **Hardness Jitter**
- **Shape Jitter**

The following dynamic adjustments are new terms. However, the idea of jitter is still consistent – it is the amount of variance:

- **Scatter X**: This refers to how the shape is distributed along the x axis, so a high jitter will create spacing between the shapes. A common application for this is in random-style artistic brushes such as debris, particles, and Bokeh.
- **Scatter Y**: This refers to how the shape is distributed along the y axis, so a high jitter will create uneven patterns up and down. A common application for this is in random-style artistic brushes such as debris, particles, and Bokeh.
- **Hue Jitter**: Hue is color, so if you turn on hue and paint in green, it will rotate around the color wheel, and the amount (%) will tell Affinity how far it can go. As an example, a value of **100%** means it can be any color of the wheel, but if **%** is set to **10%**, then it will stick close to green. A common application of this is in Bokeh brushes, where we want to use color information as the thing that shifts (that is, blue shifts to green and over to purple).
- **Saturation Jitter**: This refers to how much saturation fluctuates. High fluctuations can make the strokes go all the way from fully saturated to grayscale as the brush is applied.
- **Luminosity Jitter**: This refers to how much luminosity shifts. A common application of this is in flame brushes and particles.

Now that know all we need to about making the basic round brush, it is time to apply what we've learned to making the simplest of editing brushes – the popular Bokeh brush.

# Making your first brush

Now that we have enough information to get started, let's make our first brush. The type of brush we will make is called a Bokeh brush. Bokeh is the process of using light and shapes to add some drama and flare to an edited photo (there is a more technical definition, but in reality, that is all they are doing). Examples of before and after using the Bokeh effect can be seen in the following figure:

Figure 9.6 – Before and after using the Bokeh brush

Notice that the effect involves adding circles over various opacities and hues. The blend mode has been changed to make it appear as if it is translucent. Notice that in making this brush, the techniques we are going to use include the following:

- **Brush shape**
- **Accumulation Jitter**
- **Luminosity Jitter**
- **Spacing**
- **Blend modes**
- **Scatter X**
- **Scatter Y**

There are other ways to make this style of brush. Since brush building will be covered later in this book, we are going to stick with the easiest most, direct way of doing this (the more advanced version consists of custom shapes and textures, which is beyond the scope of this introductory chapter).

## Setting up the category

Create a new category (as covered in the previous sections). I am going to call this category Packt Book Brushes. We will be using this category to hold all the brushes for this and later chapters (they will be available as a download for this book and can be found in *Chapter 17*). If you would rather call them Bokeh or something else, find a name that works for you. This will be the category where you will create your Bokeh brush.

## Setting up your test space

When testing brushes, it is advisable to initially set up a square space (I like mine to be 1,500 x 1,500 px) so that you can try to test different brush behaviors without having to worry about the image variations underneath throwing it off.

I also typically use a black background for Bokeh (to do this, simply drag out a rectangle and fill it with black).

## Building the brush

Follow these steps to create a brush:

1.  Create a new round brush by going to **Brushes | New Round Brush** (see *Figure 9.7*).

    We will go through the different types of brushes when we cover advanced brushwork in the artistic brush chapters (see *Chapter 15, Creative Effects and Specialty Brushes in Affinity Photo*). For now, we are only dealing with a round brush. Why? Simply because the round shape is the best shape for the Bokeh style. Once you click on the **New Round Brush** option, a new brush will be added to the category you created. See *Figure 9.7* for the **64** px round brush:

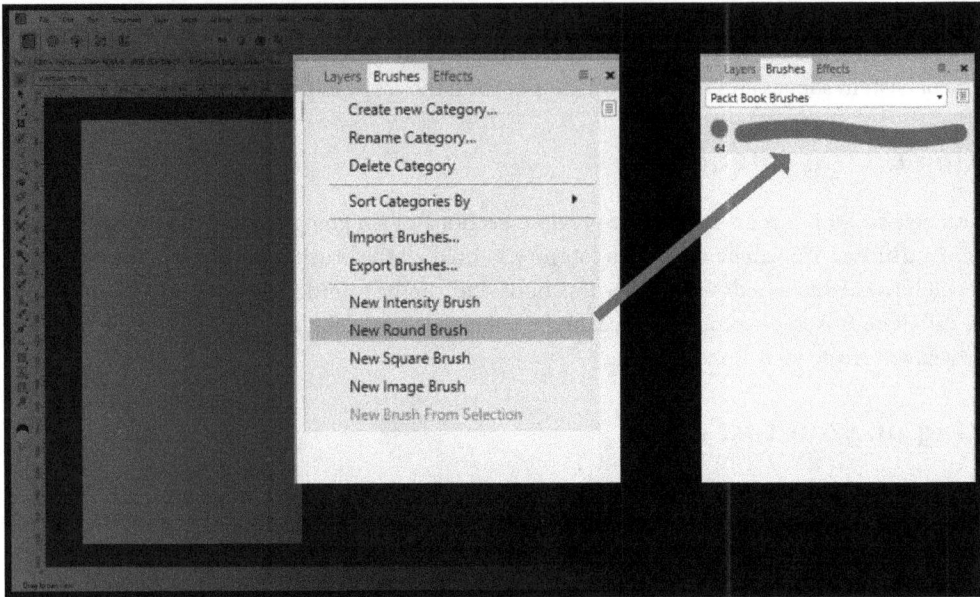

Figure 9.7 – Setting up the initial brush

2.  Adjust the **General** settings.

    Adjust the overall settings first – I have included mine in *Figure 9.8* so that you can set the same ones. I am trying to make a small twinkle light-style Bokeh, so I will most likely have a smaller brush with more spacing:

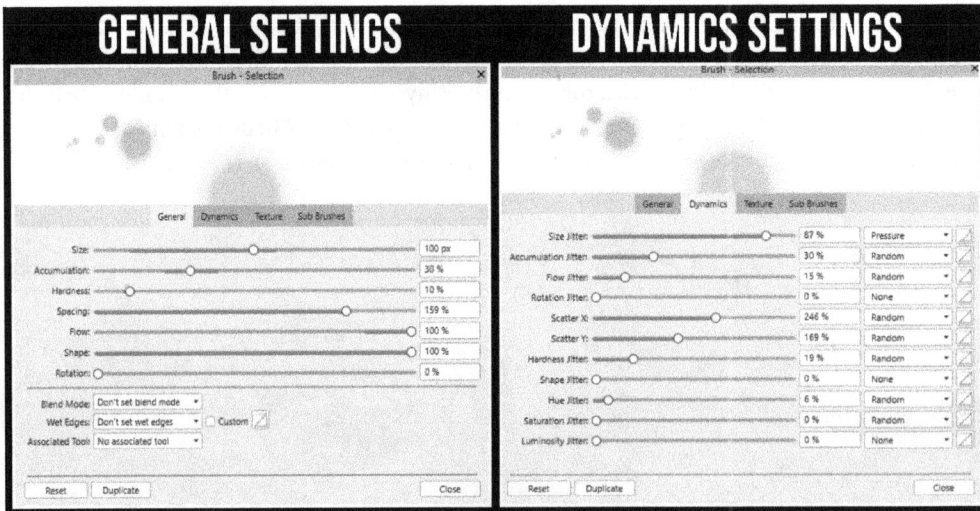

Figure 9.8 – The General and Dynamics settings for the Bokeh brush

Here are some of the highlights shown in *Figure 9.8* on the **General** tab:

- I have set **Size** to **100 px** (this is a good size for what I do; you can always adjust later)

- **Accumulation** is way down to give it a more transparent look

- I have scaled **Spacing** up so that the circles aren't on top of one another

3.  Adjust the **Dynamics** settings.

    Adjusting the **Dynamics** settings is where the trial space is used. You will want to play with this a bit to find the right mix for your style. I have included my settings in *Figure 9.8* under **DYNAMICS SETTINGS**. Here are a few things you should note:

    - Because I work with a tablet, I have enabled **Size Jitter** with a **Pressure** controller. If you are only using a mouse, you could try **Speed** or **Random**.

    - Notice that the controllers for the rest of my **Dynamics** settings are set to **Random** if they are engaged. **Random** means the program will randomly fluctuate within the limits you set (as an example, when **Accumulation** is set to **30%**, the program will fluctuate 30% of the accumulation randomly).

    - **Scatter X** and **Scatter Y** are engaged because I want the circles to be moving around.

    - There is a slight **Hue Jitter** amount, so if I use, say, yellow, it will move slightly off yellow, but not to green or anything weird.

Now that we have created a prototype brush, we need to test it on a garbage canvas to check its behavior.

## Trialing and duplicating the brush

Once the brush settings have been created, you will need to trial your brush. I adjust several times on a solid background before I move to images and then test those images. *Figure 9.9* shows an example of my trial for the brush. Notice that I used a white color on a black background (this I because I typically create the Bokeh effect in a lighter color so that I am testing close to how I will use it):

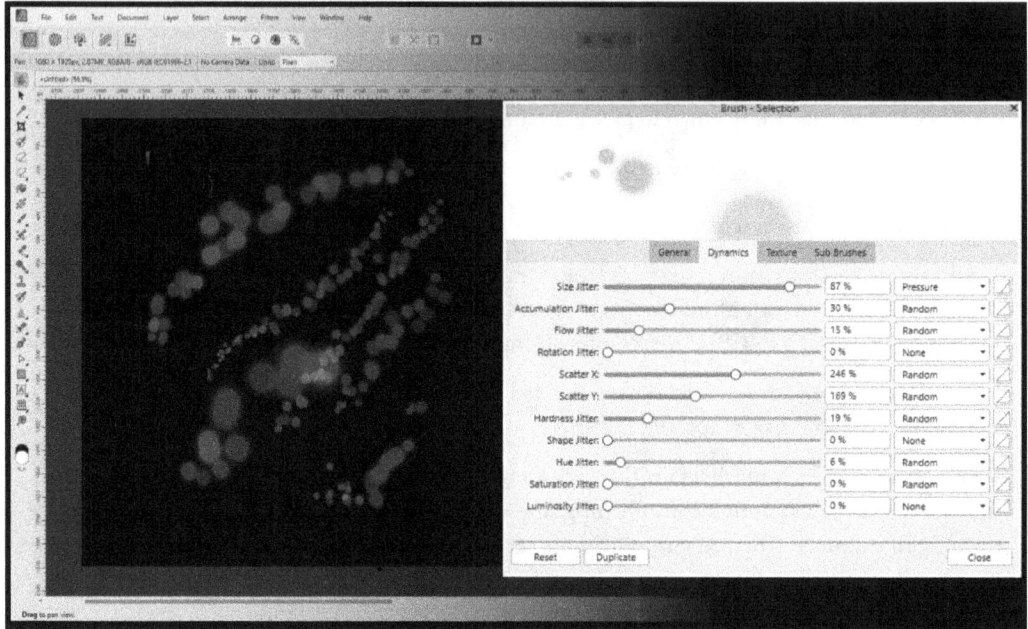

Figure 9.9 – The final dynamic settings of the brush and its trial space

Once you are done with the brush and you are happy, you can duplicate the brush as a starting point for other Bokeh-style brushes. To duplicate the brush, follow these steps:

1. Right-click on the brush preview in the brush studio.

2. Select **Duplicate Brush**.

   A new brush will be created.

3. Right-click the brush to rename and adjust it:

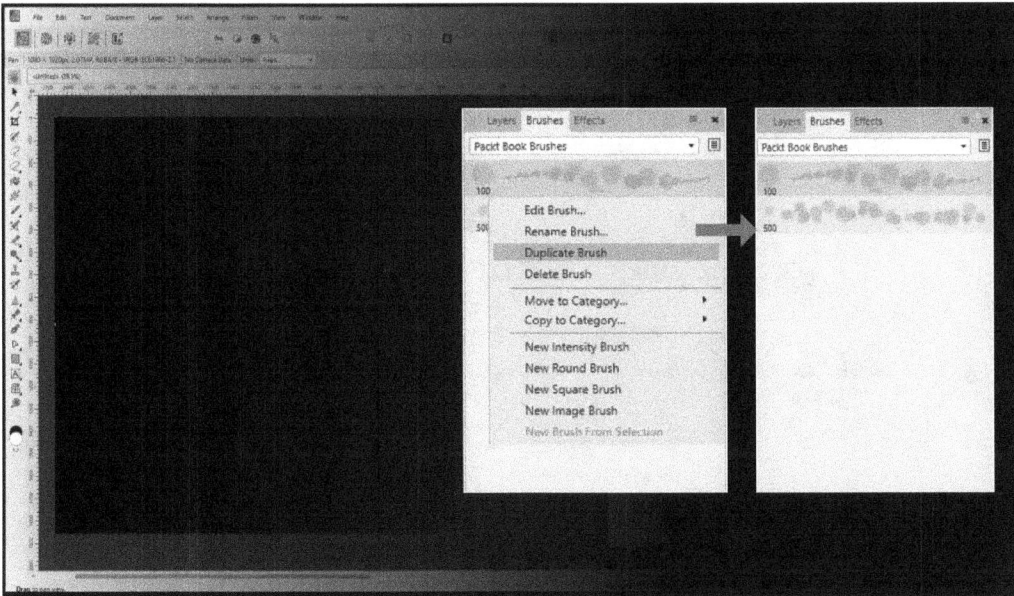

Figure 9.10 – Duplicating a brush for different types of Bokeh styles and variants

There are two brushes for Bokeh in the Packt book download – one we made and another that's a duplicate with its **Size** set to **Random** and made larger so that it can be used without a tablet.

With that, we know how to see the native options of the brush (that is, the default values). But what about when we are working on a piece? How do we adjust on the fly for that particular task? This is done through the context toolbar for the brush tool. We will tackle this in the next section.

## Adjusting brush settings using the context toolbar

So far, we have covered how to adjust brush settings inside the **Brushes** panel. However, in an individual project, you can adjust the brush settings using the context toolbar. This toolbar comes up when you select the brush tool. It is highly recommended that you do not adjust the individual brushes and instead use the context bar. This is because, for each job, an adjustment will need to be made. If you adjust the brushes constantly inside the **Brushes** panel, you won't have a predictable baseline for behavior.

### Reading the context toolbar for brushes

*Figure 9.11* shows the context toolbar for the brush tool. This is where you can adjust the settings of your brush without impacting the settings in the **Brushes** panel. So, on a piece-by-piece basis, this is where you will make your adjustments. If you adjust the brush settings in the brush itself, the changes will be permanent from there on out, so I do not recommend that approach:

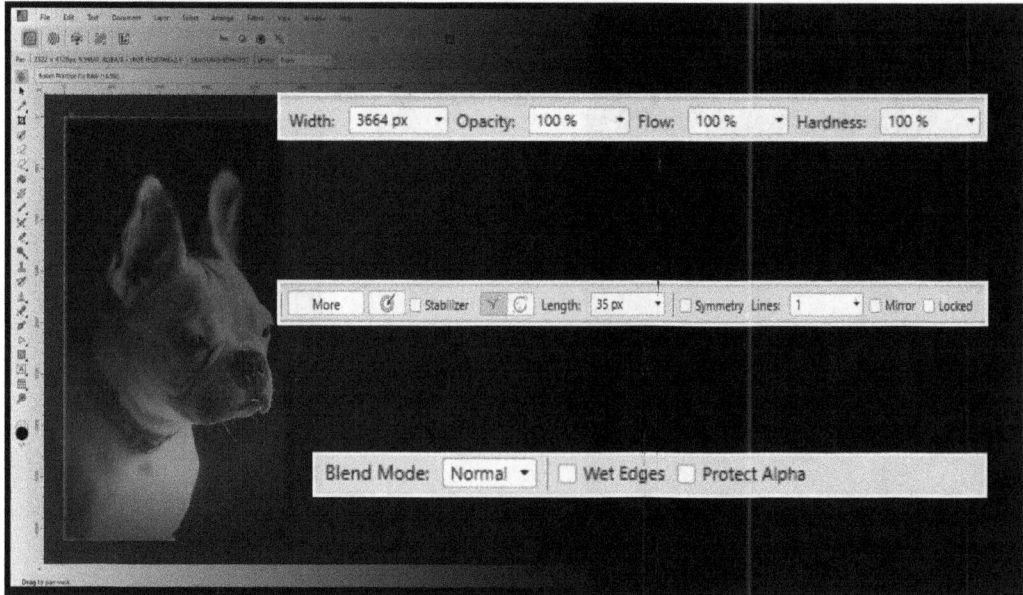

Figure 9.11 – Examples of the context toolbar for the brush tool

## Describing the options in the context toolbar

Let's go over the brush options in the context toolbar:

- **Width**: This is where you can set the brush size without changing the default size.

> **Pro tip**
> To change the brush size quicker without going to the slider, use the *[* and *]* keys on your keyboard.

- **Opacity**: **Opacity** tells you how opaque the effect will be at its maximum application. There is a difference between **Flow** and **Opacity**. If **Opacity** is set to, say, 30%, then it doesn't matter how many times you stroke over the area – it will never get past 30%.

- **Flow**: If **Opacity** is set to 100% and **Flow** is set to 25%, then the effect will be at 100% strength after four strokes of the brush.

> **Pro tip**
> I typically set my brushes between 20% and 40% flow for more control.

- **Hardness**: This specifies how hard the brush's edges are.

- **More**: This option pulls up information about the brush – the same as double-clicking.

- **Controller icon**: Depressing this tab engages your pressure sensitivity. If it is not engaged, then the default brush behavior is still in play (things such as dynamics using **Random**).

- **Stabilizer**: To assist in straight lines, the stabilizer helps you decrease the shakes and variations in your brush strokes. It is useful for pulling long lines.

- **Length**: This option specifies how long the line of the stabilizer is. I typically have mine set to **35 px**. Length is not an important value compared to the control it provides, so don't get too held up on a particular length. Instead, find a length that gives you good, steady control.

- **Symmetry Lines**: In the event you want symmetry to be engaged, click this option to engage it. Here, you have the following options:

  - **Mirror**: This is a subset of symmetry, and it means that the left-hand side will be the same as the right. This is different from the default behavior, which has symmetry working from the top and bottom.

  - **Locked**: This option locks the symmetry line so that you cannot adjust it accidentally.

- **Blend Mode**: Sets the blend mode for this application of the brush.

- **Wet Edges**: This option lets the edges of the stroke build up and is only used for digital painting. I have never turned this option on during editing.

- **Protect Alpha**: This is a feature that will not allow you to paint where there is no pixel information, So, if the space is empty, it will not put anything in there. I use this option frequently while painting but not so much for editing.

With that, we have made the brush and we know how to adjust it once it's been created. Now, it is time to apply it to an edit.

## Applying the basics of brushes in an edit

In this section, we will use the brushes we created in an edit. However, the last part of successfully using brushes is being able to adjust them inside the image.

> **Quick editor's note**
> There is no perfect brush that is 100% applicable out of the box for every image. You will always have to adjust and tweak the brush so that it fits the image.

To work with the context toolbar for the brush tool, we will be using the image shown in *Figure 9.12*. We have edited the image using brushes so that you can follow along and have also included the working file in the downloads for this chapter (see the Bokeh practice file RAW). This image has been edited a significant amount, so it is a good example of how adjustment layers, masks, and selection can be stacked. Notice in *Figure 9.12* that we have used some of the techniques from earlier in this book.

The settings for the color are shown in *Figure 9.12* so that you can use the same color for your work:

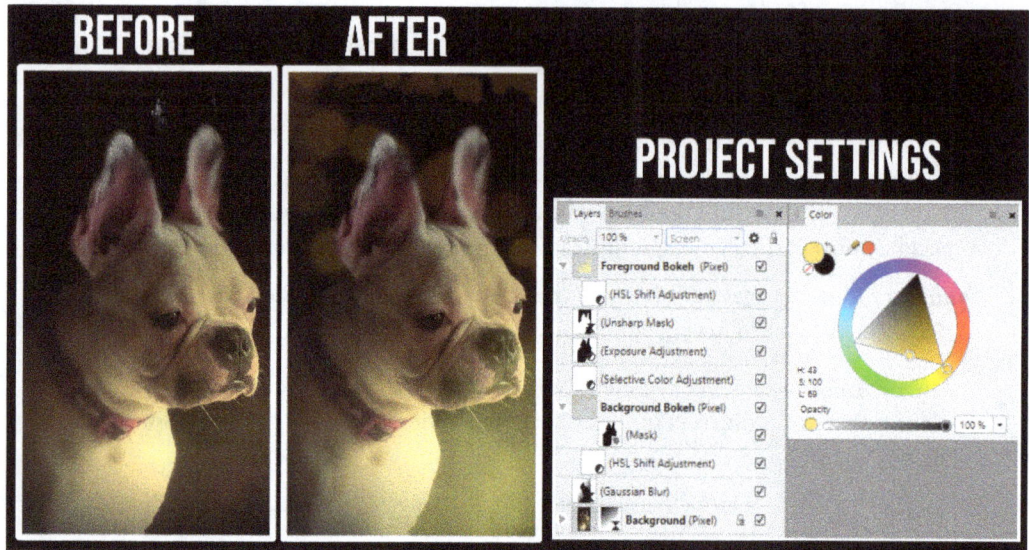

Figure 9.12 – Before and after

Let's look at an overview of the file before applying the Bokeh effect. As an editor, you must learn to read the layer stack so that you know what is happening. Doing this forensically with other editors' work is a great way for you to develop this skill:

- We used selection and spare channels to create the masks and separate them into background and foreground

- We used **Gaussian Blur** to make the background blurry and fade out

- We added **Unsharp Mask** to sharpen the foreground of the dog (remember what you learned about atmospheric perspective – objects in the foreground are sharper)

- We added **Exposure Adjustment** to drop only the background (back to the atmospheric perspective)

We will only be using the Bokeh brushes on the following layers since that is the intent of this lesson:

- **Foreground Bokeh** (see *Figure 9.12* for its location in the layer's structure)

- **Background Bokeh** (see *Figure 9.12* for its location in the layer's structure)

I have included this flow in a video if you would like to see it in action (the link can be found in the *Technical requirements* section).

## Steps in the edit

Follow these steps to edit the photo:

1.   Select the **Background Bokeh** layer. Make sure you are on the layer and not the mask.

2.   Select the Bokeh brush and a color.

     I am using the smaller of the two brushes, which has pressure sensitivity. If you do not have a tablet, go into the brush and change the controller to **Random** – it will work similarly. Do not be afraid to adjust the settings in the context toolbar for things such as **Size** and **Opacity**.

3.   Paint the Bokeh the way you like it.

4.   Change the blend mode for the layer and adjust its **Opacity** setting.

5.   Select the **Foreground Bokeh** layer.

6.   Select the Bokeh brush and color you want to use. I am using a large brush in the **Packt** category, without pressure sensitivity.

7.   Paint the Bokeh. I am doing single clocks to control the Bokeh effect with each dot.

8.   Adjust the blend mode of the layer and its **Opacity** setting to your taste.

While this edit was 100% about the use of brushes, the structure of this edit shows a far more advanced flow, including how to use spare channels to mask and how to use various blend modes to create effects. All this has helped you develop as an editor and build on the previous chapters.

# Professional tips, tricks, and important points

Here are some pro tips for you:

*   When editing and masking, rarely if ever will I use a hard round brush. Most of my editing work is done with soft round brushes with various levels of softness – it depends on how aggressively I want the edge to fade away.

*   If you want to see how an artist created their brushes for any of the brushes that you have bought or come natively loaded into the program, click on one of them – all the settings will be there for you to try and replicate. This will help you learn how the professionals build their tools.

*   I keep my **Flow** low (usually at 30% or so) and my **Opacity** high so that I have more control over strokes. This also allows me to build up more gradual effects.

*   Always duplicate brushes you want to change permanently and place them in a new category. This helps you avoid making irreversible changes to the brush and forcing you to reinstall it.

Here are some important points for you to note:

- Make sure you are on the right layer; I cannot tell you how many times I have painted on a past layer and not realized until it was too late

- Lock the layers you are happy with to avoid painting on the wrong layer

## Summary

We made a ton of progress in this chapter. We started by identifying the panel for brushes and then learned a bit about brush management. By doing this, you can now read a brush preview, as well as dig into it and take on the basic panels that control the general and dynamic settings. However, the star of the show was you creating a Bokeh brush and applying it to an edit.

In the next chapter, we will cover the last truly fundamental thing you need to understand in Affinity Photo, or any art application for that matter – the idea of color. So far, we have survived by recognizing that there are panels for certain things, and we have chosen some basic colors. In the next chapter, we will explore color as an editor to add a signature to our edits. See you soon!

# 10
# Working with Color in Affinity Photo

In this chapter, we will cover how to utilize color in photo editing, including the tools used and an in-depth discussion of the software models of color. An artist's color choices reflect the mood of the image and help define the artist's style. The same image, given a different color scheme, will convey a completely different mood to the viewer, so it is a strategic decision by the editor to use color based on the intended feeling they are trying to convey.

In this chapter, we will start with the basics of color and look at the panels that influence color, including the color panel and the swatches panel, both of which will be used to create that perfect edit. From there, we will cover the three most used color spaces in editing to create radically different effects. After that, we will cover various tools and adjustment layers that we use in daily edits to influence color. We will wrap up this chapter with a practical edit that will use color and the tools discussed throughout this chapter.

In this chapter, we will cover the following topics:

- Working with the **Color** panel and the **Swatches** panel
- Color spaces in photo editing
- Tools and adjustments used in color
- Practical editing using color grading
- Professional tips, tricks, and important points

## Technical requirements

The CiA video of this chapter can be found on `https://packt.link/YrEtF`.

You will also need the `Chapter 10` project to follow along with me.

# Working with the Color panel and the Swatches panel

In this section, we will explore the visualizations of color in Affinity Photo, how we see color in the program, and how to choose a color for our images. We will also explore the use of pallets, which is how we can save colors for later and also share them with others. This section is not about color theory but rather about the mechanics of how the Affinity Photo program shows us color.

## Color studio panel

The studio panel for color can be found in the **View** menu in the studio. If there is a checkmark beside **Color**, then it is already on the right-hand side of the interface – we just have to drag it out.

This was covered earlier in this book, so it should be easy to find. However, be aware that the look of the panel may be significantly different than what I have shown in *Figure 10.1*. This is because, depending on the version of the program and your settings, it may be showing a wheel or a slider. So, if yours looks different, don't panic. The same information on how to read the panel is still applicable, and we will cover the different visualizations of the color panel in the following sub-section.

### Reading the Color panel

This is what the **Color** panel looks like:

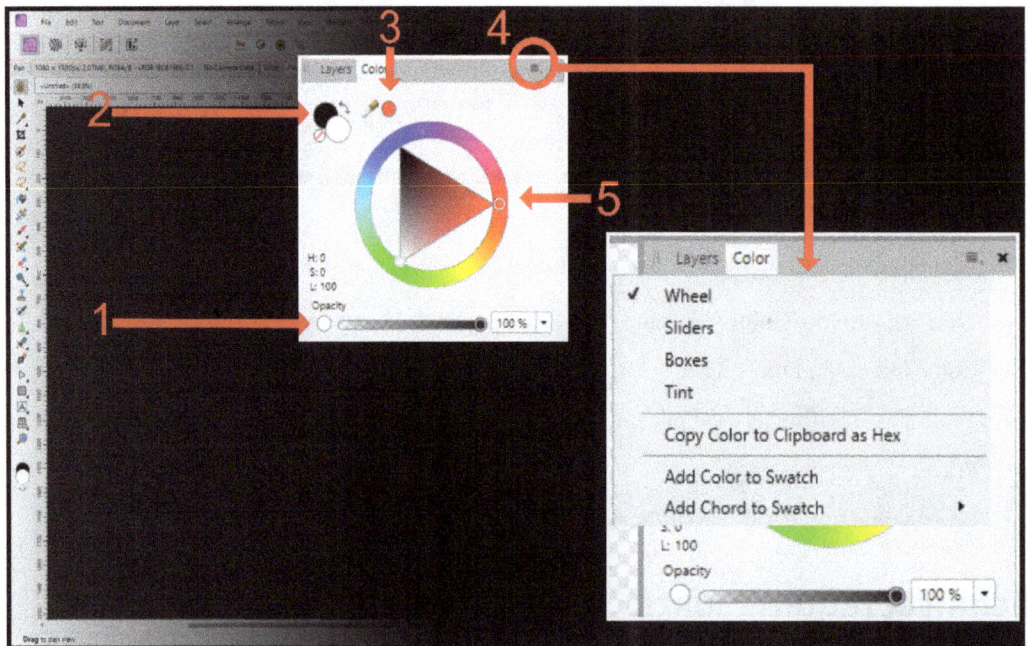

Figure 10.1 – Reading the Color studio panel

Referring to the preceding figure, let's go over the different options in the **Color** panel:

- **Opacity (1)**: This impacts the opacity of the color. By default, the color is at **100%**, so if you are looking for a lighter opacity, adjust the value here.

- **Primary and secondary colors (2)**: This shows the last two colors you used. By default, the circle in the foreground is the one that will be applied. To select the other color (the one behind), simply click it – that color will become the one in front.

- **Color picker (3)**: This tool allows you to choose a color in the image and match it. To use it click, hold, and drag over to the color you want to choose. Then, when you release, the color will be present in the circle next to the dropper. To get the color over to your area, simply click on the color next to the dropper.

- **The hamburger menu (4)**: This is where you can choose the view, chords, and other options from the menu.

- **The color visualization (5)**: This is how you can see the color you are working with. Various views can be selected through the hamburger menu; they will be covered in the following subsection.

### Adjusting the panel for various views

When it comes to color visualization, there are several views we can use, and it depends on you, as the artist, which one you like. I rarely stray from the wheel, but it is up to you. The following figure shows the various visualizations present in Affinity Photo:

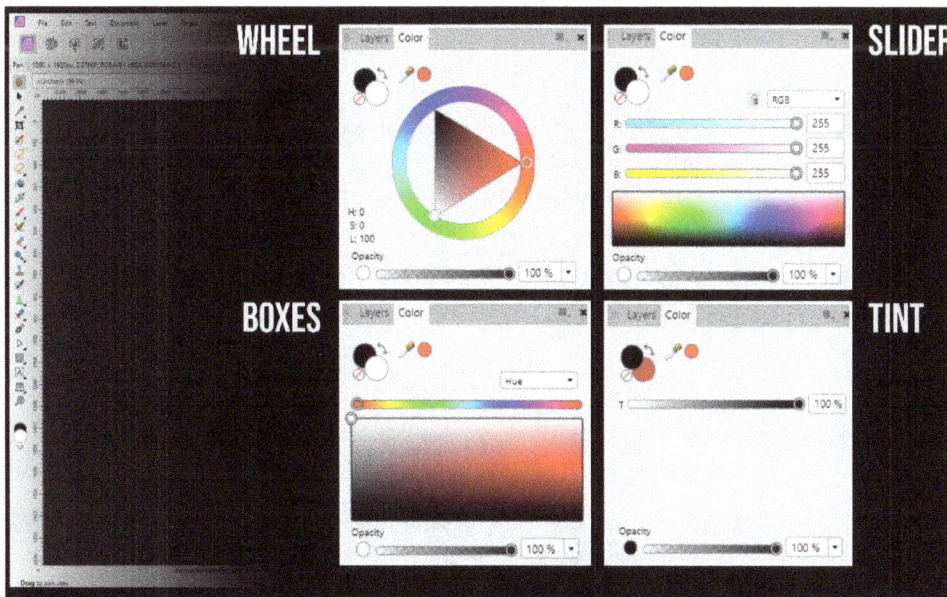

Figure 10.2 – Different views for color in Affinity Photo

As mentioned previously, there is no perfect way to describe exactly when to use a certain view and I use the wheel for 90% of my work. However, if I need various tints for a color in a monochromatic piece, I may use the **Tint** view after selecting my color. While it is impossible to tell you all the ways to use these visualizations, I have tried to describe the strengths and limitations of each in the following subsections.

### Wheel

I use the wheel 90% of the time when I edit because it is simple to understand. However, a real limitation of the wheel is getting consistent colors and variations of color. Because you can adjust it so freely, while it may look like you are using the same shade of green as last time, you might not be. Also, if you need a similar light shade of green and then later a yellow shade, it is tricky to find a similar lightness value.

If you are going to use the wheel, use swatches so that your color pallets are the same all the time.

### Sliders

Sliders allow you to set numeric values in the color space. In the example shown in *Figure 10.2*, notice the top right-hand corner, specified as **SLIDER**. The color space is RGB, so the options for the sliders are red, green, and blue. I use this visualization when I need repeatable colors (I write down the values on scratch paper), especially if I am not quite sure of the color pallet for my edit yet.

A limitation of this view is that the saturation and lightness control is not as intuitive as other visualizations, so you end up with very saturated colors.

### Boxes

This visualization gives you the option to slide between **Hue**, **Saturation**, and **Lightness** to make selections, so this is awesome when you want the color you choose to be similar in saturation or lightness if you choose another one. To switch between these options, simply go to the top right of the panel (shown as **Hue** in *Figure 10.2*) and change the selection.

A limitation of this is the repeatability of selection after the fact. Let's say you chose magenta – you may not get the same magenta again, so please use swatches and pallets to aid in repeatability.

### Tints

Once you've chosen a color, the **Tint** view allows you to carefully control the tint of that color. I typically use this with black to get consistent values of gray in my work.

The only limitation of this method is that there are no saturation or hue options and we have to set them separately by picking the color first using the dropper.

Now that we know how to view color, let's dig into how to save the colors we choose using the **Swatches** panel.

# The Swatches panel

When you begin working with color, it is best to pair it with the **Swatches** panel. This panel can be found alongside the rest of the panels in the **View** studio panel; make sure it is selected. The **Swatches** panel got its name from how you go to the store to pick out paint and leave with small cards that show the finished color, which is called a swatch of color. Swatches are contained inside the panel in a container called a **pallet**.

Swatches live inside pallets. The **Swatches** panel is essential in the workflow of color, and it serves as a repository for the colors you have used.

We have included an image I edited with a very blue, somber type of color grade (the image is included in the downloads for this section; see `Swatch panel reference image`). We will use this image to make our first pallets and review the world of swatches and color management.

## *Reading the Swatches panel*

In the following figure, I have broken down the **Swatches** panel into a few distinct areas:

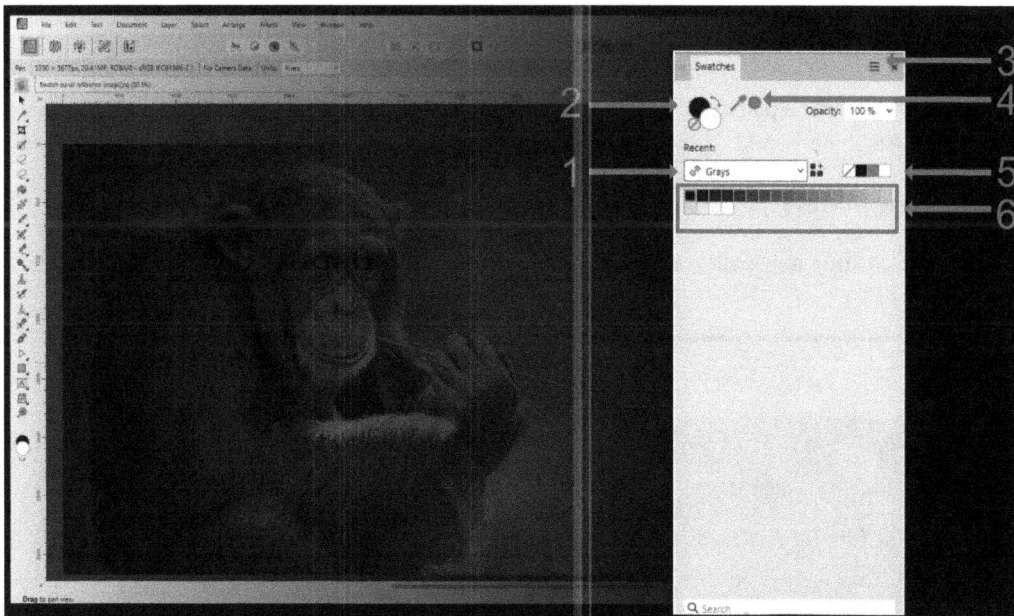

Figure 10.3 – The Swatches panel

Let's explore these options:

- **Pallet name (pallet selected) (1)**: This is the pallet we are working with. In *Figure 10.3*, it is called **Grays**. The Affinity symbol next to the title denotes whether it is an application or a document-specific pallet (these terms will be described later in the Creating new pallets section).

- **Color (current and alternative) (2)**: This is the same as it is in the **Color** panel. The circle at the front shows the currently selected color, while the one at the back represents an alternative. To switch, just click the circle in the rear to bring it forward.

- **The hamburger menu (3)**: This is where all the options are set for the panel, including pallet creation.

- **Color picker (4)**: This is the same as it is in the **Color** panel described in the previous subsection.

- **Recently used colors (5)**: This panel shows the recent colors you have used. In this case, pay close attention to the white square with the red line; this means no color. No color is an option you can set if you would like to remove the color.

- **Swatches included in the pallet (6)**: This area shows the swatches included in the pallet. Adding and subtracting these will be covered in the Adding and subtracting swatches to/from a pallet section.

Now that we know what a palette is and where to find it, let's take a look at some options we have to create new pallets in our work.

### Creating new pallets

The process of creating new pallets is simple. Go to the hamburger menu; you will see two options (see *Figure 10.4 (1)*):

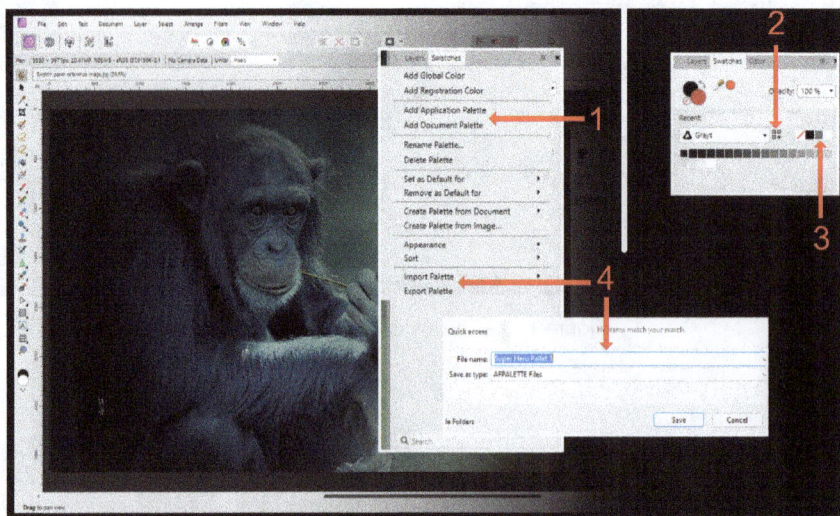

Figure 10.4 – Adding pallets, adding swatches to pallets, and exporting pallets in Affinity Photo

Let's go over these options:

- **Add Application Pallet**: When you select this option, the pallet will be available in Affinity Photo for every document
- **Add Document Pallet**: When you select this option, the pallet will only appear for a particular document and not for any other document you open

I use document pallets for project-specific needs. However, if there are themed pallets such as a "superhero color pallet" that colors images as Marvel does, it is an application pallet since I use it in various projects.

So far, you have a pallet. However, you need color to fill that pallet. How do you get color onto your pallet, and what happens if you have to remove a color from your pallet? All of this will be covered in the next section.

### Adding and subtracting swatches to/from a pallet

There are two ways to add swatches to a pallet:

- **Add the current fill to the pallet**: With the color you want in the foreground circle, click on the add to pallet icon, shown in *Figure 10.4* (**2**). This will place the swatch color in your pallet.
- **Add a recent fill to the pallet**: If you forgot to grab the color and have switched it off, chances are that you can recover it from the recently used colors (*Figure 10.4* (**3**)). Simply click on the color; it will move into the foreground circle. Then, click on the add to pallet icon.

  If you add a swatch and do not like it, simply right-click on the swatch and hit **Delete Fill**.

---

**Important note**

Gradients can also be added to swatch pallets, but it will differentiate whether or not it is a linear or a radial gradient (these terms will be defined in the *Gradient tool* section of this chapter).

---

Since we have come to the end of this section on pallet creation, I have included a CiA video to show how to create, rename, and add swatches to the pallets (the link to the CiA video is mentioned in the *Technical requirements* section).

### Importing and exporting a pallet

Importing and exporting a pallet allows you to share or load pallets others have created into Affinity Photo. This is shown in *Figure 10.4* (**4**). Here, you will see options in the menu for importing and exporting:

- Click on the **Import** or **Export** option in the menu for the **Swatches** studio panel.
- A screen will appear, asking you where you want to save an export, or where the import is located.
- The file will be saved as an Affinity pallet file; see *Figure 10.4* (**4**) for the file extension.

We have included one of our superhero pallets in the downloads. We will not use it in the chapter, but it does give you a professional pallet that you can use.

Now, what if you admire other pieces of art from other editors, or you have been given a flat image by a client and you have to pull a pallet off from the flat image? Well, we can make that happen too, and that is exactly what I will show you in the next section.

## Getting a pallet from a piece of art

This is one of the coolest features of digital art in my opinion. Affinity has two functions to allow you to analyze and utilize the colors from another image, creating a pallet based on your favorite editors and artists. You will want to do this because of the following reasons:

- To pay homage or find inspiration in others' work. I am into comic books and how comic companies do their color grading on their movie posters; heroes have always inspired me.

- If you are retouching, creating a pallet of colors that have been used allows you to match very closely and create elements that are similar in tone.

To do this, there are two options:

- **Create Pallet from Document**

- **Create Pallet from Image…**

I have included the chimpanzee image I color-graded. Please follow along; you will find the image in the downloads for this chapter.

### Creating a pallet from a document

With the document open in Affinity Photo, select the **Create Pallet from Document** option from the **Swatches** panel:

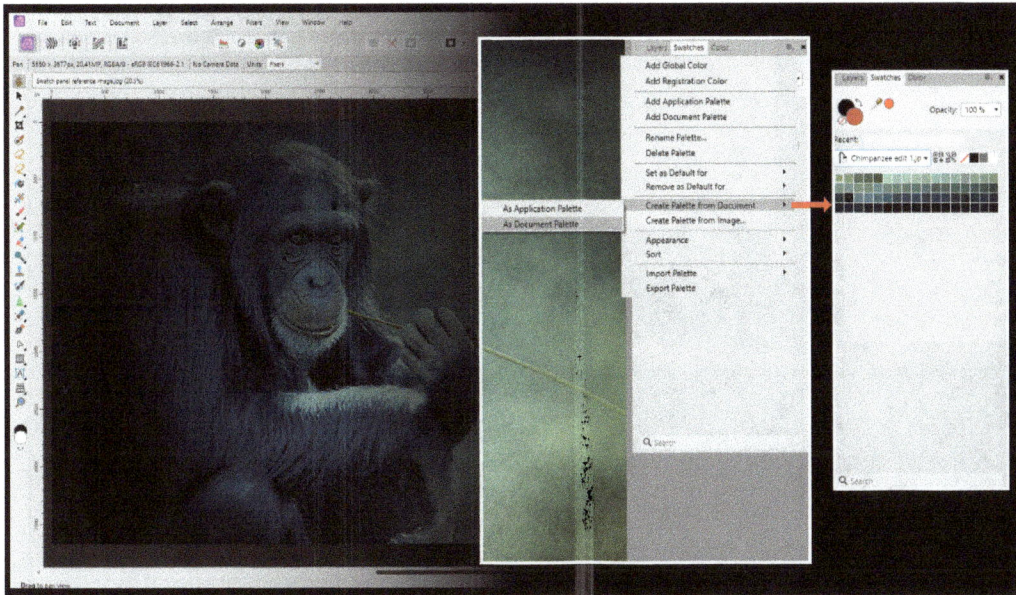

Figure 10.5 – Grabbing a pallet from a document

This will create a pallet from the open document. My only critique of this method is the sheer number of different colors it creates in the pallet. Typically there are far too many to be usable in an actual edit.

We have included this color pallet also in the downloads for this chapter in the book (import it just like we talked about previously if you want to use it).

## Creating a pallet from an image

In this section, you will be creating a pallet from another image, not the image in your document. To accomplish this, follow these steps:

1. Select **Create Pallet from Image** from the **Swatches** panel (shown on the left-hand side of *Figure 10.6*).
2. Select the image from the location on your computer; see *Figure 10.6* (**1**).
3. Select the number of colors you want Affinity to create (in this case, **5**). See *Figure 10.6* (**2**).
4. Select whether you want an **Application** pallet or a **Document** pallet:

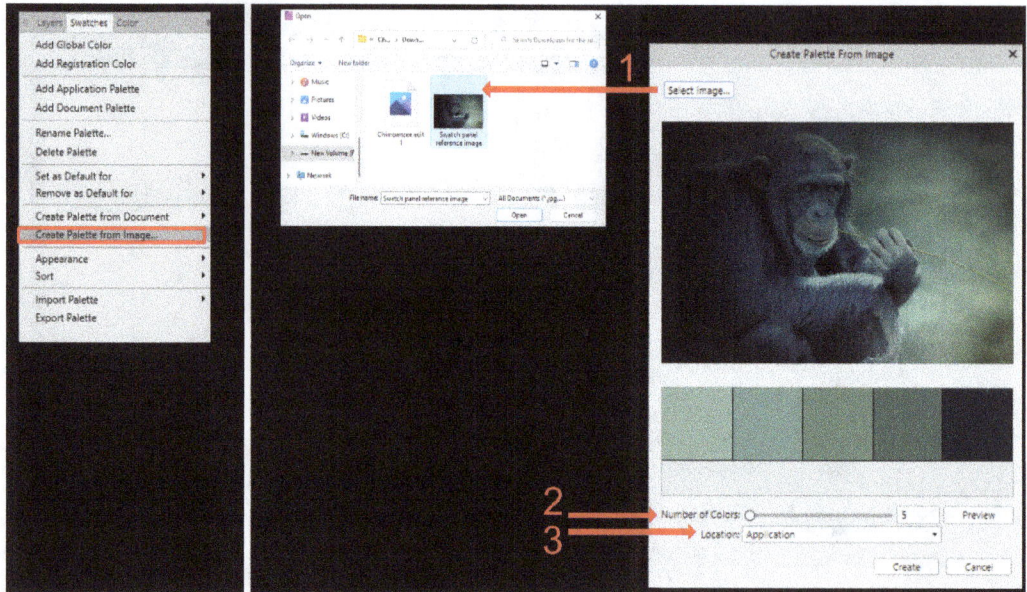

Figure 10.6 – Adding a pallet from an image

As we close this section on pallets, I have included a CiA video (link mentioned in the *Technical requirements* section) showing you how to create pallets and add color to them. This video will show the following techniques:

- Creating pallets from a document
- Creating pallets from an image

Now that we know how to work with color from a technical viewpoint in Affinity, it is time to discuss the various color spaces. This is sort of a theoretical discussion around the three most popular color models in digital art, along with a technical explanation of how to switch your art between spaces using Affinity.

An example of this would be if you started an RGB color and then decided to print halfway through and your printer used a CMYK color space (these terms will be defined in the next section). So, enough with the preview – let's get into the feature.

# The three most used color spaces in photo editing

You are going to hear the terms color spaces and color profiles when moving into your editing career. I am going to give you a simple, non-technical breakdown of them as I use them; how deep you want to go after that is up to you:

- **Color spaces**: Color spaces are different ways color is represented using math – for example, RGB, CMYK, and LAB. Each of these models uses math to represent color differently.

- **Color profiles**: Color profiles are what you assign or download. As an example, if you were to print your image on a Canon printer, you would download the Canon printer color profile; it would convert your colors into something the printer can match as an equivalent.

We will go over the three most common color profiles in this section. While it is outside the scope of this book to describe the technical terms of each in detail (people have entire manuals on color theory and grading), we have included what you need to know – that is, what they do and when you should use them.

The following figure shows the three most popular color spaces available and how those spaces create the various channels in Affinity Photo:

Figure 10.7 – The three color spaces available in Affinity Photo

We have included a color tab and a channel tab for each of the color profiles. While the print copy of this book may not show the subtle differences, notice the channels that were created for each in the **Channels** menus shown in *Figure 10.7*.

## The three color spaces

The three color spaces we will be discussing are RGB, CMYK, and LAB; each is favored by various artists for various reasons. I use almost exclusively RGB and sometimes venture into LAB for portrait edits that have high light values. That being said, let's explore them so that you have a better understanding of them.

### RGB

**RGB** stands for **Red, Green, and Blue**, which are the colors that are available in electronic devices. So, the color you see is a combination of a certain amount of red, green, and blue mixed to form the color your eyes see.

It is worth noting that RGB is additive. But what does that mean?

If the values of red, green, or blue are zero, then you have the deepest darkest black. However, if you crank the RGB values to the extreme, you get the brightest white. This is different than CMYK, which is subtractive (that is, no color is bright white and as you add colors, it gets blacker).

To try this, I have included a CiA video, where we adjust the sliders to prove this for the **RGB** space).

99% of the work I do is in the **RGB** color space, and all the printers I have worked with convert these colors just fine. However, if you want to be technical, you will want to use this color space for digital-based work that will most likely be shown on screens.

### CMYK

**CMYK** stands for **Cyan, Magenta, Yellow, and Black** and mirrors the colors of the cartridges in your printer. These colors create every color you see. Notice in *Figure 10.7* that the wheel visualization is subtly different than the **RGB** or **LAB** space. This is because it is using CMYK as physical print inks.

CMYK is subtractive, which means when all the colors are present, you get perfect black and as you remove them, the absence of color is white. To illustrate this, I have included a CiA video, where I begin with all colors and then work my way to no colors.

CMYK is typically used when you are planning on using a printed medium. Always check your printer for its color requirements as many of them will prefer RGB in the days of digital.

### LAB

**LAB** stands for **Lightness, A channel, and B channel**. In this model, we create a 3D color space with red on one end and green on the other (this will be referred to as A):

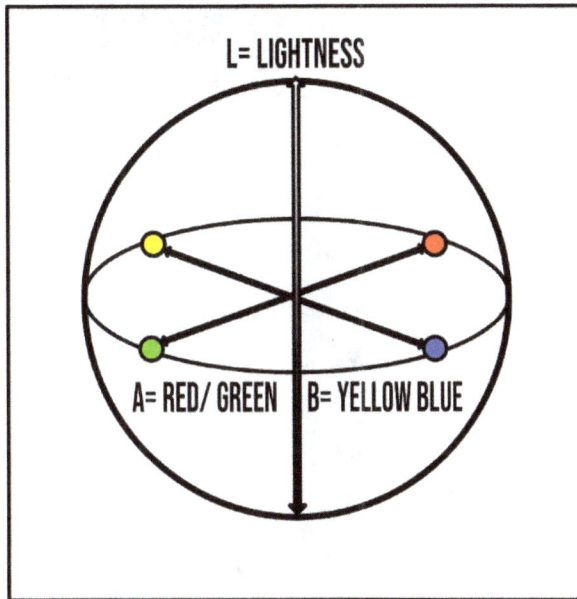

Figure 10.8 – The LAB color model

Notice in the CiA video that the last color space we adjust is the LAB. Moving the lightness from white to black makes the color absolute. Then, as we move away from the true center on either axis, we get pure color values.

The LAB color model is a three-dimensional model space. There are two competing colors on each axis (on the A channel – red and green), and as you move toward one end or the other, the color becomes a different hue – either more red or more green. However, if we add a horizontal lightness axis, that red or green hue becomes darker or lighter.

I know this might be hard to wrap your head around, but if you want to experience it, open an image in a LAB space and adjust the channels using the Curves Adjustment layer. This will make everything clearer.

Now that we know about color spaces, we will learn where they are in the menu and how to adjust them. Affinity uses the term *profile* to describe the switch between the various color spaces.

## Setting the color profile in the options menu

To set a color profile and choose between the various spaces available, follow these steps:

1.  With the document you want open, go to the **Document** tab.

2.  Choose **Convert Format / ICC Profile…**:

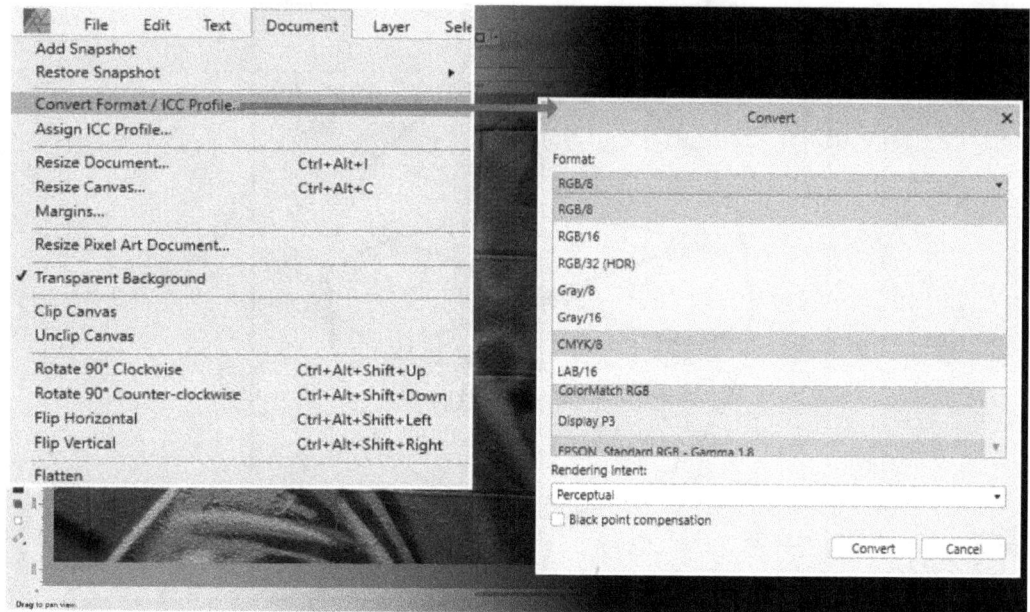

Figure 10.9 – Convert Format/ ICC Profile… in the Document tab

3.    Choose the profile you are looking for.

> **Note**
>
> We will not address the 8- versus 16-bit differentiation here. If you are interested in that topic, please refer to *Chapter 13, Advanced Color Concepts and Grading*, for a discussion on the differences.

With that, we have finished exploring different models for color. After the LAB discussion, your head may be swimming. So, let's take our foot off the gas and look at the tools that are used to apply color in Affinity Photo.

# Tools and adjustments used in color

Now that we know all about color, I wanted to provide you with a quick list of the most common tools that I use daily when performing edits. I have not covered every tool that deals with color, but rather the tools that I use most frequently, so if you see a tool you use that I don't cover here, it may be elsewhere.

# The fill bucket tool

This tool is the simple workhorse of the color world and is the tool you use to fill things. When I use the term *"fill things,"* I am referring to pixel-based items. Here are some examples of what you can use the fill tool on:

- Pixel layers
- Selections
- Adjustment layers
- Mask layers

You can't use this tool to fill shapes, text, or anything else that is not pixel-based.

## *Using the fill bucket tool*

With the tool selected, simply choose the color you want from the color menu, then click on the object you want to fill. The layer or selection will be filled with that color.

The tool will read the color of the layer and fill it with colors of a similar value. If you get bad results, up the **Tolerance** value in the context toolbar.

Remember that mask layers work in black and white, so you will not see any color. However, it will take the value to create an opacity that's consistent with that value of gray.

Let's look at the fill bucket tool's options to see how it can help us:

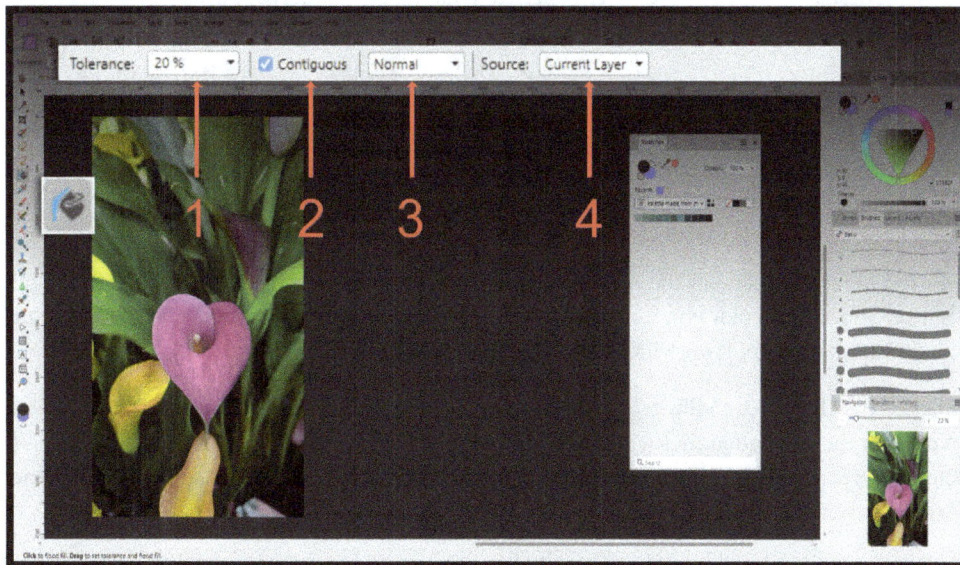

Figure 10.10 – Fill bucket tool

Referring to the preceding figure, let's go over the different options for this tool:

- **Tolerance (1)**: Tolerance tells Affinity how much to fill certain colors close to the clicked color to apply the fill to. To cover more area, increase the **Tolerance** value.

- **Contiguous (2)**: Contiguous means *touching*, so if this option is selected, it will only fill those values that are touching. If you want it to look at the entire image, uncheck this box.

- **Blend mode (3)**: You can change the blend mode for the fill if you want.

- **Source (4)**: This specifies what layer Affinity is looking at to see what value to replace with the fill.

### *Using the fill bucket tool for editing*

We have included a practice file for this tool so that you can follow along. It can be found in the downloads for this chapter (see the `Fill bucket tool practice file` Affinity Photo file). Here are a few takeaway points:

- There is a solid blue background layer. This will allow you to see the effect of the tool on the **Mask** layer.

- There is a spare channel titled **Flower** in the **Channels** panel. This will allow you to fill in the selection.

- I added a **Pixel** layer at the top so that you can see how to fill a full layer.

- There is **Lens Filter Adjustment**, which we will use to show the effect on the **Adjustment** layer.

This file is not designed to be an edit – it is designed to have a mechanical application for each of the preceding methods, so do not expect this to be a great image. We'll edit the image using these techniques:

1. To use this tool in a Pixel layer, simply choose a color. Then, with the tool selected, click on the layer.

2. To use this and fill a selection, you must make a selection (this is the spare channel with the flower in it for the working file), choose a color, and click inside the selection area. This area will flood with the chosen color.

3. To use the bucket tool on the **Adjustment** or **Live Filter** layers to mask, in the working file, simply choose the color you want to use. Affinity will use the lightness value of the color and mask the adjustment layer with the appropriate **Opacity** value.

I have included a CiA video to illustrate this concept (see the *Technical requirements* section for the link). Notice in the CiA video that when I clicked on the flower selection, it did not fill the flower completely. This is because the **Tolerance** value was too low. We will need to increase the **Tolerance** value to get a better fill.

Now that we know how the fill bucket tool works, let's move on to the next commonly used tool – that is, the gradient tool.

# The gradient tool

When we think of color, one of the most powerful tools is the gradient tool. This tool allows us to create smooth transitions from one color to another, or from an opaque color to a transparent color. It works on the same sort of layers as the fill tool:

- Pixel layers

- Selections

- Adjustment layers

- Mask layers

You can't use this tool to fill shapes, text, or anything that is not pixel-based.

## Using the gradient tool

We have included a practice file to show you how to apply the technical aspect of gradients (see the `Gradient tool practice` file). The working file has two layers:

- A red upper layer

- A blue lower layer

With the gradient fill tool selected, choose the red upper layer and simply click and drag – this will create a line similar to the one shown in *Figure 10.11*. I have a black and gray gradient selected, so your colors may vary, but the line is the same:

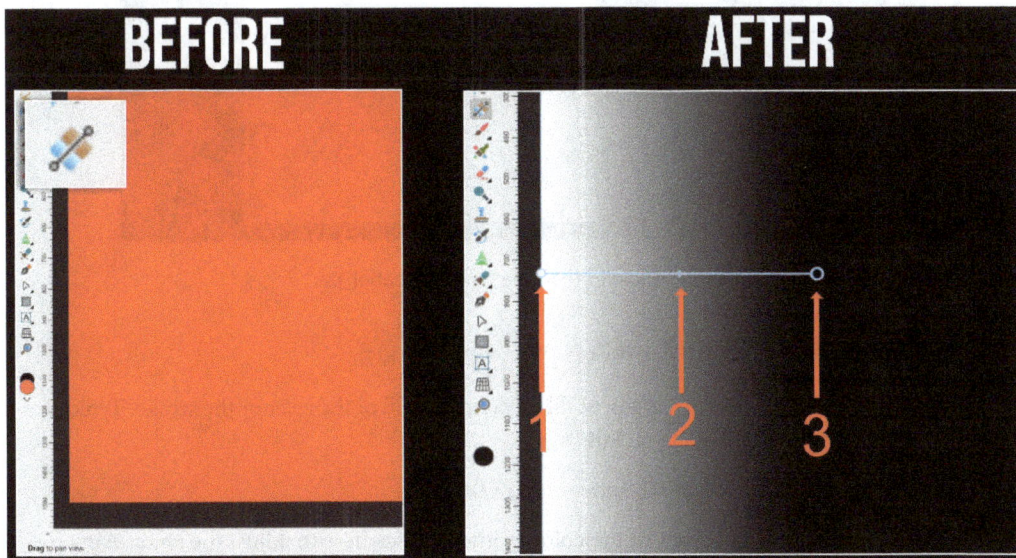

Figure 10.11 – How to read the gradient tool, part 1

Let's understand what is going on in the preceding figure:

- **(1)**: This white circle is the origin circle and is the first color. This is where you started the click and drag.

- **(2)**: This is the mid-point. Click and drag it to adjust the position from the first color to the second color.

- **(3)**: This is the second color or where you ended the drag.

### Reading the context toolbar of the gradient tool

The context toolbar of the gradient tool is shown in the following figure:

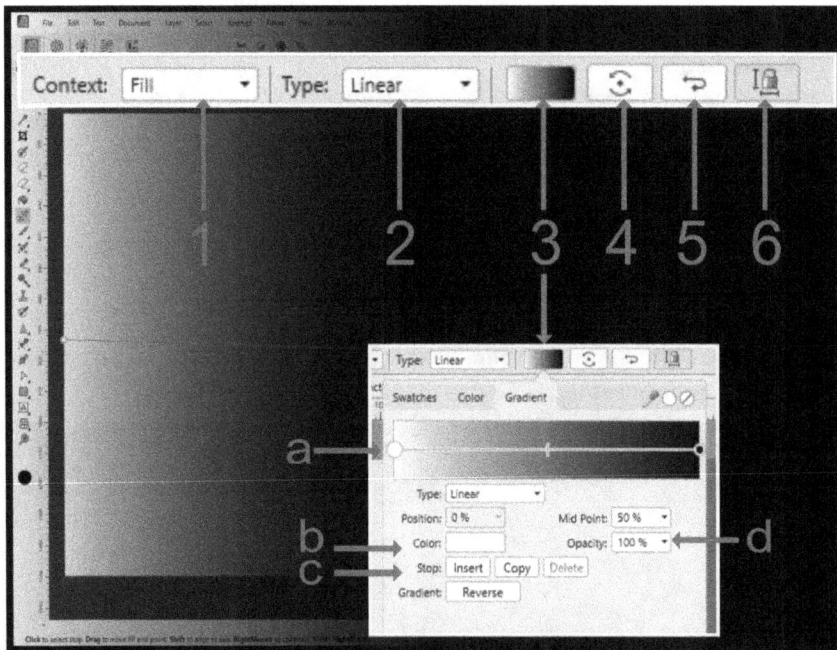

Figure 10.12 – How to read the gradient panel, part 2

Let's go over this toolbar and learn what each option does:

- **Context (1)**: You can perform a gradient fill on either the fill of the layer or the stroke. Typically, for photo editing, it will usually be **Fill**.

- **Type (2)**: Options include **Radial**, **Linear**, and **Elliptical**.

- **Color (3)**: This shows a preview of the color gradient. Click here to adjust the color using the instructions provided in the next section – that is, *Adjusting colors*.

- **Rotation symbol** (**4**): Rotates the gradient's direction.

- **Reverse symbol** (**5**): Reverses the gradient's direction. I use this all the time as I typically drag it out backward. So, instead of redoing the gradient, I just flip it.

- **Aspect ratio** (**6**): If this is checked while using the **Radial** gradient, the colors will shift in equal proportion from the *X* and *Y* directions. Make sure this is unchecked when you're using the **Elliptical** gradient.

Next, let's learn how to adjust colors from this toolbar.

## Adjusting colors

The properties for adjusting the colors in the gradient can be seen in *Figure 10.12* (**3a**). Let's look at this box in more detail:

- (**a**): This is the preview of the gradient. Click on the dots to change the color of the beginning and end. Click on the line to add stops (opportunities to add new colors). This will create complex gradients.

- **Color** (**b**): Changes the color of the selected circle or stop.

- **Stop** (**c**): This adds color to the middle of the gradient. Here, you can add stops by simply clicking on the line or remove stops by clicking on them and selecting **Delete** in the dialog box.

- **Opacity** (**d**): Adjusting the **Opacity** value can make a gradient go from fully opaque to transparent. We will use this in the editing example later in this chapter.

Reading a gradient can seem daunting, but the easiest way to learn is by using the tool and trying it in an application. We'll learn how to use the gradient tool in editing in the next section.

## Using the gradient tool in editing

To show you how to use the gradient tool in editing, I have added a simple sky color grading activity that utilizes a simple color linear gradient. In the following edits, we will be using a radial gradient, but for this one, we'll keep it simple. I have left the pixel layer as-is and switched **Gradient** to **On** or **Visible**. So, if you want to copy what I've done, you can find it in the Linear gradient editing project file in the downloads of this chapter.

This image was taken by the Atlantic Ocean during a family trip. As you can see, it lacks pop, so using a linear color gradient, we will make a simple adjustment that makes a ton of difference. Here are the steps:

1. Create a **Linear** gradient on a new **Pixel** layer, starting at the top and dragging down about halfway.
2. Using the context toolbar for the gradient tool, click on the **Color** box (**3b** in *Figure 10.12*) and select a deeper blue for the origin color and a yellowish color for the end color.
3. Change the blend mode of the **Pixel** layer to **Overlay**.

4.    Reduce **Opacity** to around **40%**.

5.    Play with the colors, blend mode, and opacity to your liking.

I have included my settings in *Figure 10.13*, including my before and after:

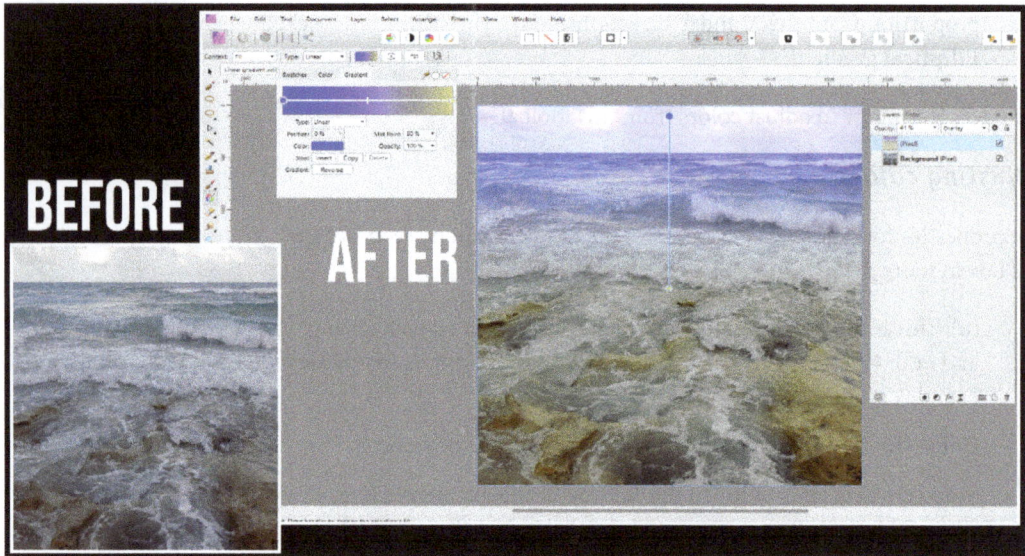

Figure 10.13 – Applying a linear gradient

With that, we know what a gradient is and practiced using a simple gradient in a real application. But now, it is time to add color grading and up the gradient game a bit by making a truly unique piece of art using our new knowledge. So, let's get into the practical part of this chapter!

## Practical editing using color grading

Now, it is time to practice how to use gradients in an actual edit. In your downloads for this section, we have added a practice file, and I have edited the file using techniques we have discussed previously to a point we can explore gradients (see the `Practical edit flower file` in the downloads). I have included a screenshot of the image in the following figure:

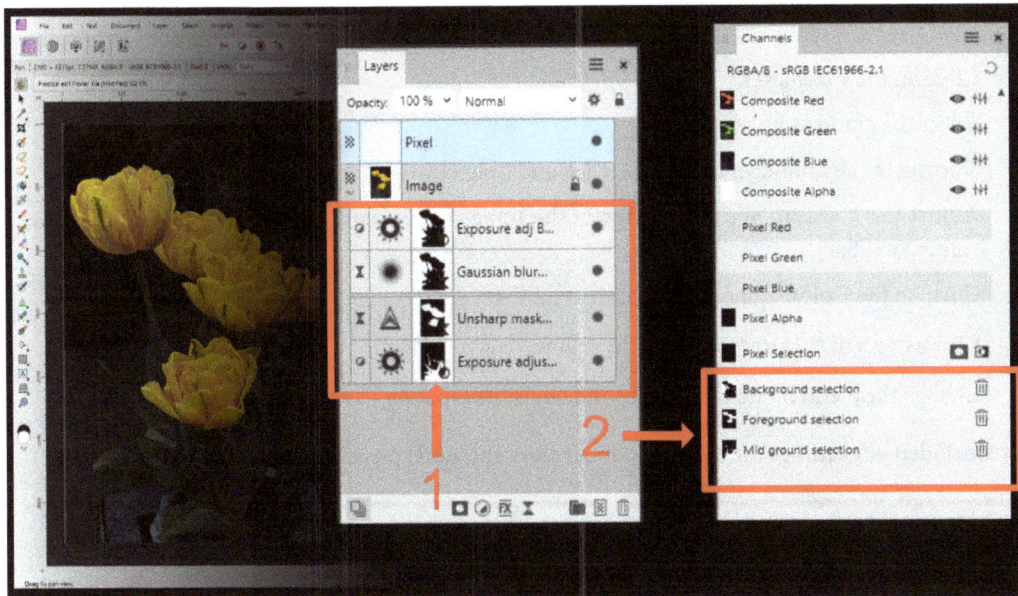

Figure 10.14 – Editing the flower file – project breakdown

Notice a few things before we get started:

- **(1)**: I edited the file with some existing edits. Here are some highlights:

  - I added exposure for the background and midground. A simple brush has been applied to the masks to make the adjustments visible in certain areas of the image (notice that the white thumbnails have black spots in them in item 1).

  - I applied an **Unsharp mask** to bring out the foreground and applied the foreground mask.

- **(2)**: I created selections for the **Foreground**, **Midground**, and **Background** areas. I have included them as spare channels for your continued use, so if you only want to apply an effect to the **Foreground** area, you can right-click on the **Foreground** channel and apply it to the **Pixel** layer.

Now that we have finished evaluating the file, it is time to add some edits to the baseline. So, let's dig in and start editing.

## Editing sequence

I have included a CiA video of us performing this operation, so if you get lost, make sure you check out the video (see the *Technical requirements* section for the link).

In the **Pixel** layer, just above the image, let's apply a color gradient using a **Radial** gradient:

1.  Click on the **Pixel** layer.

2.  Select the gradient fill tool.

3.  Starting in the middle of the image, click and drag outward.

4.  Adjust the *Y* axis to get to the edge of the image (make sure the aspect ratio is off in the context toolbar).

5.  Click on the **Color** square in the context toolbar.

    I am using a dark purple with a light yellow.

6.  Change the blend mode to **Overlay** and adjust its **Opacity** to your taste.

I have included our settings for this, including colors and node positions, in *Figure 10.15*:

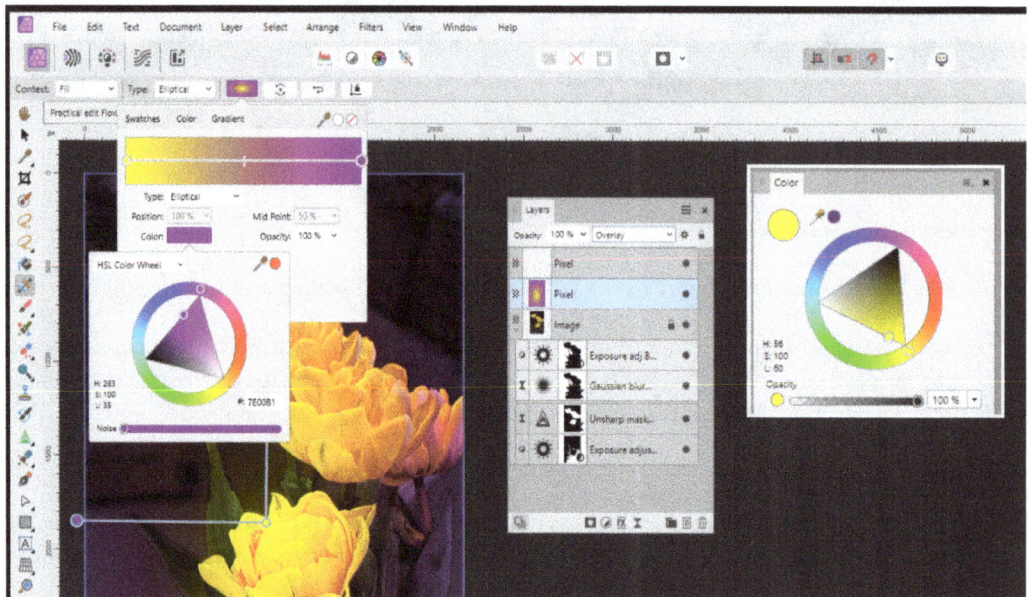

Figure 10.15 – Applying the gradient to the edit

Now, let's add some flash orbs to the image.

## Creating flash orbs

Now that we have the initial color figured out, let's add some magical lights to the flowers so that the lighter value makes sense. To do this, we will create the effect once, duplicate it, and resize it. Let's get started:

1. Select the **Pixel** layer at the top of the image.

2. Select the gradient fill tool.

3. Select the **Radial** gradient and then drag outward from the center a bit.

4. Choose a light white-ish yellow for the origin and a more saturated yellow for the outer node using the **Color** box in the context toolbar.

5. Select the outer node and reduce its **Opacity**. This will create a nice fade outward that will blend well.

6. Change the blend mode to **Screen**.

7. Now, duplicate the flares and add them wherever it works for your artistic vision.

8. I placed two of the flares under the **Pixel** layer.

I have included my gradient settings in the following figure:

Figure 10.16 – Applying the flower lights

I have also included the finished file in the downloads (`Practical edit flower file_ Complete`) so that you can see exactly how this was done.

Lastly, I have included this in the CiA video if you want to follow along.

# Professional tips, tricks, and important points

Here are some takeaway points for you:

- Save your colors in a pallet, even if you are simply working on a sketch project. Matching colors will slow your workflow, and you may not get the same ones. So, even though it takes a few seconds, it is well worth it.

- Utilize artists that you like to gain inspiration from their color pallets. This will allow you to explore and grow at a faster rate than fumbling around on your own.

- When using gradients, never underestimate the transparency fade out. There are so many ways to draw focus and creatively color-grade using gradients, which makes it a tool I use for 90% of my edits.

# Summary

We did a lot in this chapter. We started by learning about pallets, swatches, and color spaces, but perhaps the greatest single thing you can take away from this chapter is the information provided about the gradient tool, which is by far the single most important tool to use to color, mask, and adjust in photo editing. Following that, we covered a real practical application where we made two gradients and created an amazing edit using lighting.

With this first half of this book out of the way, we can finally discuss more advanced compositions. Now that we have adequately explored color and the use of tools, we will shift to what I consider the purest creative expression of photo editing: the idea of composition. When we look back at our journey, we have undertaken all of the fundamentals, such as selection, masking, coloring blend modes, and more. All these aspects will come into play when we cover composition in the next chapter. I will see you there, but for now, go outside, get some Sun, and shoot some photos – you will need them for when we cover composition.

# Part 3 :
# The Practical Applications
# of Affinity Photo

Now that we have our technical side well in hand, we will look at the actual applications for this wonderful program and apply them to very common tasks. We will be creating multi-layer compositions, photo restoration, and retouching, as well as making our own special effects brushes to give our composition and design work an extra boost with a unique flair.

This part comprises the following chapters:

# 11

# Compositing in Affinity Photo

In this chapter, we will cover what I consider to be one of the most organic expressions of freedom when it comes to photo editing, and that is the area of compositing. Compositing means taking various images and combining them. It allows artists to expand past the world that exists and create whatever they want to make in their minds. This is a specialty of mine in photography, and I am excited to share the fundamentals of what I know with you.

In this chapter, we'll take a short break from the technical aspect and begin to create, and explore. However, we will use every skill we learned about in the first half of this book, so all the technical knowledge we've covered will come in handy. To explore compositing, we will look at some fundamental concepts in composition photography, followed by exploring my workflow for composites and where to get awesome images for your composites. This will lean heavily on the composition discussions we had during our cropping and atmospheric perspective discussions earlier in this book as all of these concepts will come into play.

Finally, we will wrap up with a longer, more in-depth composition where you'll learn about some of the Live Filter layers and adjustments we can use to achieve unified harmony when using multiple images.

In this chapter, we will cover the following topics:

- Fundamentals of compositing
- Basic compositing workflow
- Filter and layers used in compositing
- Practical composition edit
- Professional tips, tricks, and important points

## Technical requirements

The CiA video for this chapter can be found at https://packt.link/4FrOo.

You will also need the Chapter 11 project to follow along.

# Fundamentals of compositing

Before we dig into the software aspects of how to composite, I wanted to give you a solid foundation of compositions and the things that need to be universally considered. At its core, compositing involves arranging and carries over to all art forms, including art, photography, design, video, and drawing. It is crafted with universal principles (different mediums and tools). Even music has similar approaches to compositions. Learning the basics covered in this section will pay dividends throughout your digital career. So, let's cover some basics and fundamental rules.

## What is a composite?

A composition is simply a collection of images that are separated and then combined or composed to create a new image that isn't there in the real world. In effect, compositions tell a story. Many composition artists sit down and plan out the story before actually coming up with *how* it gets done.

Examples of compositions can range from a few images with a few layers (as shown in *Figure 11.1*) to several images and over 100 individual layers (as shown in *Figure 11.2*):

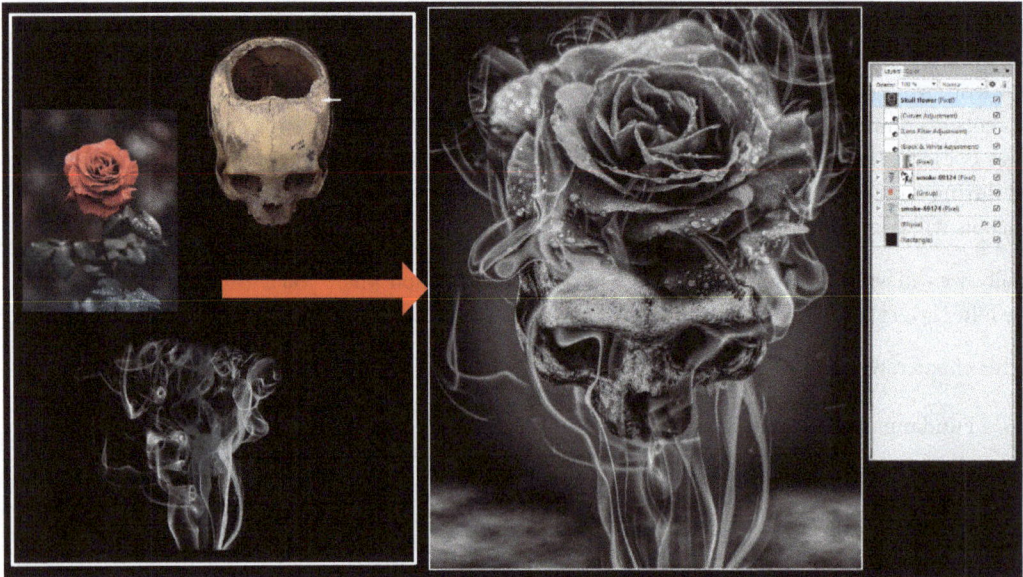

Figure 11.1 – A simple composition using a few assets and layers

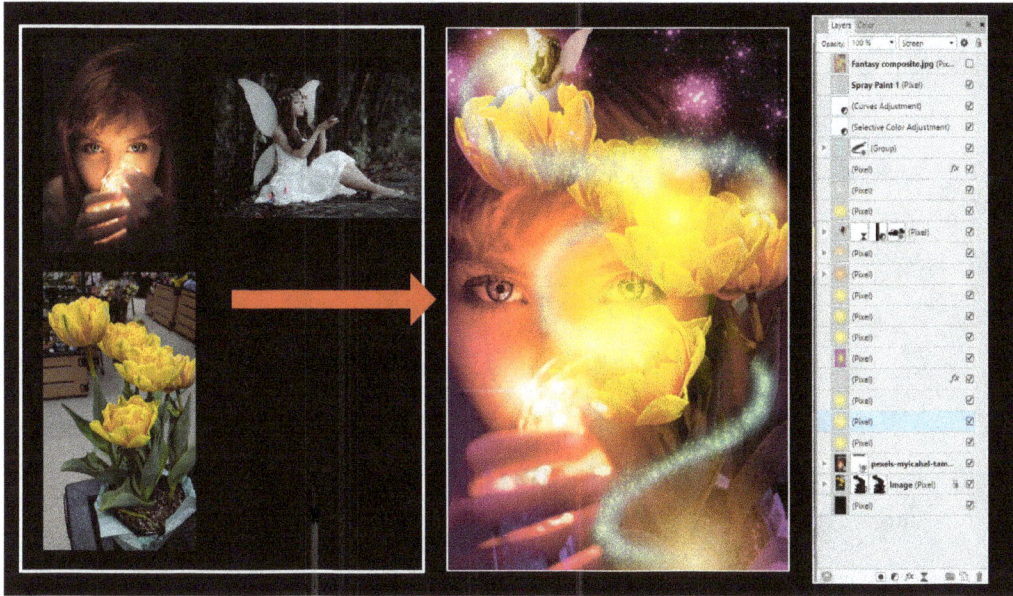

Figure 11.2 – A complex composition using many layers

Can you spot the techniques we dealt with previously in this book? This is why composition appears again later in this book – it relies heavily on the fundamentals you learned earlier in this book.

## The basics of perspective for compositing multiple images

While the fundamentals are important, there is one fundamental skill that will tank your composites before they even get out the gate, and that is the idea of perspective. For this discussion on perspective, we will use a simple example of a coffee cup.

I have included a background image and three different isolated coffee cup images for you to play with in the downloads for this example, so if you want to follow along, simply open the file and turn the layers you want to experiment with off and on.

In *Figure 11.3*, in the **AS SHOT** image on the left, we have the angle running from left to right:

Figure 11.3 – The angle and vanishing point for multiple images

Also, notice where the red line hits the coffee cup. For this to match between images, we have to either find an image with a similar angle or shoot our own. In this case, I shot the table image from the top and angled it. Notice how the red line in the composited image for *Figure 11.3* follows a similar angle. This means this coffee cup will work from a perspective in the image. You must find images that can work together from a perspective standpoint when doing a composition. Now, take a look at the following figure:

Figure 11.4 – Correct placement of the solo image in an existing environment

In *Figure 11.4*, I shot the coffee cup from various angles and placed them on the same table. You will see that one of them looks better than the others. This is because of three very important points that you must keep in mind:

- The horizon line in *Figure 11.3* should strike the object in the same place in the composition as it did in the original shot.

- In the **SORT OF CORRECT** image in *Figure 11.4*, notice how we are seeing far more of the coffee cup circle than we are the lid on the to-go cup. This is because the shot was taken at a higher vantage point. To compensate, we moved it close to the viewer. This looks plausible but weird because the image does not match the composite background angle.

- Notice in the **INCORRECT** image that the cup was shot directly on the side and that the table was shot from the top. This is evident because the ellipse on the to-go cup is drastically different than the cup from the side, and it is sticking out from the composition.

Now that we understand perspective, let's apply it to a project to make sure you understand how it works.

## Perspective – practical application

*Figure 11.5* shows a picture of my wife that I took during our trip to Milwaukee. I wanted to place a few different instances of her at various places in the image. If you wish to follow along, I have included a working file (`Perspective person file`) in the downloads of this chapter. You will see a few things in this file:

- There's a simple street scene as the background

- I have isolated her as a subject for you to place

- At the lowest layers is a group where I have included my solution, so if you want to see how I did it, it is all in there

Notice in *Figure 11.5* that the horizon line strikes her right around the glasses, so any new instance we place must strike in the same area (notice that all three instances in *Figure 11.5* strike the same area):

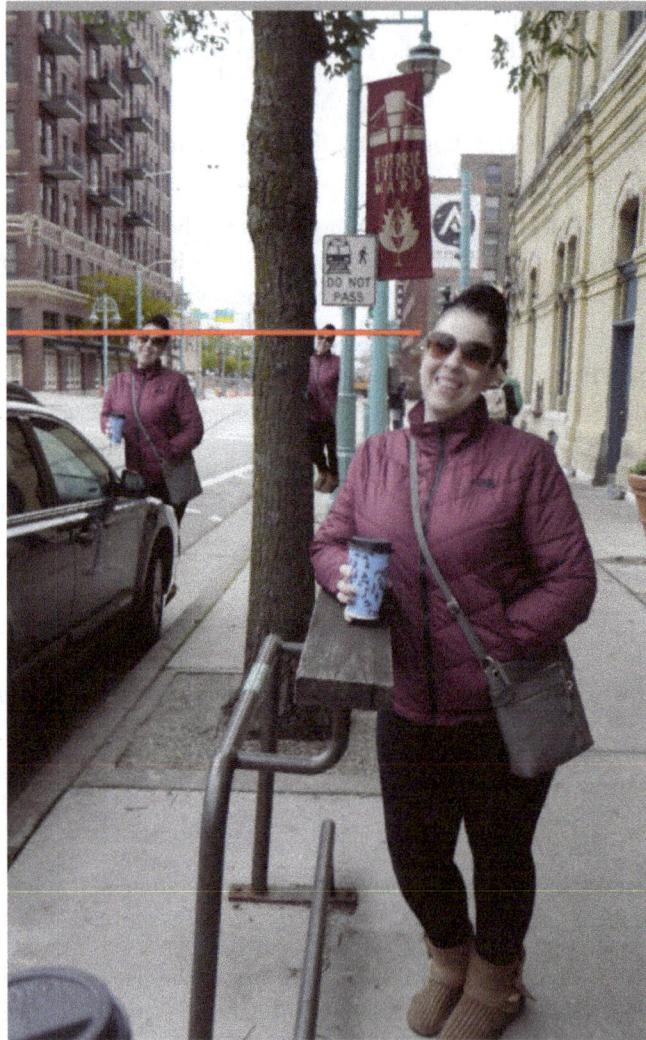

Figure 11.5 – The horizon line consistent in duplicate clones

Also, remember that the further a person gets, the smaller they look, but we still have to put them in the same area relative to the horizon.

Notice that I have not adjusted the three factors of atmospheric perspective – none of these instances have been adjusted for exposure, saturation, or detail. This is simply to explain how you should place your instances.

Now that we know a bit about perspective and how to place multiple images, the question is, *Where can we find images?* We'll explore some of my favorite places to find images in the next section.

# Finding images

We should start with the panel that Affinity Photo has for finding images, and then expand to the external world of the internet to see where else we can go.

## The studio panel

Within both Affinity Designer and Affinity Photo is a studio panel completely dedicated to stock photographs. This is called the **Stock** studio panel. To access the **Stock** panel, simply go to **View** | **Studio** | **Stock**.

There are two stock sites available in the panel:

- **Pexels**
- **Pixabay**

You will have to agree to their terms of use. Please read them carefully as most of the terms for these sites at the time of writing are CC0 licenses, meaning you can do anything with them except repackage them and sell them in collections. However, that is fine as we are changing and compositing them. You can create commercial work but you have to change it and make it yours.

To use the Stock tab, follow these steps:

1.  Type a search term into the box (I have chosen dogs):

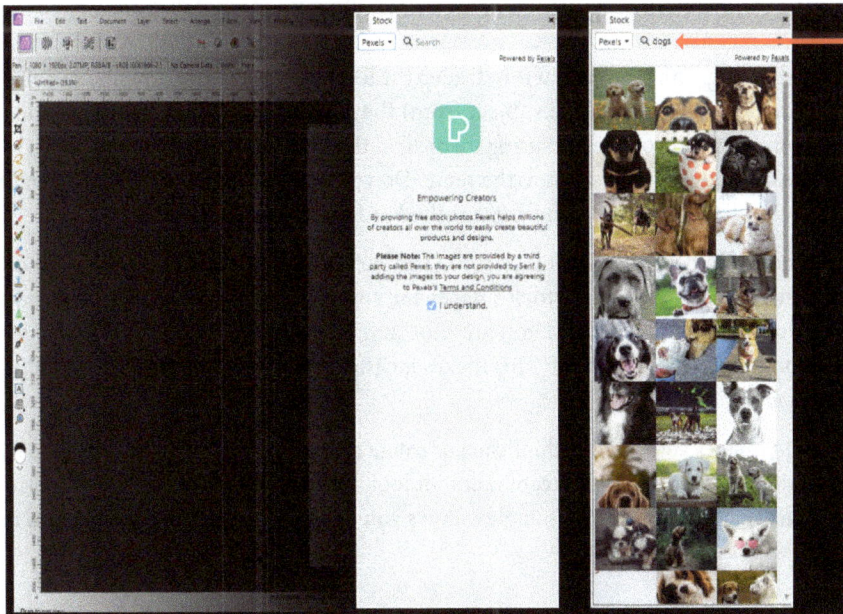

Figure 11.6 – Studio panel showing our search

2.    Click and drag an image onto the canvas.

3.    Double-click on the image to be taken to the site and download it from there.

### External sites

The studio **Stock** panel gives you access to a few sites, but I wanted to type them here as external sites so that the list is complete:

- Pexels

- Pixabay

- Unsplash

As noted previously, refer to the site's licenses for its current restrictions and permissions.

## Shooting images for your library

As you develop a style, you will find that there are elements that you use repeatedly and like to work with. For me, these are textures, graffiti styles, and different lighting setups. As you mature and create more and more compositions, it may be advantageous to begin shooting references. I do this constantly and even make an entire day of going to a location just to shoot references.

Here, I have included some tips I learned about shooting references to make sure you have what you need. This will ensure that when you get back to your workstation and your composition, it turns out as you expect:

- If you are shooting with a project in mind, keep the level (where the horizon line hits the image and what angle you approach the subject from) the same in each shot. If you are shooting a collection of items and you plan on using them all in the same shot, I recommend using a tripod to make sure that the height is always the same. Do you remember the coffee cup from the first activity in the chapter, and how the angle shifted as I got higher and higher?

- Shoot in more of a flat style, and make sure your surroundings aren't over-saturated or super bright. When you are shooting stock images for your composition library, keep the profile simple – do not use crazy filters if you are shooting on your phone. Also, if you are working with a camera, keep the colors flat. This makes it easier to color-grade in post-production and unify the piece.

- Think about images you can flip, duplicate, or rotate to create interest. I typically shoot images that I can adjust and modify to create different looks; a good example is a leaf. Shooting four or five leaves that fall from various angles allows you to create some cool effects. I also use this technique with debris and dirt.

- Shoot small items on a solid color background to make it easy to mask out. Wherever possible, I shoot indoors for small objects, and I always shoot on a solid color background – even a solid color wall is good. This makes selection easier.

- Make sure the lighting is consistent. When you shoot, make sure you minimize lighting hot spots or strong light wherever possible. If you do not, you may have three objects all with different lighting sources… and that is difficult to fix in compositing.

Now that we know where to find images, it is time to look at a sample workflow you can use to make this idea a reality. We will cover that in the next section.

# Basic compositing workflow

Now that we know what the act of composition is, and we have a handful of images to work with, it is time to look at the workflow differences between what we have done so far and the unique challenges of composition. So far, the only workflow we proposed (in *Chapter 7*) for standard single-image editing was as follows:

1. Crop the image to get the composition correct.
2. Adjust the levels for the image.
3. Adjust the focus to tell the story of what to look at.
4. Apply the color grade.
5. Balance and finish.

Now, we will get a bit more mature and increase the complexity of the workflow since we will be harvesting multiple images to create a unified image.

## The composition workflow

I have included an example composition that I have done to illustrate these concepts (see the `Composition breakdown` file in the downloads for this chapter). However, in each step, I have broken down the image into layers so that you can see the image come together. Hence, the images for each of the following steps have been deconstructed. However, note that the working file included in the downloads is complete in case you wanted to see how it is assembled.

The purpose of the file is not to show you how I isolated the images step by step (we will do an edit in the next section), but rather to show you how the workflow stacks on one another to achieve each step.

This is a composition I did for the famous Tupac Shakur poem *The Rose That Grew from Concrete*, so all the images I have chosen were taken to support the story I wanted to tell.

## Image selection

The first step is image selection. It doesn't matter whether you plan on using stock images or shooting your own – the pitfalls are always the same. Here are the best practices for this step:

- Make sure the images have similar perspectives. Images with drastically different perspectives will never form a believable composition.

- Make sure the images are not drastically over-edited – these may never line up later.

Take a look at *Figure 11.7*:

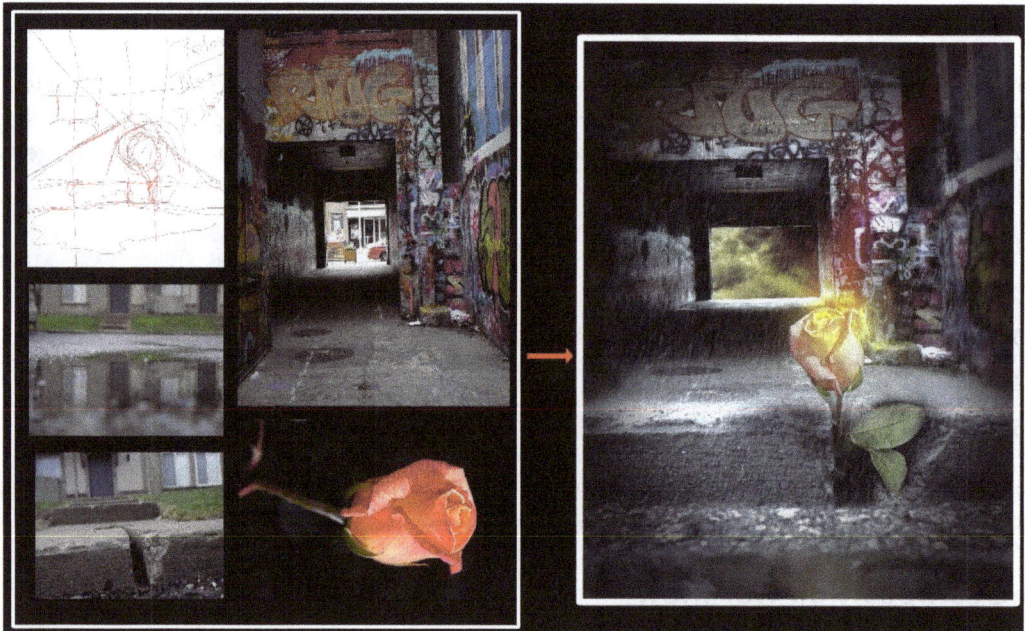

Figure 11.7 – Selecting the images for the composition

Notice the following things:

- The rose is on a dark background to aid in the selection process

- The curb I shot outside my apartment – notice that the perspective and horizon are consistent

- Notice the story sketch in the top left-hand corner that was created before I searched for images

Now that we have selected our images, let's move on to the next step.

## Image isolation

In image isolation, we extract the image pieces we want from their original images. Here are the typical tools that are used during this portion:

- The selection brush

- Drawing masks with the pen tool and then placing the image in a clipping mask

- Using mask layers and then painting out the background image using the brush

In the following figure, I have pointed out the different elements and how I isolated them:

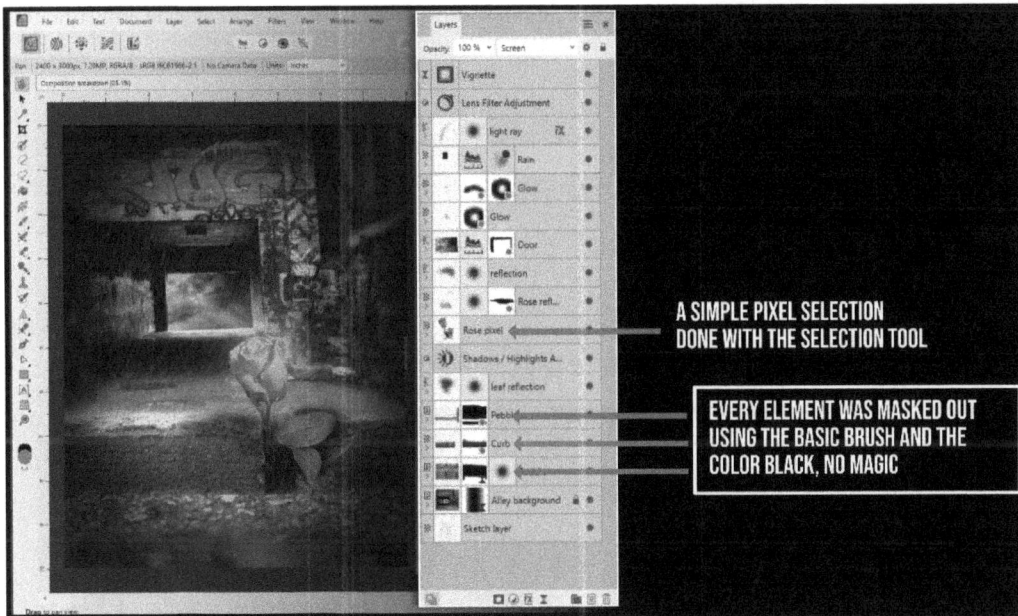

Figure 11.8 – Image isolation and techniques from earlier sections in this book

If you want to see how these things work, simply turn the mask in the project file on and off – it will show you exactly what has been masked out.

> **Pro tip**
> I highly recommend taking these project files and deconstructing them as they will show you exactly how we assembled the image. Think of it like an autopsy. You are evaluating how it was created.

## *Image unification*

During unification, we lay out the composition and place the image where we want it to be. This process involves using horizon line matching, perspective matching, and more. Also, remember the following regarding atmospheric perspective:

- Objects in the foreground tend to be brighter, sharper, and more saturated
- Objects in the background are darker, more blurry, and less saturated

Typically, this can be accomplished by utilizing the adjustment and filter layers and **nesting** them in the individual objects. We dealt with these layers in the workflow from *Chapter 7* – we are now just applying these to the individual elements, rather than the entire image. Here are the typical adjustment and Live Filter layers I use in this phase:

- HSL adjustment
- Exposure adjustment
- Curves adjustment
- Gaussian blur live filter layer
- Unsharp mask live filter

This step is also where we align light sources to make multiple pieces carry the same light source:

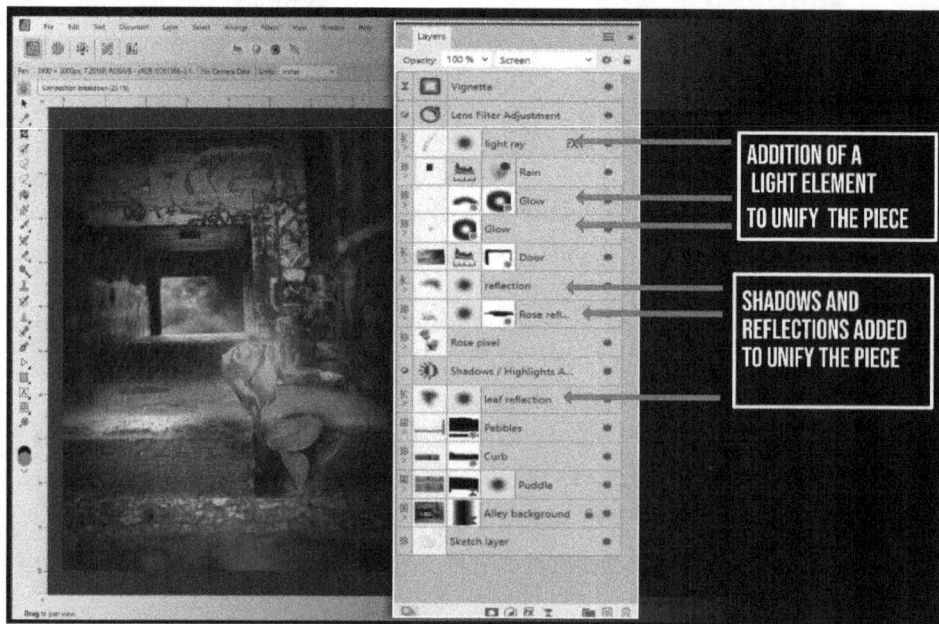

Figure 11.9 – The modified file highlighting where lighting is applied

Notice the creation of the lighting element on the rose – we are grabbing the viewer's attention and saying *this is where the focus is*.

In *Figure 11.9*, notice that I added shadows for the leaf to cast along the curb, and added the reflection of the rose in the pond.

Also, notice that the alley was made darker as it was in the background.

These elements help make the image a unified picture and make it look plausible.

> **Pro tip**
>
> Apply a black-and-white adjustment layer over the top of the image when you think you are done and check whether the image looks plausible and unified or whether you have light and dark elements that stand out. This is a great way to check your unification, but always verify that the color version agrees. This is a tool, not a rule.

## Color grading

In this step, we will apply color grading layers over the top to give the composition that wow factor – that feeling you want as an editor. We will be using some different layers in our composition. However, note that color grading has a dedicated chapter in this book where we'll explore more options (see *Chapter 13* on advanced color grading techniques).

The following are some common adjustment layers that are used in color grading:

- Selective color
- Curves
- Gradient maps
- Lens filters:

VIGNETTE CONTROLS THE VIEWERS GAZE

GIVES THE IMAGE A NICE BLUE TONE
THIS WILL MATCH THE RAIN LAYER

Figure 11.10 – Exploration file showing the color grading layers

## Texturing

Lastly, once the composition has the color we want, we need to texture it. Texturing adds the last bit of pop to the image and can vary from image to image and artist to artist. For gritty work, I may use cement; for fine art, I may use a canvas texture; if I am working in a rainy environment and want it to look like a window, I will use a **Rain** layer as a texture (*Chapter 15* is dedicated to brushes and layers that you can use to create this rain technique). Here are some common best practices for textures:

- Try to use a desaturated texture so that you do not ruin your color grading with it.

- Adjust the blend mode and **Opacity** to your taste.

- Remove the texture around focal points such as faces or eyes. This allows those elements to stand out.

- Do not let the texture overpower the composition, but rather add to it – it is like adding salt when cooking:

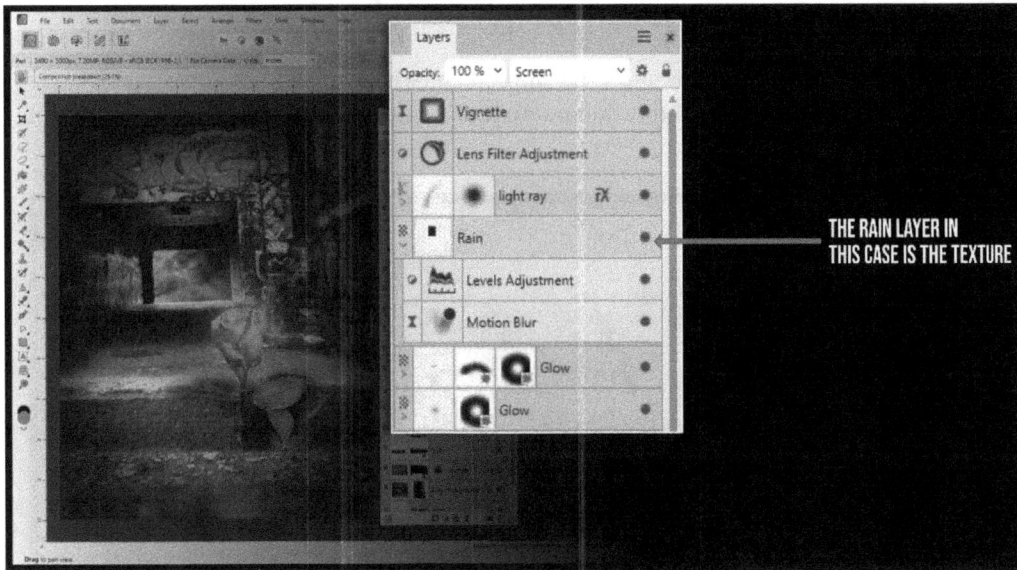

THE RAIN LAYER IN
THIS CASE IS THE TEXTURE

Figure 11.11 – Addition of the Rain layer

The **Rain** layer was made as an artistic element. In *Chapter 15*, we will make one of these and show you the process.

Now that we've walked through the workflow associated with the act of composition, we need to turn our attention to some of the filters and layer types that are used in this process. This is the meat of the process as composition will challenge every technical skill you have to take these unrelated images and make something brand new.

# Filter and layers used in compositing

While this book would be 1,000 pages long if we covered every filter that can be used in compositing, I have decided to narrow the scope of the filters and adjustments we use constantly for this task. While there are certainly an unlimited number of combinations and uses, I have attempted to share how we use each in the most common work areas.

## The Liquify filter

The Liquify filter allows you to nudge and adjust objects while specifying some pixels not to touch. This is the non-destructive version of the Liquify persona (this persona will be covered later in this book). Think of the Liquify filter as the tool you may use to make a caricature portrait of someone or show ice cream melting. In fashion photography, the adjustment layer is used to tuck in certain parts of the body that the model may possess. This is a Live Filter layer and can be found by going to **Layer | Live Filter Layers | Distortion**.

### *Reading the filter interface*

I have included a working file for this tool in the downloads (see `Liquify filter practice file`). I have taken the following figure and numerically matched the numbers with the areas of the interface:

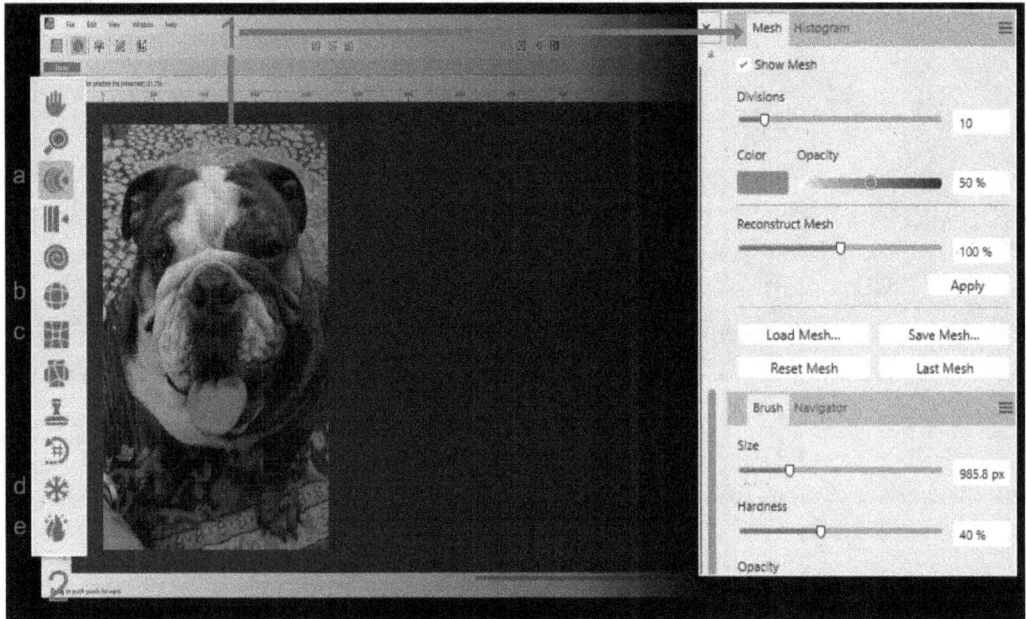

Figure 11.12 – The Liquify filter

Let's take a look at these in more detail:

- **Mesh (1)**: The Liquify filter works on a mesh. The mesh divides the image into sections. The finer the mesh, the finer the adjustment. I have only ever modified the mesh a few times, so do not worry about it too much.

- **Tools (2)**: These tools are *only* available in the Liquify filter and live Liquify persona. I have included a few of my most used tools here to get you started, but feel free to explore further:

  - The push forward tool (**a**): This nudges the pixels forward.

  - The pinch tool (**b**): This bloats the pixels, similar to a fisheye lens.

  - The punch tool (**c**): This shrinks the pixels, similar to how anime characters have smaller noses.

  - Liquify freeze (**d**): If you paint on a pixel with this tool, it freezes the pixel and doesn't affect it. Think of this as masking.

  - Liquify thaw (**e**): This tool erases your mask.

There are more tools and tabs than just these. In *Chapter 17*, we will look at them in more depth, but this is a frequently used filter we use during composition to make images match.

## Applying the tool

Using the reference photo I gave you in the downloads (see `Liquify filter practice file`), please follow along with this demonstration. This is *not* a complete tutorial on the finer points of liquifying, but rather an exploration of the tools in a practical application.

I have included six steps in *Figure 11.13* and *Figure 11.14*, as well as a CiA video (see the *Technical requirements* section for the link):

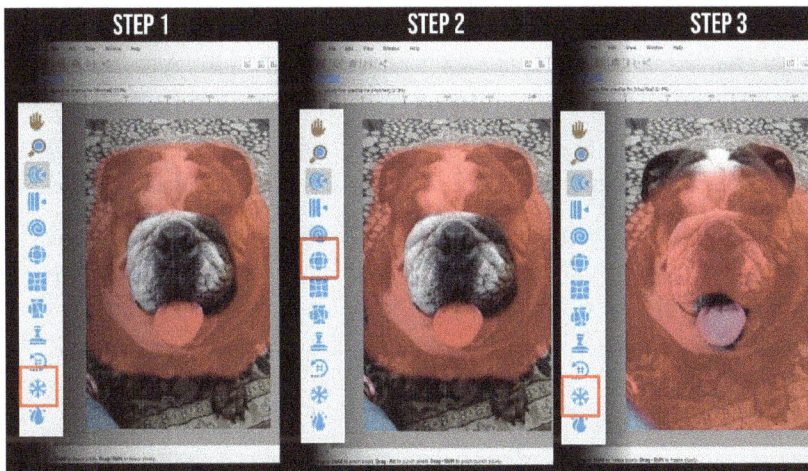

Figure 11.13 – Adjusting the snout (steps 1-3)

Figure 11.14 – Adjusting the tongue and the ears (steps 4-6)

Let's go over the steps:

1.  Freeze the pixels around the snout of the dog (this includes the nose and jowl area).
2.  Use the pinch tool to blow it up to a good size.
3.  Thaw the selection you made using the freeze tool, and then refreeze while isolating the tongue.
4.  Pinch the tongue to expand it – after all, the main characteristic of this bulldog is its tongue.
5.  Clear the freeze and, with the punch tool selected, drop the brush size and apply it to the bulldog's eyes.
6.  Freeze the pixels in the bulldog's head and push the ears out slightly.

We will see the change when we hit **Done**. You can toggle this filter on and off to see the before and after. Now, you may not like this adjustment, but at least you know how to apply the filter. As a bonus, you can just remove the filter so that the image goes back to normal – it is 100% non-destructive.

Typically, I liquify to make things fit or blend. In this case, I use techniques like this so that I have a heart start when creating an illustrated character. I create the distortions and sketch them out afterward.

Liquification is one of those essential things that makes different images sometimes seem as though they match. We can nudge, shift, and more to make things look like they fit together. However, the most important aspect is the HSL adjustment layer, a tool that makes images look like they all exist in the same environment.

## The HSL Saturation Adjustment layer

We have used the HSL a few times, and in compositions, we use the tool to *match* these values from image to image. Many times, multiple photos taken at various times can have different values and this is a good way to make sure your images have consistent values with *each other*. I have attached an image for you to follow along with in the downloads of this chapter. (See the HSL example folder. Inside this folder, you will find a file named HSL adjustment follow along file so that you can follow along.) The purpose of this file is to show you how the HSL adjustment layer can be used to get a few images to match, *not* to show a completed composite. I have not finished it nor unified it – this process is simply me getting the different images to match.

### Reading the panel

I have added an image from the aforementioned file in *Figure 11.15*:

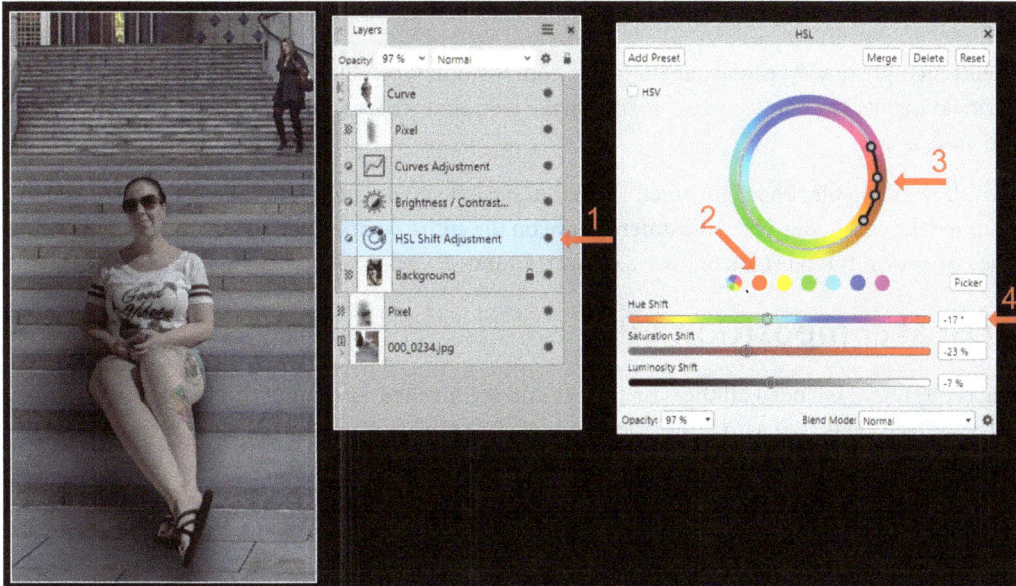

Figure 11.15 – How to read the HSL Shift Adjustment panel

Take note of the following items so that you know how to use the panel, and feel free to adjust the sliders and play with the values. After all, if you don't like what you've done, just close the file and reopen it:

- **(1)**: Notice that the HSL adjustment layer is *inside* the isolated image. Also, notice that we made a pen-based curve mask for her, so *all* the layers are nested inside the curve. In my compositions, all images have an HSL layer. Lastly, notice that if you flip open that background layer, there is an **HSL Shift Adjustment** layer in the background as well. It's a best practice is to have one for every raw image layer.

- **(2)**: In the HSL layer, you have the option of targeting certain colors. Because the figure in the image was shot with bright light originally, we had extreme reds in the subject. To target a certain color, just select it here (notice the white circle around the red). If you want to change the entire image, simply adjust the sliders in section **4** without making a targeted selection.

- **(3)**: If you select a targeted color, these four dots will come up. You can adjust the range of the values you target by clicking and dragging. To select a shorter range of red (in this case), pull the small circles around the hue closer together; to select a wider range, pull them apart.

- **(4)**: In this section, the hue, saturation, and lightness can be adjusted for the selection. Notice that we reduced the saturation of the woman after we changed the reds so that they match the bluish-gray color of the image.

*Using the tool*

As mentioned previously, each element in my composition typically has an HSL layer attached. It is a critical part of the composition flow to make sure that we have each image at similar points before unification.

In the follow-along file, I have included a black-and-white adjustment layer above the rest so that you can see the effect your HSL adjustments have on the entire composition. If any value stands out, perhaps because it is too bright, too dark, or over-saturated, click on it and amend it.

## Practical composition edit

For this edit, I have included all the base images and the finished working file in the downloads for this chapter (see the `Practical composition file` folder in the downloads for this section). Where applicable, I have included CiA videos to assist in your understanding.

*Image selection*

For this edit, I developed an idea of island life, so I grabbed a photo of a wine-style glass, some ocean, and a flower:

Figure 11.16 – Before and after

Before we get into isolation, let's take a moment to see where these images came from so that we can appreciate the fact we can find great images anywhere and capture them with our cell phones:

- The wine glass was shot at a local store.

- I used images of the Atlantic Ocean shot on vacation.

- The splashes came from a mall fountain where I captured water rolling off the feature.

- Lastly, the flowers came from the checkout line at a grocery store. There is no limit to where you can find stock images – you just have to be willing to go out there and get them.

## Image isolation

In the isolation stage, a ton of different techniques are used. I have broken them down by image and into a series of steps. *Figure 11.17* shows the respective steps:

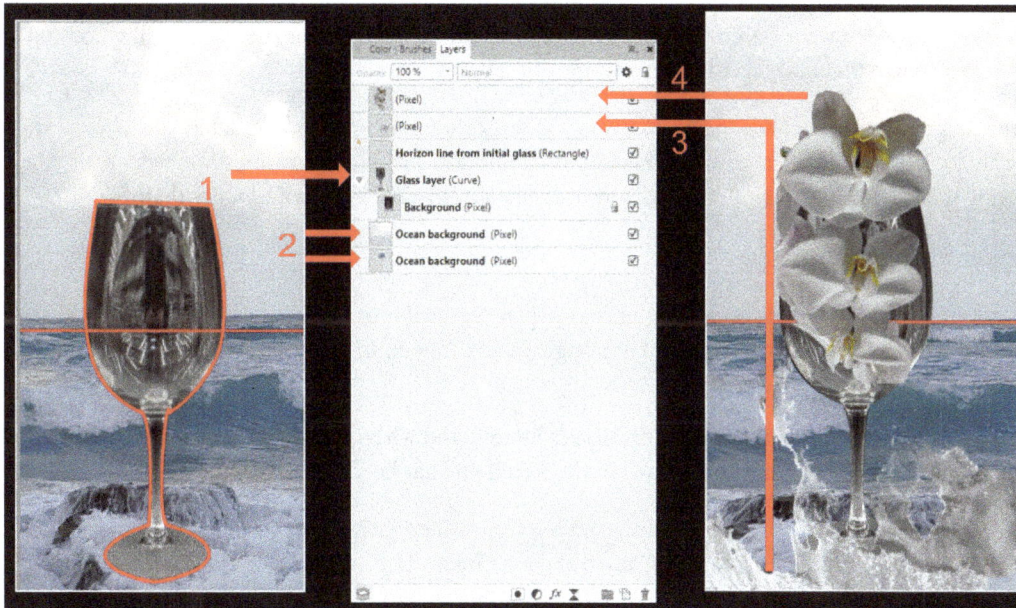

Figure 11.17 – Image isolation

Let's take a look at these steps in more detail:

- (**1**): I utilized a pen mask for the glass. I did this because we have a smooth object. We would have to use a tool such as the selection brush if we had jagged edges. Refer to *Chapter 4* on masking for the technical aspects of the pen and node tools. I have included a CiA video showing the selection of the glass (see the *Technical requirements* section for the link).

- **(2)**: No selection was required for the background image. However, in the working file, I have included an element called the horizon line. It is important to make sure the background hits the glass in the same area. You may have to zoom into the background image to achieve this.

  If you zoom in, this will leave a blank area at the top of the image. To fix this, select the rectangle selection tool, making sure you are on the background layer. Select the background, create a new layer by copying and pasting, and then stretch it so that it covers the canvas. This process is covered in the CiA video (see the *Technical requirements* section for the link).

- **(3)**: For the splashes, I used the Selection brush because the splashes are all over the image. Select the splashes you want and then export them as a new layer. Cut and paste this new layer into the composition and rotate it to place them.

  This process was covered in more detail in *Chapter 5*, but I have included a CiA video showing my selection process and placement. Notice in *Figure 11.17* that the splashes are different in color than the water, which is fine as we have not unified these two yet.

- **(4)**: For the flowers, I used the Selection brush again, isolated the flower stack, and then copied and pasted that new layer into the composition.

At this stage, you should have the basic layers you are going to work with – isolated and stacked. It may not look like much right now, but this is the beginning of making the composition work – simply stacking the images is just the first step. Now, we must arrange the images and unify them.

## Image unification

For image unification, I perform a series of steps where the goal is to make the image look believable. Later, we will worry about coloring and lighting and all of the finishing parts, but this step is about making the image look believable.

I have included the output of the steps in various figures, and wherever necessary, I have also added a CiA video to assist (it will be mentioned in the step if one has been included):

1. Glass is a reflective property, so we need to create a sandwich effect. I will have a background layer of glass and then a reflective layer of glass (*Figure 11.18* shows how I labeled them in the working file):

   - **(1a)**: Duplicate the glass layer and label the upper one as `Reflective` and the lower one as `Background glass`.

   - Change the blend mode of the `Reflective` layer to **Overlay** and change its **Opacity** value to **56%**. Change the blend mode of the `Background glass` layer to **Exclusion** with an **Opacity** value of **79%**.

2.  Now, we need to move the flower layer into the sandwich and make the splashes work.

    (**2ab**): Move the flowers between the `Reflective` and `Background glass` layers. This will give the illusion that the flowers are inside the glass. On the splash layer, label the layer `Foreground splashes` (this is how it is labeled in *Figure 11.18*). Change the blend mode to **Luminosity**.

3.  With the reflection done, let's liquify to make the flowers appear in the glass and wrap the splashes around the glass (shown in *Figure 11.18*).

    (**3ab**): Using the Liquify Live Filter, apply it to the `Foreground splashes` layer and work with the liquefication until it looks good to you. Using the Liquify filter on the **Flower** layer, make the flowers look like they are confined within the glass, tucking them in and then pinching the middle of the image to show the curvature (link is mentioned in the *Technical requirements*):

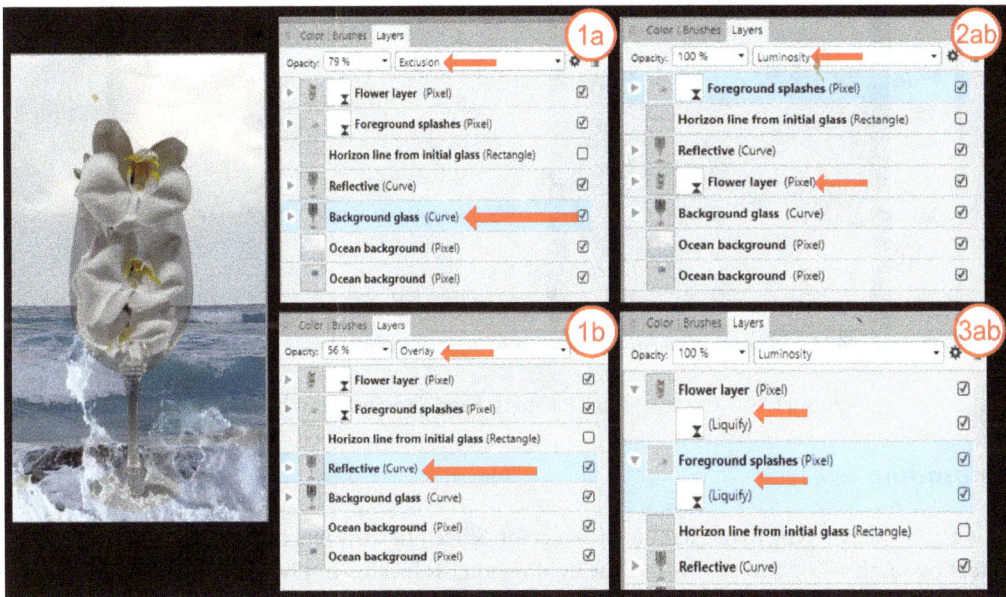

Figure 11.18 – The layer stacks for Steps 1-3 for image unification

4.  Let's add another splash inside the sandwich to make it work (shown in *Figure 11.19*).

    **4(abc)**: Copy your splash layer and place the copy inside the glass sandwich. Rotate it 90 degrees and erase the outside splashes with a hard brush. Liquify to make the water fill the glass, making sure it does not exceed the glass.

5.    Finish the unification by embedding the glass in the waves of the background and breaking up the splashes (this is shown in a CiA video):

- **5(a)**: Use a **Mask** layer to erase the glass base into the ocean

- **5(bc)**: Use a texture brush on the eraser tool to break up the splashes and add some variation and break the hard edges:

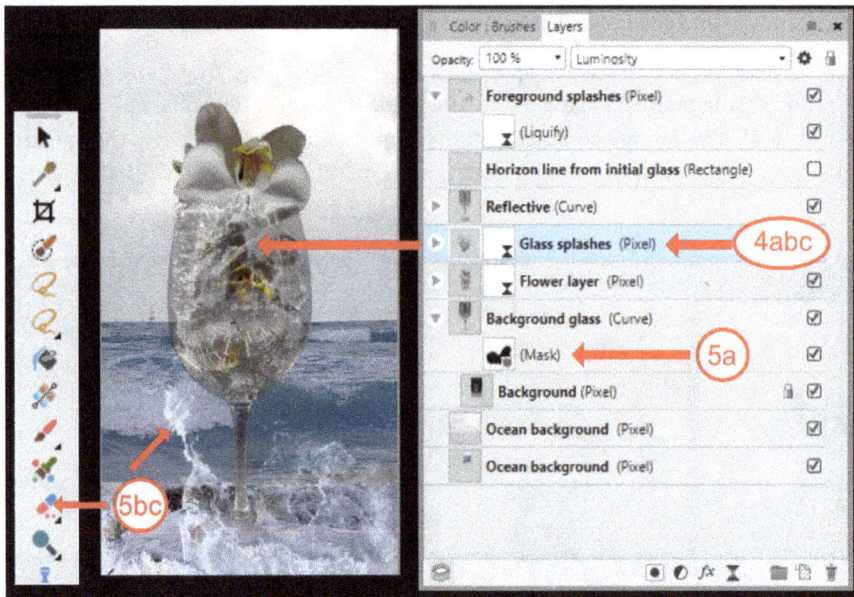

Figure 11.19 – The layer stack for steps 4 and 5 for image unification

## Color grading

For this activity, I am going to use light to create color. Since I like the feel of the piece, I am going to put attention on the glass and the flower. The lighting filter is located in **Live filter layers | Lighting**.

I have included a CiA video so that you can follow along (see the *Technical requirements* section for the link), but I have also described the process in *Figure 11.20*.

### Using a lighting layer

Lighting layers are one of my favorite ways to add visual interest to compositions. While it may seem overwhelming, once you understand how a lighting layer can light up an image, it becomes very intuitive. As shown in *Figure 11.20*, let's go over the settings that can be found within the Lighting layer:

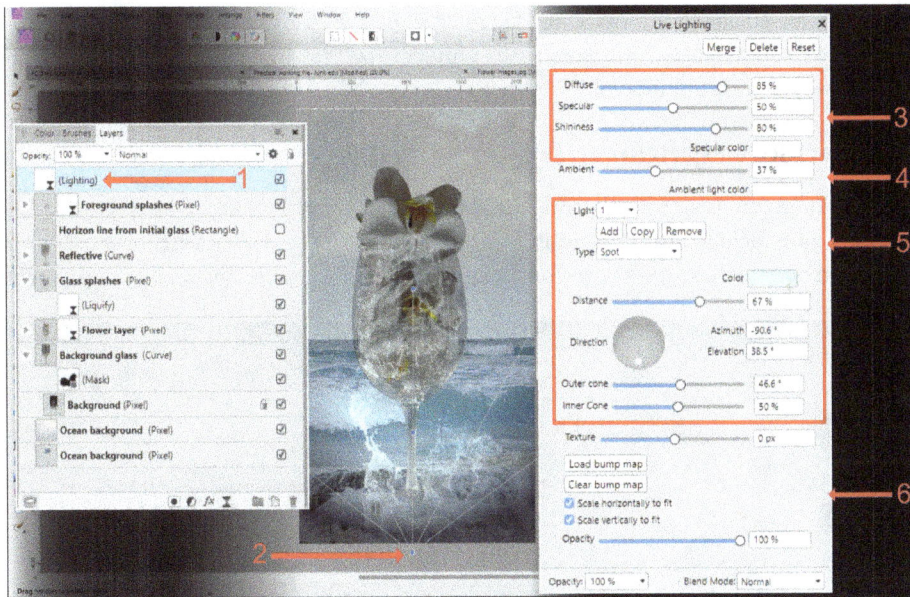

Figure 11.20 – Adding a Lighting layer

You can find the following settings for the Lighting layer:

- **(1)**: I have positioned the Lighting layer above every other layer. You can attach it to just one layer, but as we are color-grading the entire image, I placed it up top.

- **(2)**: By default, the **Lighting** layer creates a node-based light, so this is where you can manipulate the nodes and adjust the position, intensity, and more of the light.

- **(3)**: This area of the lighting controls the specular light (reflected light on shiny objects). I sometimes like to change the color of the light to get a very different look to the ambient environment.

- **(4)**: Ambient light is natural light, so this is where you adjust its intensity and color. Think of it like the Sun.

- **(5)**: This is where you can add, remove, and change lamps, as well as adjust the lamp color. It is possible to set up multiple lamps to light a scene.

- **(6)**: Lastly, light can imply texture. I don't use this often, but a good application of this is a brisk wall where you want to show the texture of the light bouncing off the wall.

Don't hesitate to play with the Lighting layer and mix colors and have a creative go at the project. Color and light convey meaning and expression, and the mood for the piece and the use of a Lighting layer can take a piece from good to great.

Now that we have unified and lit the image, the next step is to adjust the **macro** image. This step cannot occur until all the smaller images have been made uniform and look like they all belong to the same image. So, now, we will add vibrance and brightness.

### Adding the vibrance and brightness layer

In this step of color grading, we will adjust the brightness and vibrance of the piece to get that island look I am after. I have included the settings in *Figure 11.21*. Note that I haven't included a CiA video since we explored both of these layers previously, and by now, you should know how to use them:

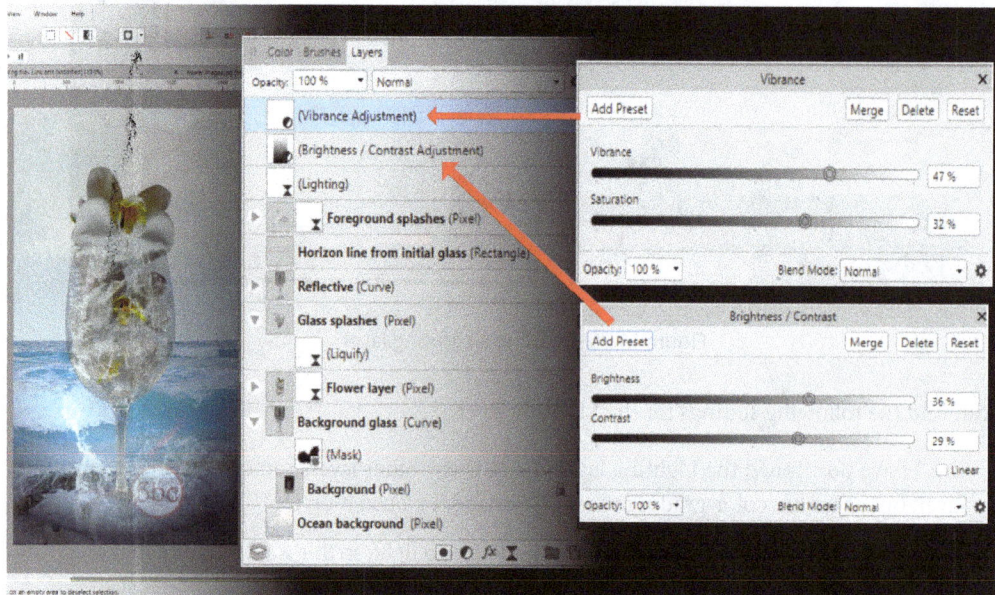

Figure 11.21 – Adding the vibrance and lightness layers

Notice that the brightness layer has a gradient mask applied to the adjustment.

I added this gradient adjustment because I wanted to make the top of the image the focal point for the composition, not the sand.

Now, it is time to finish up. This is the last 10% of your work and involves balancing texture, contrast, brightness, and more.

### *Texturing and finishing*

To finish, we will adjust the sharpness and overall contrast of the image and its elements.

## Finishing the splashes

To finish the splashes, we must follow these steps:

1.  Add **Unsharp Mask** to the splashes and then **HSL Shift Adjustment** (my values are shown in *Figure 11.22*):

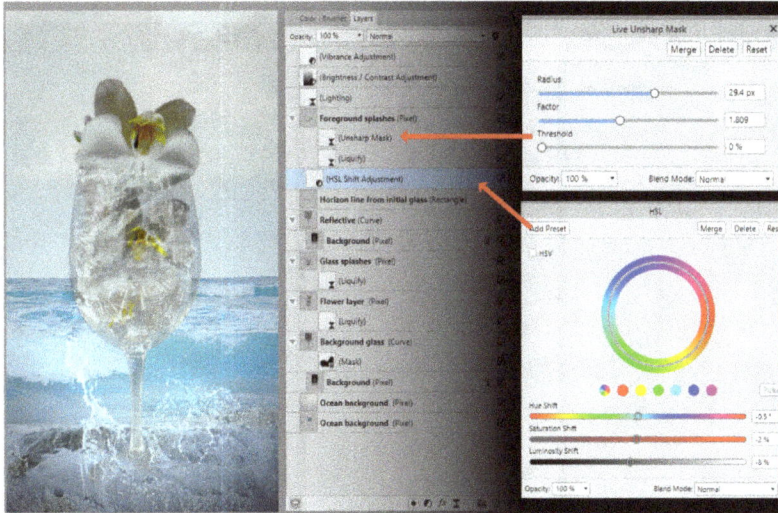

Figure 11.22 – Adding a texture and an HSL shift to the image

2.  Lastly, duplicate the splashes to bring them out more. Then, inside the glass splashes, add **Unsharp Mask**. This brings out the detail in the bubbles. My adjustments and the existence of the additional splash layer are shown in *Figure 11.23*:

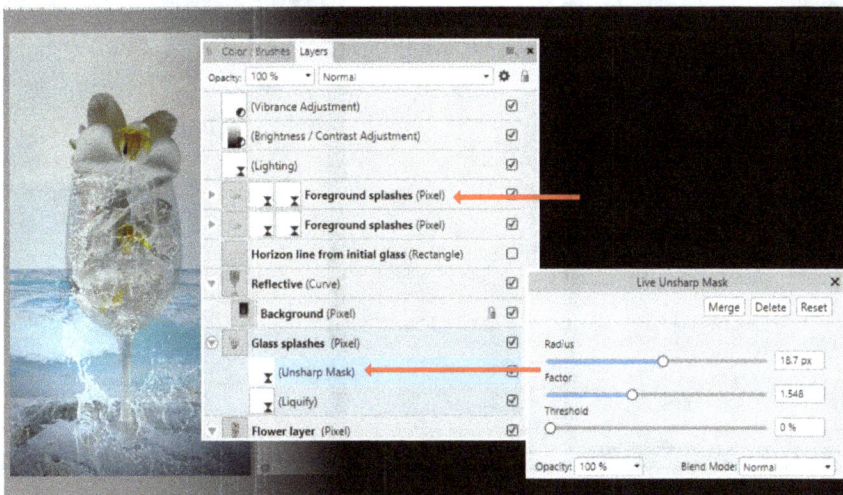

Figure 11.23 – Adding a duplicate layer of splashes to make them pop

## Adding curves and balancing to the glass

Lastly, we will add a few curves to balance out the glass. These curves and their adjustment values are shown in *Figure 11.24*.

I have included various curves here, although there is one that is special:

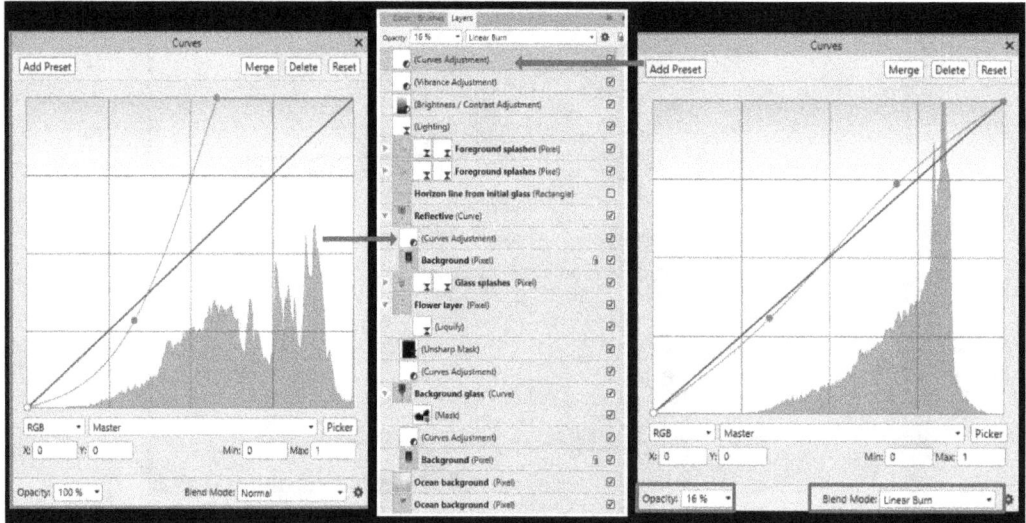

Figure 11.24 – Curves for the top 2 layers (pay attention to what layers the arrows point to)

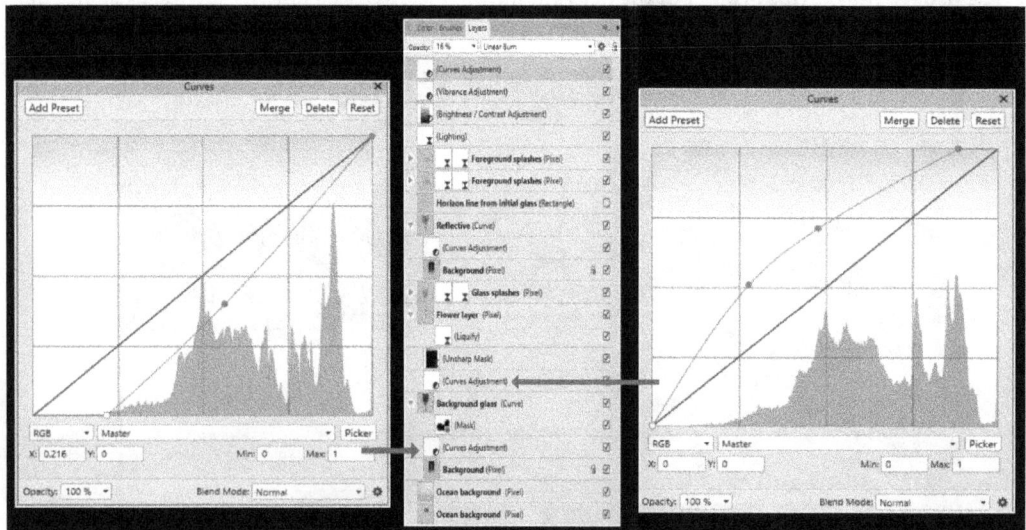

Figure 11.25 – Curves for the bottom 2 layers (pay attention to what layers the arrows point to)

Take a look at the top curve, where I adjusted the entire image contrast (top right of *Figure 11.24*). I adjusted the blend mode for the curve, as well as the opacity of the adjustment. This shows the power of combining adjustments, blend modes, and opacity.

## Professional tips, tricks, and important points

To build your stock images, just take your phone to the store and start snapping images. You can develop a large library of modern styles by simply photographing products.

Always work non-destructively. Notice that in this chapter, I did not make any decisions about where I would need to walk back. As you add more layers and more complexity to your compositions, you want to be able to make adjustments.

Depending on the composition, you may want to alter your workflow. Some elements may require some balancing before you can adjust other elements, so do not be too attached to one workflow.

Develop *stacks* of adjustments that you use all the time. Coming back to the cooking metaphor, these are similar to the bases you make beforehand that make your dish uniquely yours.

## Summary

You now have the most freeing artistic aspect of photo editing under your belt – compositing! You did a lot in this chapter, from learning the fundamentals of perspective, vanishing points, and the workflow to making it all happen – you are now mature in your journey into photo editing.

In this chapter, we covered the idea of what a composition is, the universal idea of perspective and placement, where we can get images, common tools such as the Liquify and HSL Shift layers, and how to use workflows and applications for composition.

In the next chapter, we will cover another sought-after skill known as portrait retouching, which we can do using the healing tools available in Affinity Photo. The reason we have included this section first is that the process of retouching photos often involves recreating pieces using the problem-solving and composition techniques you learned about in this chapter.

# 12

# Photo Restoration and Portrait Retouching in Affinity Photo

When it comes to studying photo editing, two other common areas of specialization are **photo restoration** and **portrait retouching**. While these are two separate specialties that deal with separate objectives, with photo restoration being used to restore older/damaged photographs and portrait retouching being used in the beauty industry, the tools are very similar. The most common tools used in both these applications are commonly referred to as **Healing** tools. This includes tools that either clone areas of desirable photos into others or remove undesirable things such as cars, scratches, and blemishes on the skin.

While it is *impossible* to give you *all* the tools, or *all* the possible workflows for these activities, I have attempted to give you a working knowledge of the most common ones I use in my daily edits. Also, much like compositions, repairing older photos and fixing portraits is, at its heart, problem-solving. How you would remove a crease is completely different from one photo to another, depending on the image. So, as you grow as an editor, the toolbox you have to solve these *problems* also grows with you.

In this chapter, we will cover the common healing tools and give you workflows for photo restoration as well as beauty retouching. Due to its use in the industry, we will be spending a fair amount of time on frequency separation as well. So, let's dive in.

In this chapter, we will cover the following topics:

- Exploring three common tools for photo retouching and restoration
- Understanding photo restoration
- What is frequency separation?
- Portrait retouching workflow and practical application
- Professional tips, tricks, and important points

By the end of the chapter, you will be able to perform a basic restoration for common issues with older photographs, as well as retouch a basic photo using techniques such as frequency separation and healing tools.

# Technical requirements

The CiA video for this chapter can be found at `https://packt.link/KvSD0`.

You will also need the `Chapter 12` project to follow along.

# Exploring three common tools for photo retouching and restoration

In this section, we will explore three of the most common tools for learning the image replacement process. When we use the term "image replacement," we are implying that we will be removing part of the image with undesirable qualities, such as objects or colors, and replacing them with a piece of the same image that is desirable. Examples of image replacement in practical editing include the following:

- Removing cars from portions of the street
- Replicating trees and foliage in landscape photography
- Removing damage to a wall in architectural photography
- Removing undesirable skin blemishes in portrait retouching

## Clone stamp tool

The Clone Brush uses a target area from an image to replace the area the brush travels over. Now, in other versions of Affinity, the context menu may change for this tool. This is the nature of software, but the overarching concepts presented in *Figure 12.1* will still stand up. Now, let's check the interface for the Clone Brush:

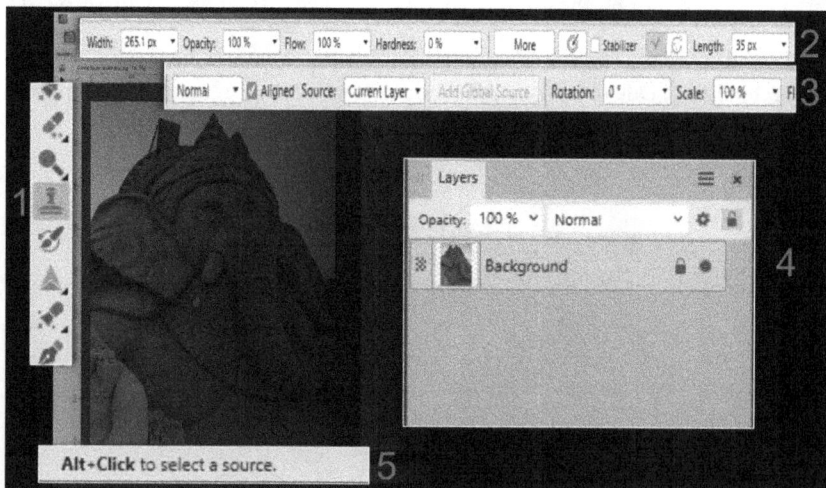

Figure 12.1 – Clone tool breakdown

Let's take a look at the interface in more detail:

- **1**: The Clone Brush can be found in the tools menu, as shown in *Figure 12.1*. If it does not appear on your toolbar, refer to the chapter on modifying the toolbar to add it (*Chapter 1*).

- **2**: Here, we have the brush settings for the Clone Brush. This is where you will see common settings that were already covered when we looked at brush basics. Typically, I do not adjust the opacity or flow for the Clone Brush, but I do modify the hardness to avoid solid edges appearing in the replacement.

- **3**: This is the area of the context toolbar that describes the selection instructions for the brush and what layer to select from. Never in my career have I changed the scale, rotation, or other properties and I typically keep the current layer selected.

- **4**: The **Layers** panel shows the *only* layer in the composition, so this is the layer we are currently on.

- **5**: This is something we have not covered before. When a tool is selected in Affinity Photo, the **instructions** for the tool are displayed at the bottom of the screen. So, while we will use the Clone Brush in an actual application, here, you can see that you can set the target value (covered in the Using the Clone Brush section) by holding *Alt* and clicking.

In this chapter, we will be removing signage from a photograph and people from in front of a building, and showing you how to remove simple blemishes in a photo.

### Using the Clone Brush

The process of using the Clone Brush involves two elements – the target area and the brush. To use the Clone Brush, follow these steps:

1. Create a selection if desired. I added a selection around the sky using the Selection brush to make sure I did not clone the elephant.

2. Press *Alt* and click on the area you want to duplicate or clone.

3. Next, using your brush, paint over the area you want to replace.

This method is demonstrated in *Figure 12.2*:

Figure 12.2 – Setting a target area in an image

## Best practices with the Clone Brush

Here are some best practices for you to follow:

- The size of your brush matters. A larger brush may pull in picture elements around the target area and place them in your cloned area. So, the idea is to keep the brush small.

- If you click and drag the brush to clone the image, it will move the target area too. So, if your target area has a bunch of imagery around it, do not click and drag – instead, click and release to clone the area. In *Figure 12.2*, there is a lot of sky around my target area. This means I can click and drag without the risk of the elephant coming through.

- You may have to reposition the target several times when cloning large areas. To minimize this, in the context toolbar, you can choose to uncheck **Align Source**.

A few wrong steps with the tool will help you gain some mastery, so do not be afraid to use that undo button.

> **Important note**
>
> I have included a CiA video to support the information in this book (see the *Technical requirements* section for the link), as well as the source image in the downloads for this chapter (see the `Clone Brush example` file). Now that we have covered the Clone Brush, you should be able to do simple sky replacements and randomized pattern replacements, such as grass on a field. Now, we will look at another proactive tool that tries to predict objects and patterns: the Inpainting brush.

## Inpainting brush

The Inpainting brush will take a selection you paint over, and it will look around the image to try and match the colors, texture, and overall feel of what is around it. Let's look at the Inpainting brush's context toolbar. We have broken down the Inpainting tool in *Figure 12.3* to show you the portions of the interface that are important:

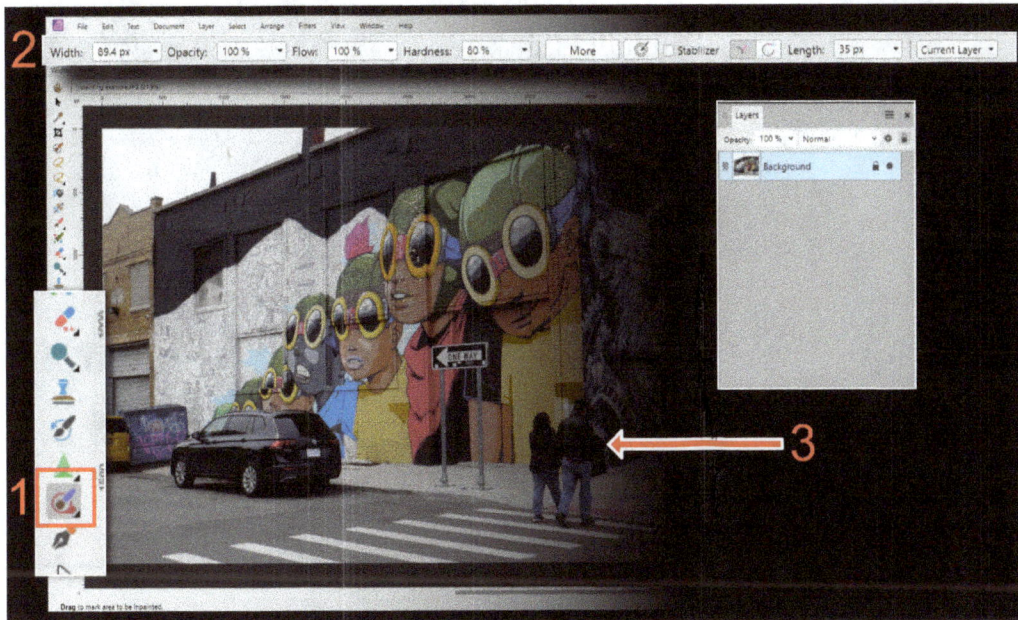

Figure 12.3 – The removal area

Let's look at these areas:

- **1**: The tool can be found with the other healing tools in the tool section.

- **2**: The context toolbar has no new sections that were not previously covered in the section on brush basics; I really have never adjusted the settings from their defaults.

- **3**: Notice the reason why the Inpainting brush will work for this image: the couple is surrounded by solid colors and multiple things (the sidewalk, the building, and the corner). So, the Clone Brush would not give us the best result due to the number of objects around them. Now, if the couple was against, say, a patterned way where the wall was the focus, then inpainting would not work. This is because Affinity can't match the pattern with inpainting.

## Using the Inpainting tool

Select the **Inpainting** brush from the menu and paint over the couple see *Figure 12.4*. Affinity will utilize the algorithm to decide the best background for replacement. As you can see in *Figure 12.4* it did a great job:

Figure 12.4 – Before and after

> **Note**
>
> We have included a follow-along image that you can work with (see the `Inpainting example` image in the downloads for this chapter).

### Best practices for using the Inpainting tool

Let's look at some best practices for using this tool:

- Adjust the brush size and the amount you attempt to inpaint at any one time.

- For the couple we did in the follow-along, it was best to paint over the couple as one big square to prevent the algorithm taking small bites. However, if we were to take the ONE WAY sign, it would be best to select the sign first, then the pole, and then the other pole. I would also take the pole in several sections that matched the color behind it so that Affinity has less variation to interpret. This is shown in the video (see the *Technical requirements* section for the link), where we performed the replacement.

Practice with various scrap images and different amounts of inpainting and brush sizes to get the best results.

With the Inpainting tool covered, and you as the editor confidently aware of when and how to use it, we can move into the simplest of tools – the Blemish Removal tool, or as I call it, "the one-click wonder."

## The Blemish Removal tool

The Blemish Removal tool is a good choice for making small repairs to a consistent area where you cannot use the clone stamp or the area is so complex that the Inpainting tool is not an option.

However, it is worth mentioning what the Blemish Removal tool does: it creates a blurred patch using data around the brush. Because of how it does this, it is not recommended for large patches that have a lot of detail; what you will be left with is a blurry mess, so this is a fine last-step style tool.

### Using the Blemish Removal tool

Using the Blemish Removal tool is simple: the context toolbar only has a size adjustment, so simply adjust the size so that it's slightly bigger than the imperfection, and then click where you want.

I have included an image of my grandmother (see the Grandmother-Blemish download file). *Figure 12.5* shows a before image and a magnified after image for one spot in the picture. The dot we are removing is indicated by red arrows:

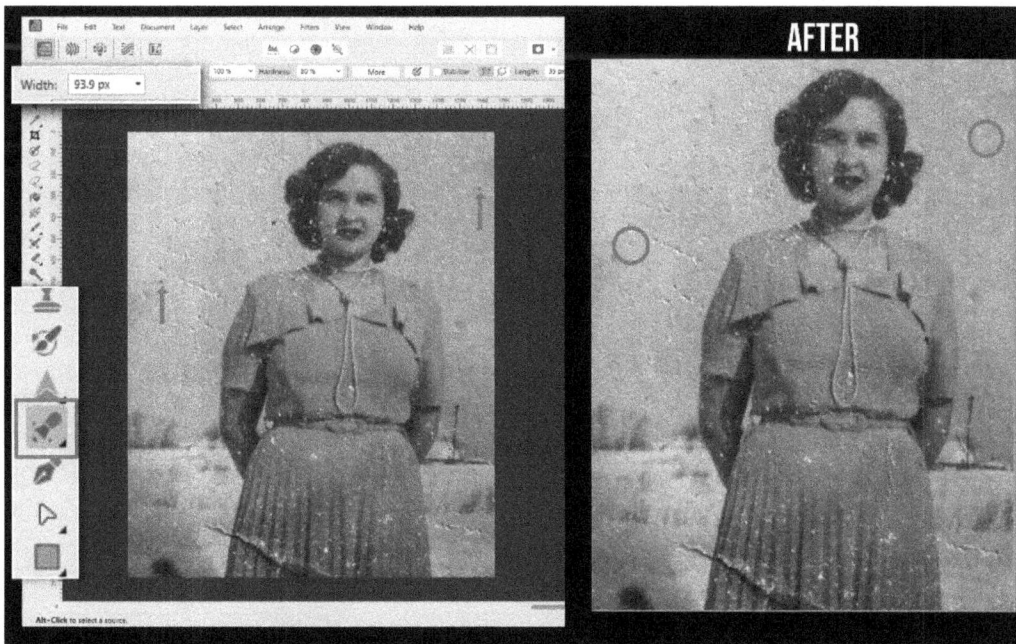

Figure 12.5 – Before and after the targeted blemishes are removed

If you want to see the video for this, I have included a CiA video (see the *Technical requirements* section for the link).

**Best practices for using the brush tool**

Here are some best practices:

- Use a brush size close to the size of the blemish you are removing
- Do not use this tool in areas where details have to be maintained
- Do not use many instances close to one another as they will begin to look weird

This concludes this section on basic tools. With that, you have documented proof of your ability to apply them, as well as an awareness of when to use each one. This means it is time for our discussion on workflows, which allow you to apply the tools you've learned so that you can achieve a more complex edit. This will be the focus of the next section.

# Understanding the photo restoration workflow and its practical application

Tool knowledge without workflow awareness is worthless – it is similar to a carpenter knowing how to wield a hammer but not having enough knowledge of carpentry to build a workbench with it. So, in this section, we are going to be starting with workflows – or the bigger picture of how to use a hammer to build your actual project.

For this application, we will be working with a photo of my grandparents that I have included in the downloads for this section (see `Original image` file inside the `Photo restoration project` folder). As you can see, it has seen better days. The reason I chose this photo is that it displays many of the common issues with photos we commonly get asked to enhance and restore. These include the following:

- Specks and flecks in the image
- Washed-out dynamics
- Wrinkles and folds

The workflow I have found most useful for photo retouching or restoration will be covered in this section.

**Structural repair:**

- Clone tool
- Inpainting tool
- Dust and scratches filter
- Blemish Removal tool

**Detail refinement:**

- **Unsharp** mask coupled with a mask layer on the filter
- Gaussian blur live filter layer with a mask on the filter
- Duplicate copies of the adjustment layer to bring out details

**Dynamic adjustment:**

- Non-destructive dodge and burn using the curves layer and a mask (covered in this example)
- Exposure live filter
- Shadows and highlights

Now, let's look at the practical application of these workflows.

## Practical application

As previously noted, we have included a practice project in this chapter. You can find it in the downloads for the section titled `Photo restoration project`. The file we will be using is labeled `Original image`.

From here on out, we will be applying the aforementioned workflow to the real world so that you have documented evidence you can apply in the context of a workflow.

### Structural repair

I have included the steps for this workflow in the CiA video (see the *Technical requirements* section for the link), so if you get lost, check the video out for step-by-step instructions.

Let's start by applying structural repairs to the `Original image` file. Follow these steps:

1. Straighten and crop the image to eliminate the scalloped sides of the image; we will be using no border for this edit. This image was not scanned from the original but is a photo of the photo.

2. Duplicate the image twice and relabel the duplicates as `working copy` and `detail`. We are doing this because the next step is destructive, meaning that we cannot change the adjustment back. To ensure we do not affect the original image, always make a copy.

3. Select the **Background** area with the **Selection** brush and add a spare channel. Label this channel **Background**. Invert the selection (found in the **Selection** menu at the top of the screen). Once the selection has been inverted, this will be the foreground. Create a spare channel and label it **Foreground**:

    A. At the end of this step (**3a**), you should have a spare channel saved for the background and the foreground parts of an image.

You can see the results of *steps 1* to *3* in *Figure 12.6*:

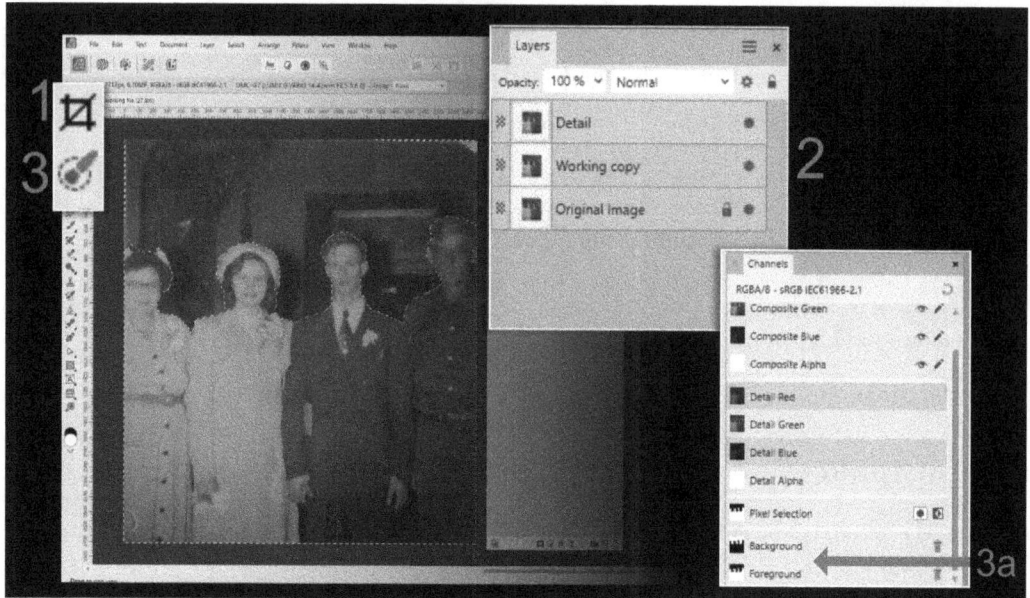

Figure 12.6 – Working image showing the layer stack for steps 1-3

4.  Turn off the "detail" and apply the Clone Stamp tool to the working layer to remove the large white blemish in the top right-hand corner of the image.

5.  On the working copy layer, apply the background channel to the layer. This will ensure the filter layer only affects the background. Run the dust and scratches filter on the **Working copy** layer. You can find this filter under **Filters | Noise | Dust & Scratches…**. Note that this is not a live filter, so once you apply it, you will not be able to change it. The strength of the background layer should be around 8 px.

6.  Now, select the foreground channel on the working layer; this will select the wedding party. Apply the dust and scratches filter at a radius of between **4** and **6** and hit **Apply**. In effect, what we just did was apply a lower level of the filter to the foreground.

7.  Now, on the **Detail** layer, add a mask layer and fill the layer with black so that it's not visible. With a soft round brush, paint white on the faces of the wedding party. This will bring out the details in the wedding party's faces, and the faces will be the focal point of the image. The use of a mask also ensures that any structural issues still present in the detail layer are not applied to *Figure 12.7*.

8.  Lastly, let's take care of the camera hotspots in the background. Above the door, there are places of discoloration, either due to the camera when the image was taken or age. For this, we will use the Clone Stamp tool and select an area close to the image – in this case, the **Working copy** layer:

    I.    Set the hardness of the clone stamp to zero.

    II.    Single-click the door frame in the image. We cannot click and drag as it is too dangerous to make the alignment right.

The following figure shows the final steps for the structural repair of the photo:

Figure 12.7 – Working files showing steps 4-8

This concludes the structural cleanup for the image. In the next step, we will tackle the detail refinement process. The reason we fixed the structure first is that the details we will bring out and the dynamics we will apply will not be right if we try to fix the structure after the fact (this means the errors in the structure will cause bigger issues in later steps, so we need to fix the structure first).

## Detail refinement

When it comes to detail refinement, the goal is to make the *important* aspects of the image more pronounced, and those *less* important details fade into the photo. When retouching, never forget that we are limited to the quality of the original photo, so while we can make improvements, we need to be honest in our expectations. Follow these steps:

> **Note**
>
> I have also included a CiA video showing the process (see the *Technical requirements* section for the link).

1. On the **Working copy** layer, select the **Foreground** channel; this will select the wedding part figures. Apply an unsharp mask to the foreground and turn the value up to where the foreground looks good. See *Figure 12.8* for the settings that I used. Notice the use of the hard light blend mode.

2. Using the **Detail** layer, add another **Unsharp** mask and increase the detail to a level that looks good. What you are doing is creating a hierarchy of focus for the eye. The hierarchy is **Background-Foreground-Faces**.

3. Notice the artifacts that are now present in the image once we add details. We will use the Blemish Removal tool to remove the image of the groom's face and the maid of honor's face under the nose (the areas I worked on are shown in *Figure 12.8*).

These adjustments are shown in the following figure:

Figure 12.8 – Blemish adjustment

Now that we've finished the blemish adjustment, it is time to move on to the dynamic adjustment process, where we give it a bit of pop.

## Dynamics adjustment

Dynamic adjustment uses some of the same techniques we have used before, and the goal is to pull the different tones out of the image and emphasize where we can. My primary focus is on the bride's suit – it is too white, and the details and contrast have been lost in the image, so I want to pull it out.

> **Note**
>
> As with the rest of the steps, I have included a CiA video showing the sequence of operations (see the *Technical requirements* section for the link).

Let's start adjusting the dynamics for the image. Follow these steps:

1.  Create a **Curves** layer and drop the midpoint *down*. This will be the *burn* layer (burning means darkening). We will strategically use it to add more shadows behind the party. Once it has been dropped (see *Figure 12.9*), select the fill bucket tool and fill the **Curve** layer with black (in effect, masking it). Now, you can paint darker shadows by using a soft round brush with white; this is a professional way to burn and dodge. Adjust the blend mode to multiply and adjust the opacity to your liking.

2.  Create a **Curves** layer and *raise* the midpoint. This will be the dodge layer (dodging means adding lighting) and we will strategically use it to add more highlights to the dress. Once it has been raised (see *Figure 12.9*), select the fill bucket tool and fill the **Curve** layer with black (in effect, masking it). Now, you can light highlights on the arms and dress by using a soft round brush with white. Change the blend mode to **Screen** and adjust the opacity to your liking:

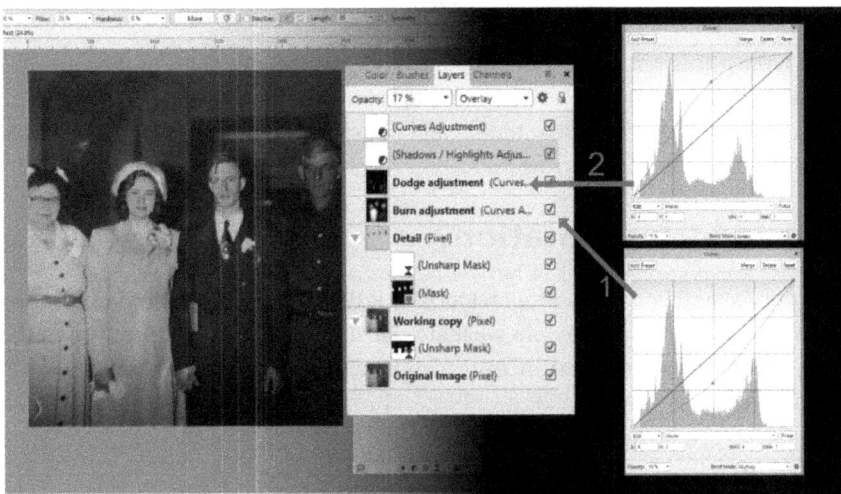

Figure 12.9 – The curve layers for dodge and burn

3.  Add a **Curves** layer to the stack and push the nodes in, as shown in *Figure 12.10*. Notice the change to the blend mode and opacity.

4.  Add a **Shadows / Highlights Adjustment**, adjusting the shadows (**17%**) and highlights (**41%**), and change the blend mode and opacity (these values are shown in *Figure 12.10*):

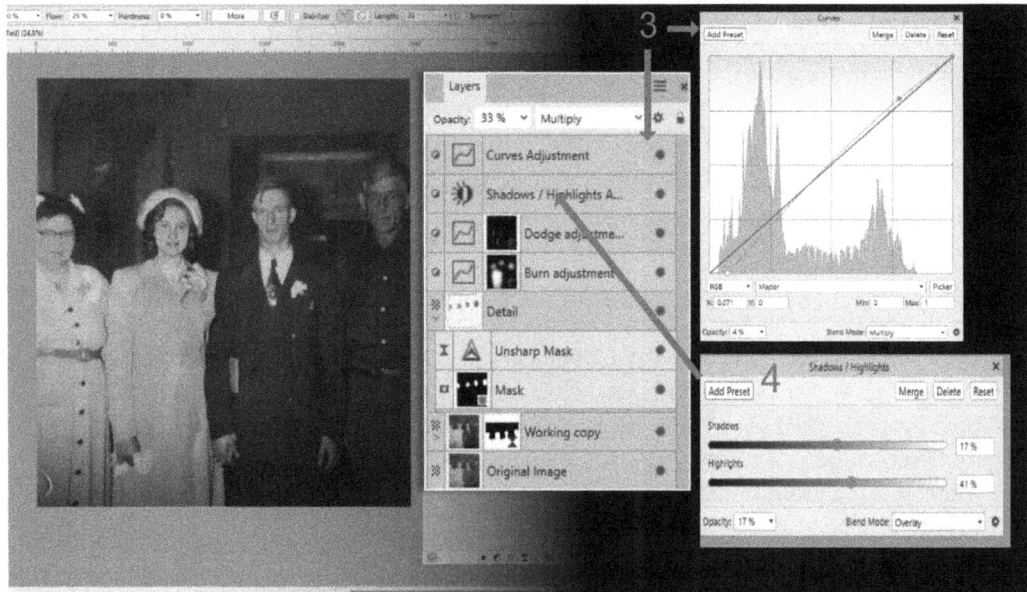

Figure 12.10 – Steps 3 and 4

With that, we have completed the edit. While there are entire volumes of courses out there on editing and retouching photos, this gives you a good workflow for getting started.

Next, we will turn the page from restoration to retouching. You will use *many* of the same tools and techniques mentioned. However, the goal here is not to bring the photo back but to make it *better*. In the next section, we will cover a very basic toolset and workflow. Note there is far more out there on retouching, so think of this as your starting point.

## What is frequency separation?

One of the most utilized techniques in portrait retouching is the act of frequency separation. Frequency separation allows you to split an image into what are called frequencies. There are two of them – a high frequency, which contains the texture information, and a low frequency, which contains the color and tone information.

Notice in *Figure 12.11* that there is a texture layer over the light red square. This is exactly how a texture works when applied to a photo. So, instead of adding a texture layer that's applied over a layer, think of frequency separation as extracting the texture from a photo and separating the photo and the texture:

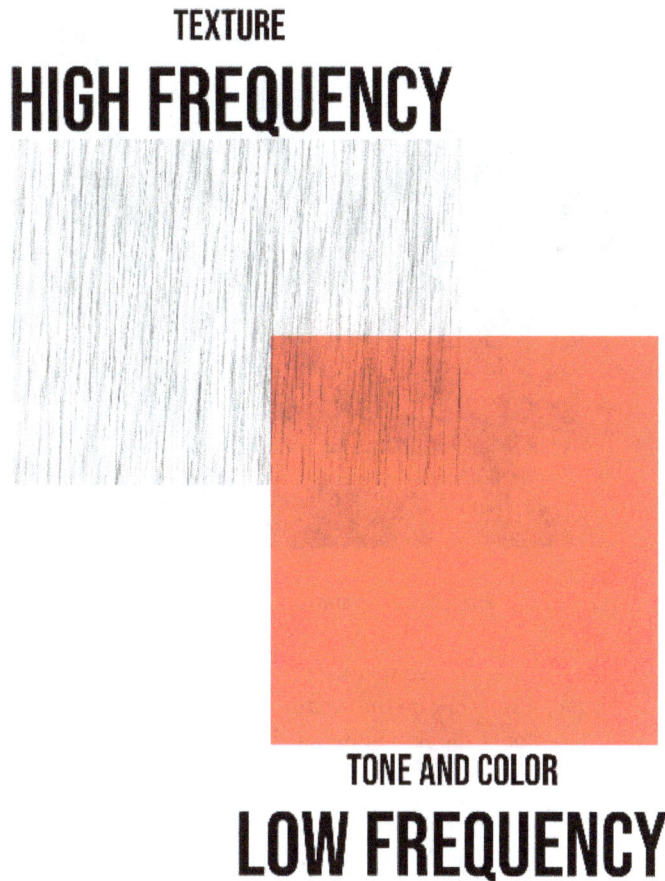

TEXTURE

# HIGH FREQUENCY

TONE AND COLOR

# LOW FREQUENCY

Figure 12.11 – The theory of frequencies – high is texture, low is tone and color

In the following sections, we will dive into the more practical uses of this theory, as well as complete a project to show the concept in practice.

## Why would you use frequency separation?

When editing, we frequently need to knock down light spots, modify dark spots, and so on, and retouching things such as skin smoothness and the consistency of color across the skin is important. However, when correcting, we need to make sure we do not impact the detail of the image and smudge out of detail.

The following image of my wife was taken on a beach in Florida. It's a perfect example of where we can apply frequency separation. The gray area shows the higher frequency (remember the texture with the red square in *Figure 12.11*), while the right-hand side (the blurry side) shows the underlying tonal information:

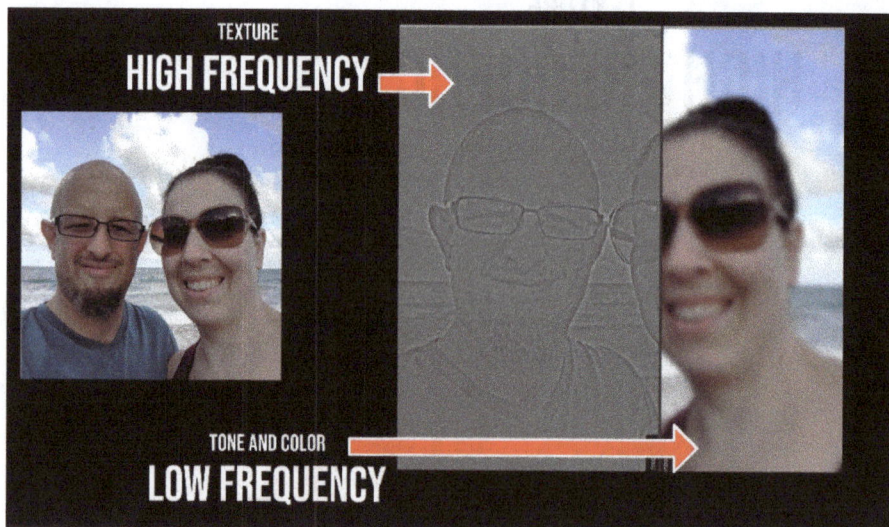

Figure 12.12 – Frequency separation applied to an image

This is why we use frequency separation – so that we can use different tools and workflows on two different layers. As an example, if we work on the **Texture (High Frequency)** layer, we can remove wrinkles without affecting the color of the skin underneath.

## How to use frequency separation

The **Frequency separation** filter is in a different place than the other filters we have discussed so far. It is not a live filter but rather a destructive filter. Destructive filters (filters that you cannot adjust or take back once they are applied) can be found in the menu at the top of the screen under **Filters | Frequency separation**.

The **Frequency separation** filter is different from destructive filters in that, once applied, they cannot be modified. So, to use them, make sure you create a duplicate of the image before you apply it. You never want to lose your base image and not be able to recover it.

We have included this image so that you can follow along if you want. It can be found in the downloads section (see `Frequency separation practice file` inside the `Frequency separation files` folder). When you apply the filter, the screen will change, as shown in *Figure 12.13*, and a preview slider will appear. The two halves of the screen represent the areas as high frequency (grayed out) and low frequency (blurred out):

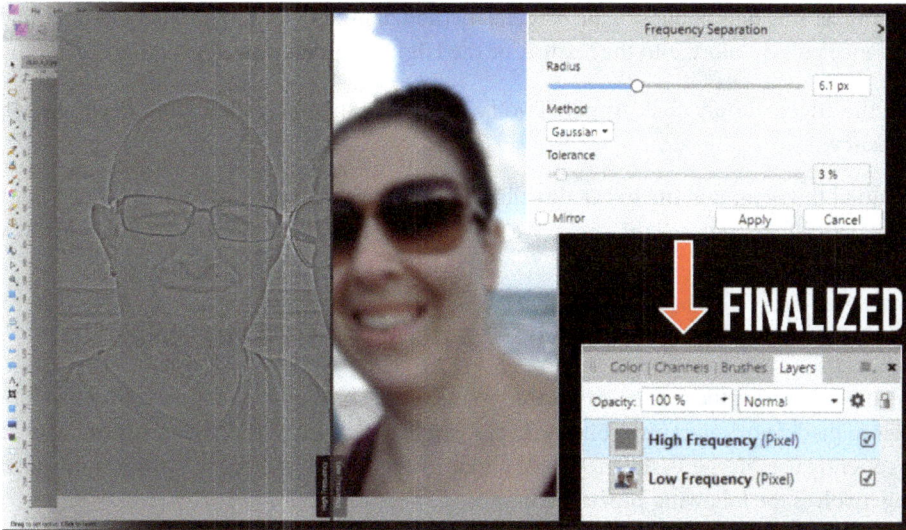

Figure 12.13 – The adjustment and separation in two distinct layers

Now that we know what the gray and blurry areas are showing us, let's learn how to adjust the image *before* finalizing it.

## Adjusting for frequency separation and finalizing the effect

Adjusting the amount of frequency separation is done on a case-by-case basis. In *Figure 12.13*, we highlighted the adjustments you can make, but it is 100% up to the editor what they choose in the image:

- **Radius**: This is the amount of separation. The higher the radius, the more detail is pulled into the high-frequency layer.

---

**Best practice 1**

Do not make the radius so high you are pulling color information in. I pull it until I see color come in and then back off a bit.

---

- **Method**: There are various methods that Affinity uses to pull the low frequency from the high, and the best method for your image can vary, so try all three and see what gives you the best result.

---

**Best practice 2**

I traditionally use **Median** as my method because I find it gives me the most detail. However, because my glasses gave *median* issues in the practice image, I have chosen to use **Gaussian** for this example.

---

- **Tolerance**: This only applies to the Bilateral method. It asks you how *much* it looks at. A higher tolerance pulls in more into the texture, or the **High Frequency** layer.

To finalize the effect, simply click **Apply**. Two layers will be created called **High Frequency** and **Low Frequency**.

I have included the separated file in the downloads (see `Frequency separated file_To follow along`) for you to see the effect. We will be using this file in the next section when we do the portrait retouch.

# Portrait retouching workflow and practical application

Now, it is time to practice the process of portrait retouching. The workflow for portrait retouching is very similar to that of portrait refurbishing, except we utilize frequency separation early in the process and we frequently use **Merge Visible** to create a new image at each stage. My workflow for portrait retouching involves the following processes:

- Structural adjustment (removal of artifacts, cropping, and so on)
- Imperfection removal – that is, skin smoothing, lighting hotspot removal, and overall blemish removal.
- Dynamics adjustment (non-destructive dodge and burning, curves adjustments)
- Colorizing and finishing

## Practical application

To practice this, I have included an image of my wife on our wedding day. This image has already been edited, so we have very little to do, but this is good as we can focus on the workflow rather than the enormous nature of a full edit. We have included this image in your downloads for you to follow along (see `Original image` inside the `Portrait retouching` folder).

I have also included this edit in a series of CiA videos, which are referenced in the steps. Let's get started.

### Structural adjustment

Let's get started with the structural adjustment. Follow these steps:

1. Start by making a copy of the image and locking and making the original not visible. Label this new image `working copy`.
2. Now, we need to fix the back area that is sticking out of the dress. To do this, we are going to make a curve mask. The reason I am choosing a curve mask is that, with the cement behind it, it will be too hard to align the back, the cement, and the grass:

   A. Using the pen tool, draw out a shape with three points – a simple triangle.

B.   Duplicate the image and nest it in the triangle. This will create a mask with a "patch" in it.

C.   Now, we have to select the image inside the triangle and push it left so that the cement and green area covers the part of the body we are trying to mask. This is shown in *Figure 12.14*.

D.   Finally, using the **Node** tool, you can adjust the curve of the triangle so that it matches the back.

3.   Finish the image by right-clicking on the layer stack and choosing **Merge visible**. This will make a flat image of your work, including the patch and lock, and make the lower layers invisible. Label the fixed flat layer `structural fixed`:

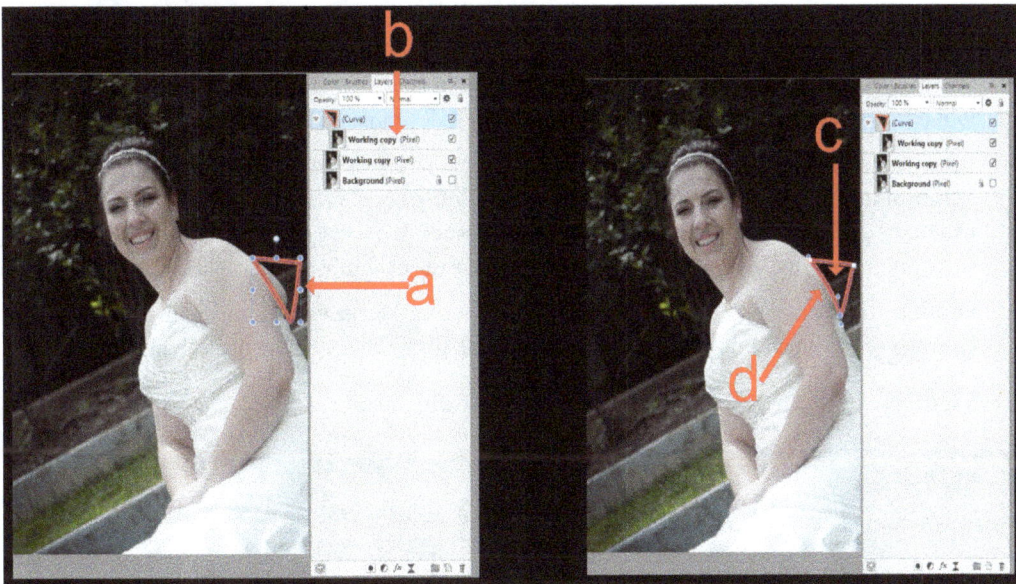

Figure 12.14 – Practical application adjustment

> **Note**
>
> This process is shown in the CiA video (see the *Technical requirements* section for the link).

Now that we have fixed the structure, it is time to move on to the next step in our workflow: imperfection removal.

## *Imperfection removal*

Let's get started with the imperfection removal technique. Follow these steps:

1.  Duplicate the flattened image (titled **structural fixed**). The next step will be destructive, so we need to make sure we maintain our work.

2.  Choose the frequency separation (refer to the section on frequency separation to remember where it is) and then choose **Median** with a 4 px level. This is just about where we start seeing some color coming in, so this is where we'll stop. This will separate the image into high frequency and low frequency layers.

3.  On the high frequency layer, we will work on areas of texture such as under the eyes, the birthmark near the shoulder, and the lines on the forehead. For this, we will use the Clone Stamp tool. My settings are shown in *Figure 12.15* in *step 3*, as well as where we targeted the tool (choose an area of the image that has a smooth desirable texture).

4.  Now, we will turn our attention to the underlying tone. To adjust the tone, we can use the Blemish Removal tool and the Healing brush. All we are doing here is removing hotspots and issues where the photographer's light may have caused issues. We will apply this to the birthmark to smooth out the tone, as well as to a few spots above the eyebrow on the right eye.

5.  Finish by right-clicking on the layer stack and choosing **Merge visible**, relabeling this post as **blemish removal**, and duplicating it by turning off all the layers below it and locking them.

You can see these steps, as well as their results, in the following figure:

Figure 12.15 – Blemish removal steps

> **Note**
> This technique is shown in the CiA video (see the *Technical requirements* section for the link).

Since we covered the techniques for dynamic processing (see photo restoration) and covered some color grading tips in previous chapters, we will not re-hash them in this section. Instead, I encourage you to return to the techniques you know for these sections and experiment, try some things, and make some mistakes. However, I will leave you with the before and after of the activity.

To see this, activate the original image at the base of the stack, and then the topmost layer. Uncheck the topmost layer – it should show the original. This is a great way to check your work.

## Professional tips, tricks, and important points

Now, let's look at some important tips:

- Photo restoration and photo retouching are all exercises in problem-solving, and each problem requires you to come at it differently. So, as you build your toolbox of solutions, develop workflows and stacks that you know work.

- When you are replacing parts of an image, make sure you read the areas around it to figure out what tool to use. For example, in the wedding image edit, we made a path because using the clone or the Inpainting tool would not give us the best curvature.

- You may not be able to save an entire image, so focusing your attention on the parts that truly matter (such as the faces of my grandparents in the restoration lesson) grabs the audience's attention and places it on the pieces that they want to see.

- When doing photo retouching, remember to keep the strength low and the hardness on the adjustments soft – humans do not have hard edges.

## Summary

Now that we know how to heal and refurbish an image, you are all set to repair and adjust the images you make. In this chapter, we covered the basics of cloning and healing images using things like the Blemish tool and the Inpainting brush. We also learned about the very common tools for image touch ups like frequency separation. Now it is time to have a deeper discussion of color. At this point, you have every basic tool you need to create anything you can think of in a photo, but it's color grading that can give an image your signature style. Even something as simple as a wedding photo can take on a completely different feel when different color harmonies are applied. So, in the next chapter, we will cover advanced color grading.

# 13
# Advanced Color Concepts and Grading

Now that we know about healing tools and fundamentals, it is time to kick our understanding of color up a notch. This involves two things to increase your skillset as an editor. The first is an increased knowledge of tools and settings for advanced color grading, and the second is the application of these tools to achieve the very popular looks that we see in modern edits. So, in this chapter, we are going to cover the tools and adjustments, as well as a discussion of **High Dynamic Range** (**HDR**) color spaces and their applications, finishing with the application of these techniques in a few very modern styles.

Now, you can get by with the basic color chapter earlier in the book; however, in my experience, as you get more mature in your editing, these techniques take you from good to great so that when people see your work, they know it from across the room. So this chapter is as technical as it is artistic, and coming back to the cooking analogy: the colors you use create your unique flavor for an otherwise common dish.

In this chapter, we will cover the following topics:

- The basics of color modes
- Adjustments and tools used in color grading
- Color grading practical applications

## Technical requirements

You will need the Chapter 13 project to follow along with me.

# The basics of color modes

You may have heard the terms *8-bit*, *16-bit*, or *HDR*, and been confused about how to choose between them and utilize them in your editing…or may not even know what they are. In this section, we will offer up a simple basic explanation of these terms and share with you the best settings based on my experience as an editor so you know *when* to use *what*.

The reason I wanted to include this in the advanced chapter is, thus far, we have been dealing with simple 8-bit color modes, and for basic screen-based edits, this is more than adequate. However, as you grow as an editor, you may want to take your images further, so let's get started.

## What is a color mode?

A **color mode** is a way of describing how the color of an image will be displayed inside Affinity Photo, or any program for that matter. I personally like the term *color depth* as this implies the richness and variation of color, but you can use either term interchangeably. Typically, you will see an option named **RGB/8**, as shown in *Figure 13.1*.

The numerical value after the / symbol in this term is **8**. This means that it is an 8-bit color mode or color depth. This value is the number of colors in each primary channel available per pixel to choose from.

We can see that this is the default setting when you open a new project in Affinity Photo:

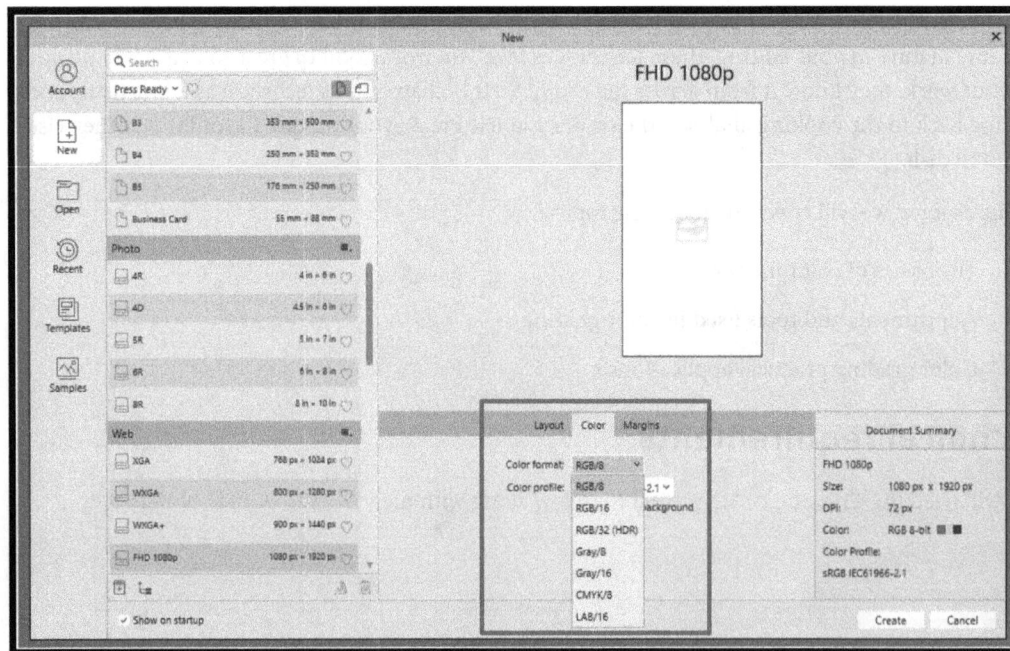

Figure 13.1 – How Affinity displays various color modes

## So, what does 8-bit mean?

To calculate the number of colors per pixel available, we take a base number of two and raise it to the power of the bit count (in this case, two to the eighth power). The result for an 8-bit color mode is 256. This means that there are 256 colors per pixel available for the software to use. We can see this in *Figure 13.2* in the **Color** studio panel.

Notice that pure red has a value of **255**; that is because it is an 8-bit format for the image.

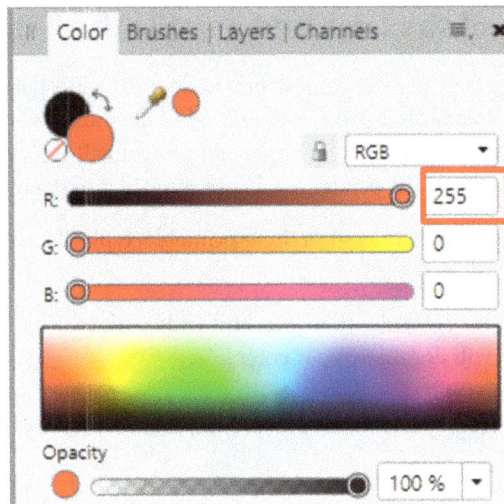

Figure 13.2 – Color mode displayed

Now, the human eye sees around 16 million colors (that's right, we can discern roughly 16 million unique colors), and with an 8-bit color mode, the image has only about 256 colors for each channel to mix and work with in a pixel. So, in reality, what I would be tempted to call this is the basic standard color mode (the most frequently used). This is the bare minimum for monitors, graphics cards, and most residential printers.

## The difference between 16- and 8-bit color modes

If we follow the logic just mentioned, the 8-bit color mode has 256 colors available per channel; however, if we work in a 16-bit color space, there are 65,000 to mix and match to display in a pixel. And as expected, this has an impact on your hardware. Computers need more powerful components to work with higher color modes such as 16-bit and even 32-bit modes. Additionally, the monitors that display these increased color depths experience a heavier pull and drag as they struggle to refresh and adjust.

> **Note**
>
> A 32-bit color mode gives you the equivalent of 4,294,967,296 variations, far beyond what the human eye can even discern, so in practical experience, this is simply wishful thinking as humans, monitors, and printers largely just cannot even comprehend this level of discernment. So, for me, I do not work in HDR and, as a basic editor, unless this is your field…most likely, you never will either.

## What do I use most of the time and what is the best practice?

Typically, I work almost exclusively in 8-bit color depth, and for the most part, this is acceptable 99% of the time and assures that, whether it is print or digital, all of the industry-standard equipment will handle this just fine. If I am working on a really important piece, I will typically utilize a 16-bit color depth, accepting that when it goes to the printer, I will lose some of it to the conversion over for print.

To convert from these high color counts to something printable, typically, you will have to tone map (we cover that in *Chapter 17*).

In my opinion, the use of 16-bit color depth begins and ends with editing; when it is displayed or printed, the 8-bit color depth will dominate your colors.

## Setting and modifying your color mode in Affinity Photo

Setting your color space initially begins with your image. If you are importing an image, the color space and color depth are set in your camera first and foremost. However, if you are setting up a new project, when you select **New**, there is an option to select **Color** options in Affinity Photo (see *Figure 13.3*).

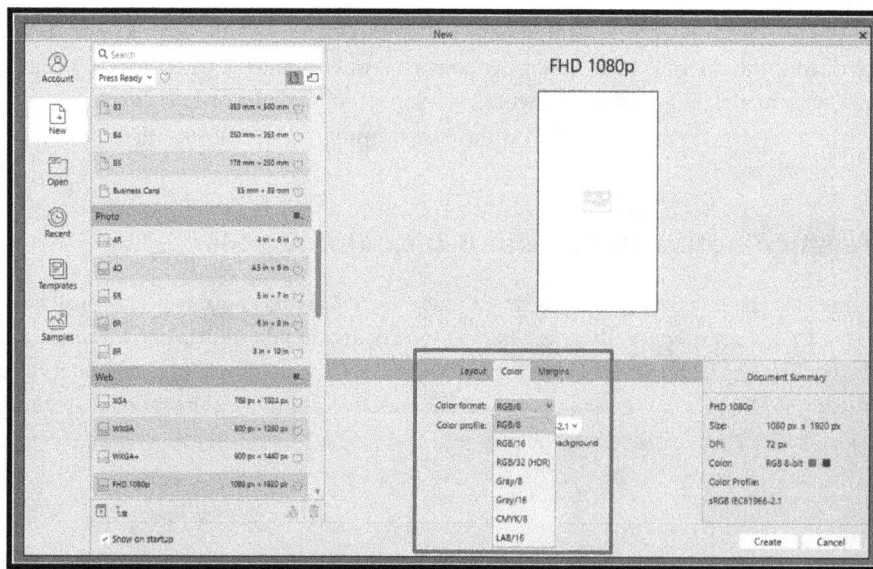

Figure 13.3 – Where to select color space in a new image

Now, if you have an imported image, the color profile from the camera will be shown in the context toolbar (see *Figure 13.4*). In order to modify this, go to **Document | Convert Format / ICC Profile...** and then choose your new profile:

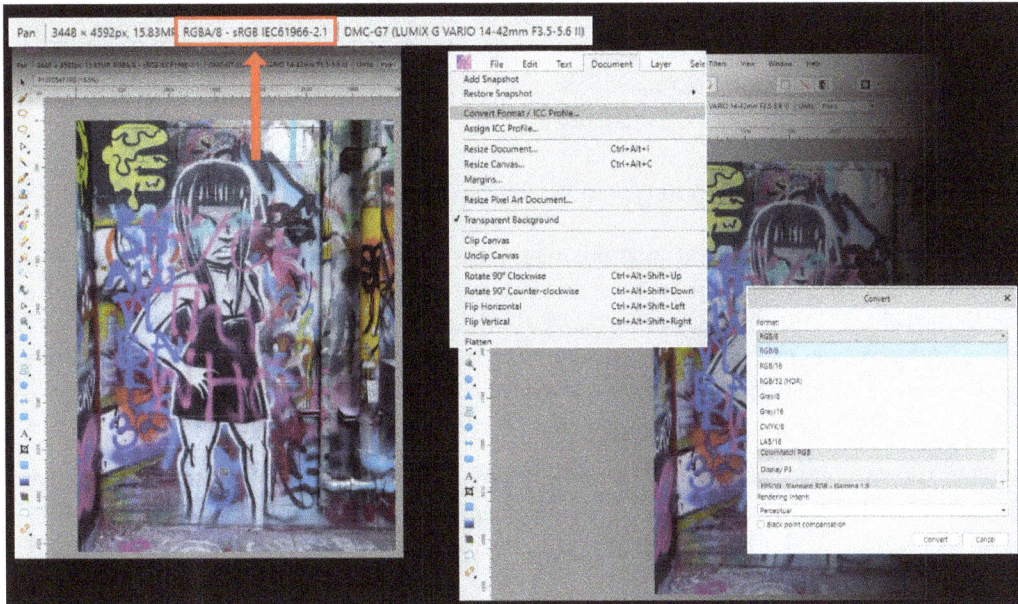

Figure 13.4 – Display/change color space

Now, it is worth noting that simply changing your format will not make colors magically appear. The image was shot with a certain amount of color data, so it cannot be created.

If you want to add more color data from the outset, many photographers shoot in RAW mode (we have this covered in *Chapter 17*). So, this is a subject for later, but typically, if you are doing high-dynamic photography, then you are shooting in RAW mode anyway.

Now that we know a bit about color spaces (8-bit versus 16-bit) and how to set and modify them, it is time to look at the tools used for color grading like a pro.

# Adjustments and tools used in color grading

There are so many adjustments and tools used in color grading, and it really is so subjective that each artist can develop their own unique style around their color grading formulas. We have chosen, in this section, to incorporate the most frequently used, dynamic tools and adjustments that give my work its signature style.

These are not all the tools, but those that I feel provide the most value and the tools that I will be utilizing in the color grading examples later in the chapter. To make sure we standardize the tools, I have included a practice image in the downloads for this section (see `Color grading base image` file inside the `Color grading tools_Example` folder).

We will be using the same image throughout this section. I have taken the image to the point where it is structurally sound, and now it is time to color grade. So, we will be using the same image to show the tools; that way, the image does not influence the tool.

## Hue, saturation, and lightness adjustment

We have covered this tool before and utilized it in previous chapters, so we know where to find it and how to use it to change the entire image; however, when it comes to color grading, we will be effectively targeting the colors. In the past chapters, we adjusted the entire image, but below the circle, we can target individual colors (see *Figure 13.5*). In *Figure 13.5*, the red colors are selected, meaning we will only affect the red colors when we make our adjustments. This is also shown by where the dots appear in the hue wheel (see the red arrows in *Figure 13.5*):

Figure 13.5 – Color grading using a simple HSL adjustment

You can set each color separately this way – the cyans, magentas, reds, yellows, and so on. So, for the flames, we will adjust the reds and yellows, and make a separate adjustment for the cool blues (see the **AFTER** photo in *Figure 13.5*).

I have included the working file in the downloads as well so you can see my exact settings. Simply click on the colors and you will see the HSL value adjustments by color.

## Selective color adjustment layer

This is one of my favorites and a definite staple in my own editing. **Selective Color Adjustment** tries to target only defined colors and change only those colors by adding or subtracting amounts of the base colors inside. I have taken a screenshot of the tool in its current state and broken it down into its parts:

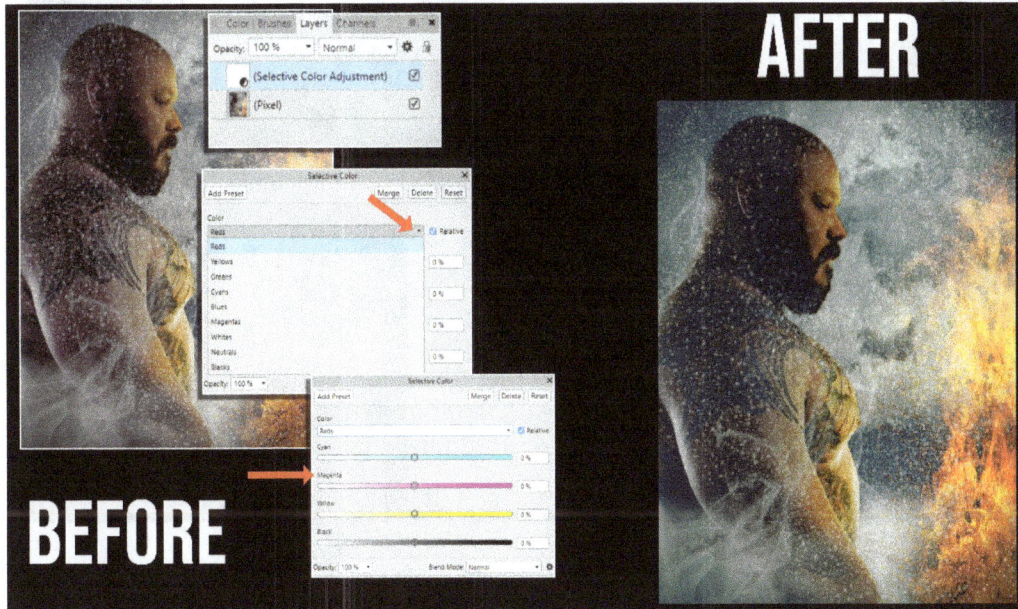

Figure 13.6 – Color grading using selective color

Follow these steps to make edits to your image:

1.  Choose the color you want to adjust. In the example, we have chosen to adjust **Reds**, which means the color adjustment will only apply to red colors.

2.  Adjust the sliders to add or subtract colors to influence the red colors. So, if you wanted to add a bit more magenta to it, slide it to the right.

We have included the color-graded image in the downloads for the chapter, so if you want to see my adjustments, simply open the file, select the adjustment layer, and then click through the colors.

## Spilt Toning Adjustment layer

In split toning, we divide the image based on **highlights** and **shadows**, and then the **hue** is adjusted for those areas. We can also adjust the balance of shadows and highlights. So, if you want more of the shadows to be affected and less of the highlights, simply move the **Balance** slider. I have broken down the windows in *Figure 13.7*:

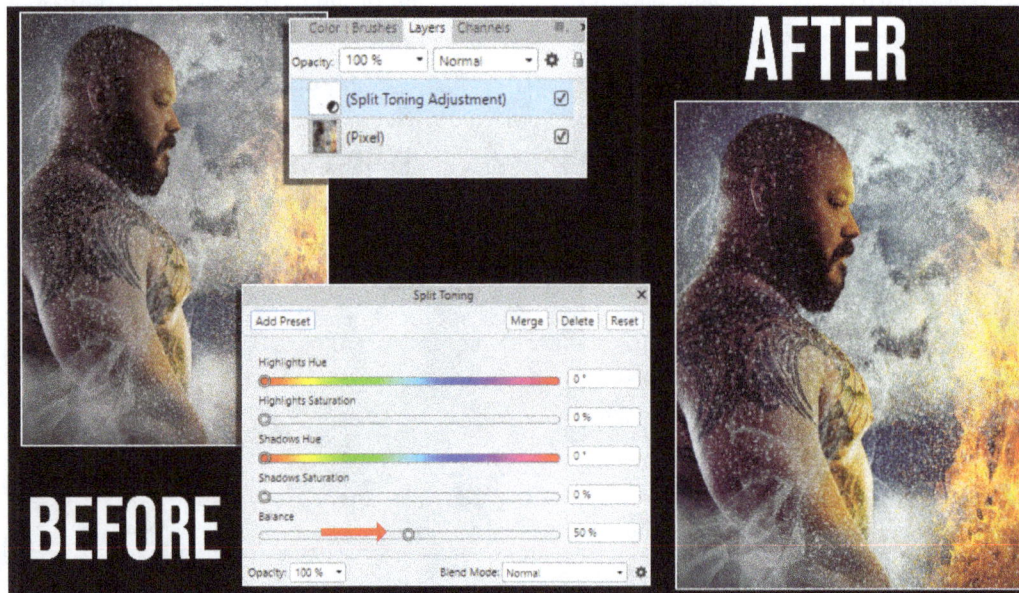

Figure 13.7 – Color grading using split toning

Notice in my working file that I changed **Blend Mode** from **Normal** to **Soft Light** (changing the blending mode is a professional trick in color grading to add a different variant). The reason it is a professional trick is many beginning editors are so interested in just the adjustment that they never think about using the blend mode in conjunction.

## Channel mixer adjustment

As we have previously covered, there are three channels in the color profile: **red, green, and blue** (**RGB**). In **Channel Mixer Adjustment**, we are adding or subtracting color to sort of blur channels. *Figure 13.8* shows how I used this tool:

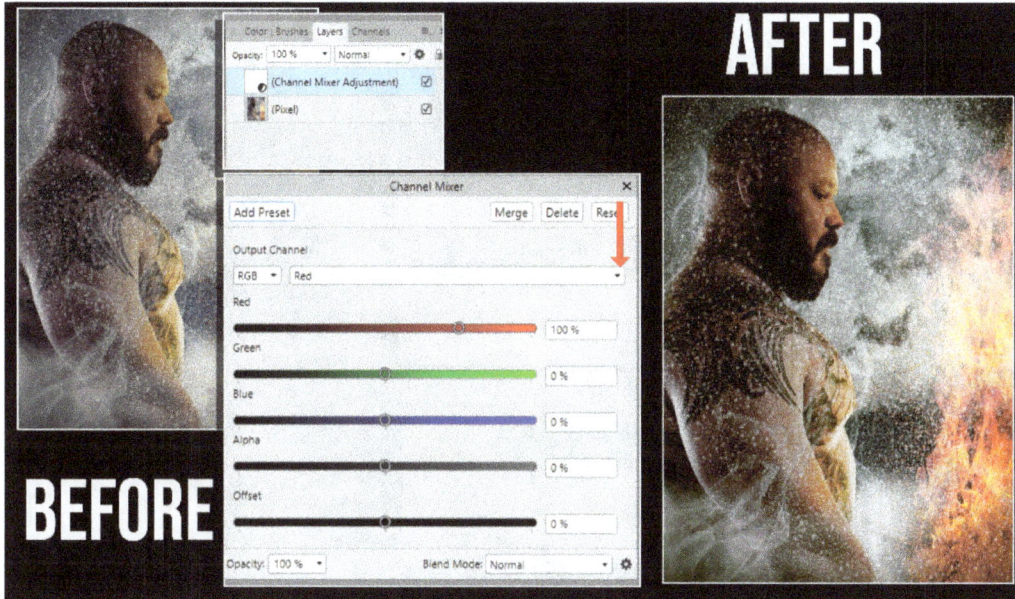

Figure 13.8 – Color grading using channel mixer

In this example, if you wanted to work on the red channel, you can add or subtract red, green, or blue. This will drastically modify the image, so subtle changes make a huge difference.

Personally, in my career, I have never adjusted the alpha or offset, so have a play with them if they speak to you; if not, they rarely (if ever) get touched. I have included my adjustment file in the downloads for you to follow along with my adjustments.

## Gradient maps

Gradient maps are a very powerful tool. They work on the idea of dark and light areas of the image. In *Figure 13.9*, the far left of the gradient map window represents the color that will be applied to the darks, and then the reds are the light part of the images. By default, in the center is the green, in the mid tones.

Now, out of the box, the default gradient map may not meet our needs, so to customize your gradient map, we can do the following:

- We can remove mid-stops (such as the green) by clicking on them and hitting *Delete*
- We can add mid-stops by clicking on the line

We can change the colors of any of the circles by clicking on them and then picking the color in the color window underneath:

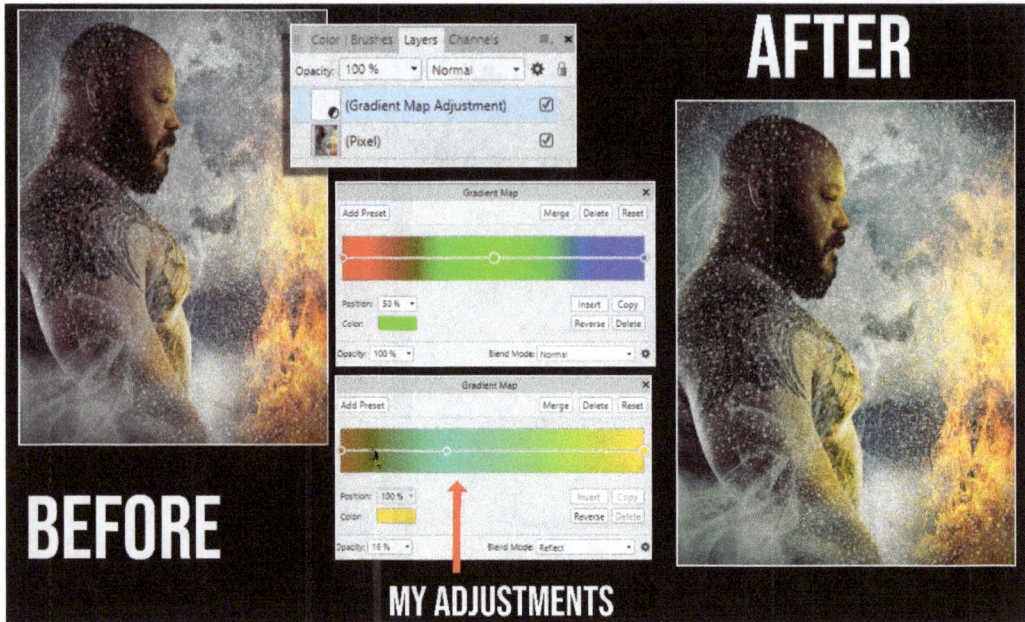

Figure 13.9 – Color grading using gradient maps

In order for the gradient map to work, we need to adjust the blend mode and opacity. By default, the **Normal** blend mode doesn't look very good, so blend modes and opacity are key adjustments when using this tool.

> **Pro tip**
> Use masks to apply the gradient map only to selected areas such as the background.

## Color balance

**Color Balance Adjustment** works on the idea that there are various diametrically opposed ends of the color spectrum (i.e., the opposite of cyan is red, the opposite of magenta is green, and the opposite of yellow is blue). To work with **Color Balance Adjustment**, simply slide the sliders around to find a balance that gives you something you like.

My adjustments for the attached color grade image are shown in *Figure 13.10*:

Figure 13.10 – Color grading using color balance

## Using curves to color grade

In *Chapter 7*, we covered the idea of the **Curves Adjustment** layer, and how it can be used to finely tune certain areas of dark and light to add contrast to an image. During those lessons, we used the **Master** channel. However, now that we are discussing the idea of color grading, we can discuss using this same functionality by using the **Red**, **Green**, and **Blue** channels (located where the **Master** channel was previously identified).

The adjustment works exactly the same: at the far left of the horizontal access are the darkest parts of the image, and at the far right are the lightest parts. However, along the vertical axis is the amount of the channel (R, G, or B). So, if we play with these, we see a dramatic difference.

In *Figure 13.11*, I have included my settings for each channel and the resulting output. I was going for an overlay dramatic book cover sort of look, so I realize this may not be for everyone, but this is color grading. Our opinion can be different depending on the intent.

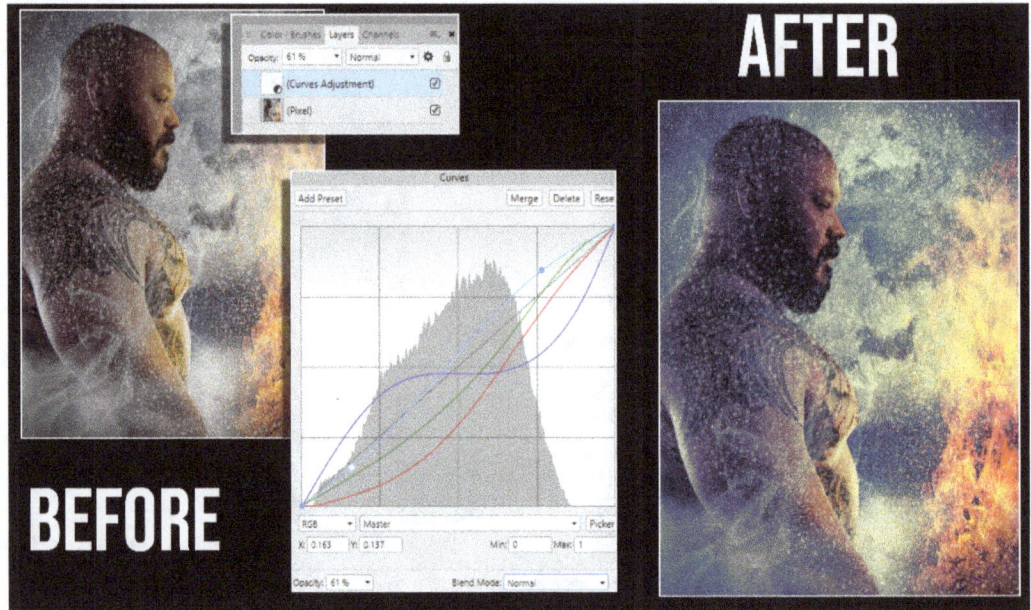

Figure 13.11 – Color grading using curve channels

## LUTs

**LUT** is a commonly used acronym for **Look Up Table**. This is a mathematical way of saying, *If I have color X and I apply a LUT adjustment, I want Affinity to change color X in the image to color Y*. Think of it like a conversion chart: you apply the LUT and Affinity looks up all the colors and, based on the LUT, changes the colors to another value according to the new table.

This adjustment layer relies on the numerical values of colors. Let's say that the green color you are working with has a certain numerical value on the HSL scale (as an example, **H-111**, **S-65**, and **L-33**) and during the edit, you want to get it darker and a little more toward the blue (as an example, **H-165**, **S-57**, and **L-24**). If you like that look for your green and want to save it, you will export a LUT.

Now, the next time you have an image and want a similar look, you can import a LUT, which will tell Affinity to convert that particular value of green to the darker, more blue value.

In short, this adjustment layer allows you to carry a signature look over to other images with one click.

The underlying images have to be similar though; you cannot apply the same LUT to two wildly different images and expect it to perform the same. In *Figure 13.12*, we have a base image and, on the right, we have a few LUT adjustments to show what is possible with this adjustment layer:

Figure 13.12 – The effect of LUTs on editing

## Exporting LUTs

We should begin by sharing how to export LUTs because, as you edit and find a style you like, you will want to export the look you like to the images you commonly shoot. I have included the screenshot for exporting LUTs in *Figure 13.13*. The option for exporting is found in the **File** menu:

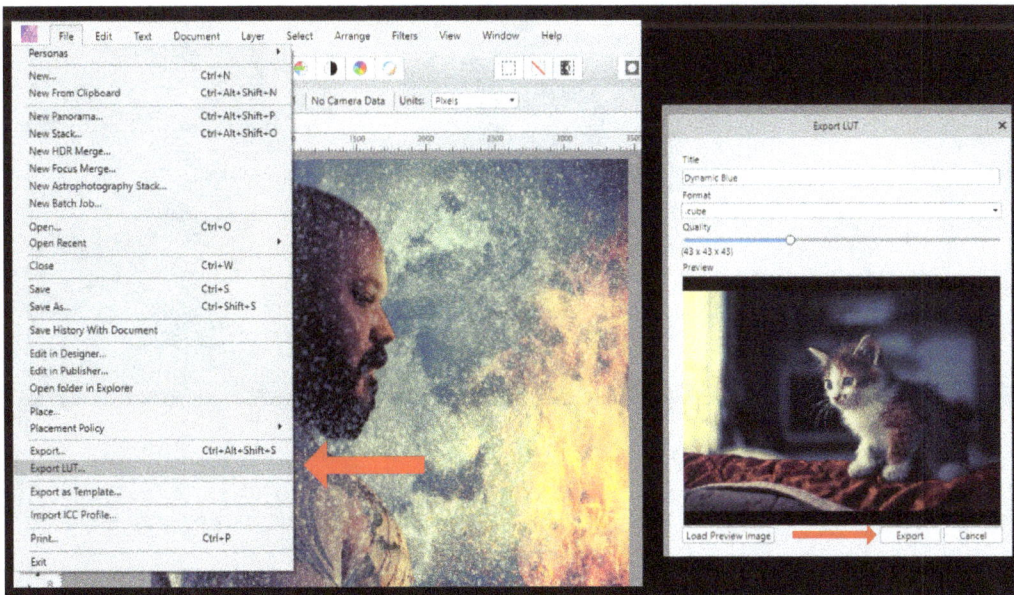

Figure 13.13 – Exporting a color grade as a LUT for later use

Notice the image of the cat comes up — this is a rough image of how the LUT would look, and simply, after that, we just name it and that's it. On the subject of file format and quality, a mid-quality will work just fine, and cubed is the general default. I rarely deviate from these settings.

Place the LUT where you want to store it because, in the next section, we will be importing it onto another image.

This **Dynamic Blue** adjustment is available in the downloads for this chapter.

## Importing LUTs

LUTs can be purchased from a variety of sources and there is no limit to the vast variations you can make from images you find. In this section, we will use the **Dynamic Blue** LUT I included in the downloads for the chapter.

Simply choose an image. In this case, I have chosen an image (I placed it in the downloads for this chapter; see `Base image for LUT practice` file inside the `LUTS` folder) and I have applied the LUT. Notice the before and after images:

Figure 13.14 – Before and after LUT adjustment

Because of the LUT, we crushed the yellows, made the blacks more pronounced, and deepened the blues.

If you do not feel like making your own LUTs, you can infer a LUT from the **Adjustment** menu and it will apply one from an image you choose by guessing the color changes. Inferring a LUT means the program will look at the image you made and infer a LUT from that image, so if you have an image you like, you can infer a LUT from that image. Think of this as being the same as the color picker... except that for the entire color scheme, you are picking up the LUT that the image used.

Now that we know the top tools for color grading, let's combine them into an application in the next section.

# Color grading practical applications

While we will be doing practical color grading, I realize that everyone's interpretation of color (what looks good and what looks bad) is different, so if any of the adjustments I make to my edits do not suit you, feel free to change them. My work is usually heavily saturated and highly dynamic (I do not do subtle very well) so make adjustments to suit you.

In this section, we will be covering some very popular adjustment styles you see in common edits, but feel free to make your own modifications,

## Color grading practical edit 1 – teal and orange film look

One of the most common and frequent adjustments in the area of films is the popular teal and orange look, and this is a simple edit accomplished with a few of the previous elements stacked together:

- Color balance
- HSL
- Channel mixer

The goal of this adjustment is to get the channels to give you an orange tone to highlight the skin, and completely offset it from the use of blue, which is a complementary color (see *Figure 13.15*). I have included a practice .jpg in the downloads for this lesson so you can follow along.

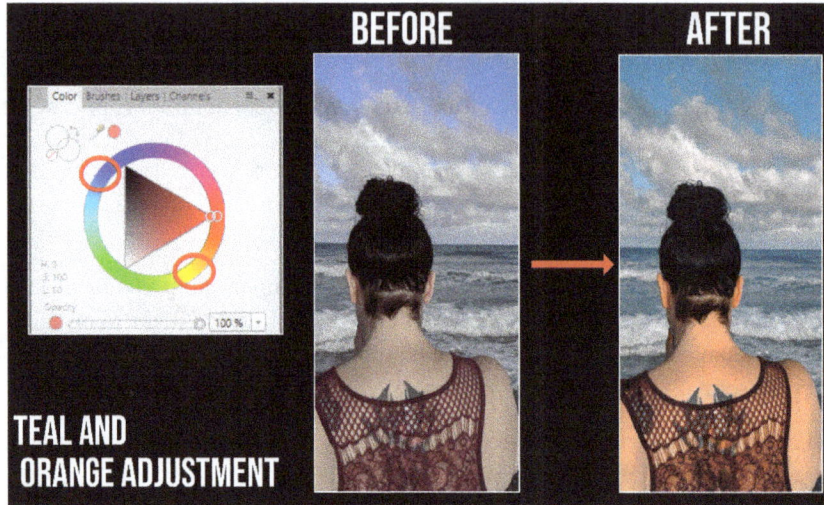

Figure 13.15 – Teal and orange showing complementary status

To make this edit, there are only three simple steps, and I have included my settings in *Figure 13.16* as well as the working file, so you can follow along:

1.  Add **HSL Shift Adjustment** and push the saturation a bit higher (around **15** is where I have it).

2.  Add **Channel Mixer Adjustment**, select the **Blue** channel, and enter the values for **Red** as **-50**, **Green** as **150**, and **Blue** as **0**.

3.  Select the **Color Balance** tool and move the shadows toward blue and the highlights toward yellow.

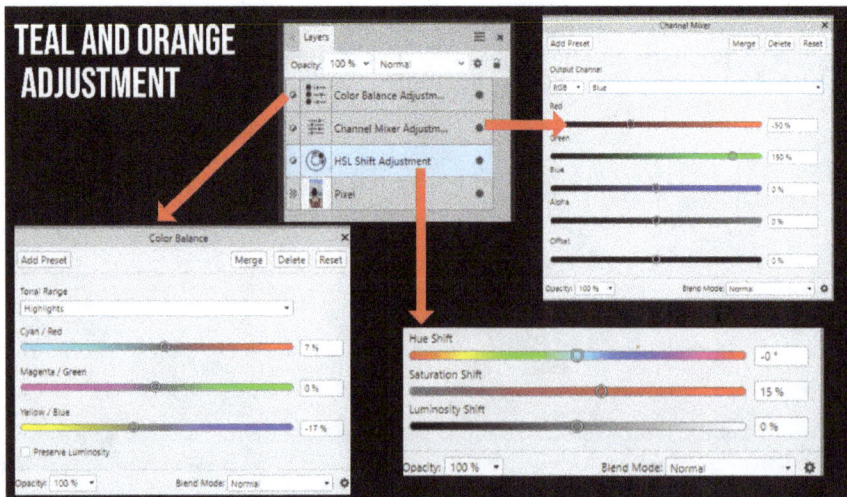

Figure 13.16 – The three layers and adjustments to achieve the teal/orange look

That is it – a simple and effective classic teal and orange color grade.

## Color grading practical edit 2 – dynamic urban

In this edit, we are replicating a very popular sort of urban look, where the red and oranges are pulled forward and the rest is desaturated to a significant amount (see *Figure 13.17*).

Figure 13.17 – Urban orange look

The first few adjustments are shown in *Figure 13.18* and the last two in *Figure 13.19*. I have included the raw .jpg for you to follow along as well. I have also included the finished file in the downloads:

1.  Add a **Color Balance Adjustment** layer making color adjustments to the shadows.
2.  Add a **Curves Adjustment** layer creating some more contrast in the image using the **Master** channel.

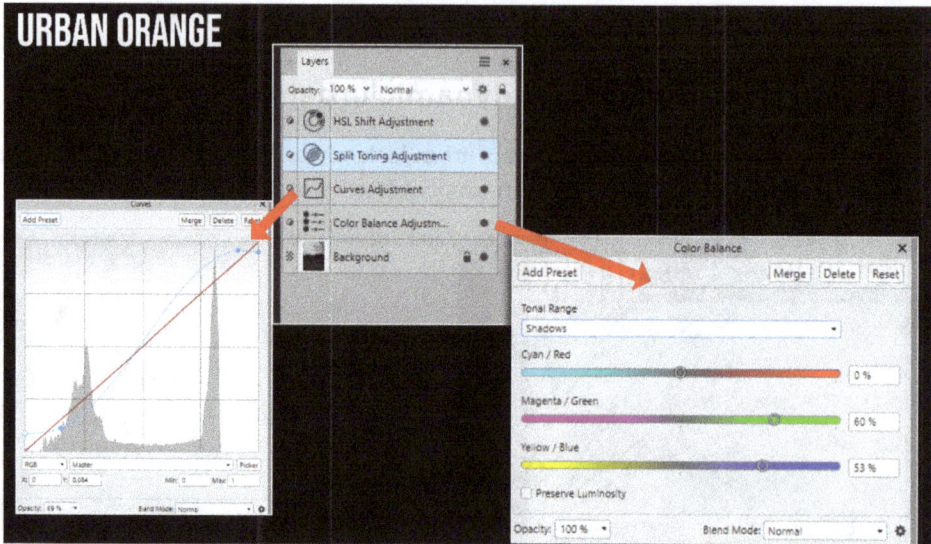

Figure 13.18 – Adjustment stack to achieve the urban orange

3.  Add a **Split Toning Adjustment** layer and adjust to taste. Make sure you do not turn it up too far in the saturation and create hot spots, or areas where the data has been wiped out by bright white.

4.  Lastly, utilize an **HSL Shift Adjustment** layer to the yellow color and increase the saturation, leave the red alone, and in the other colors, reduce the saturation to **0**.

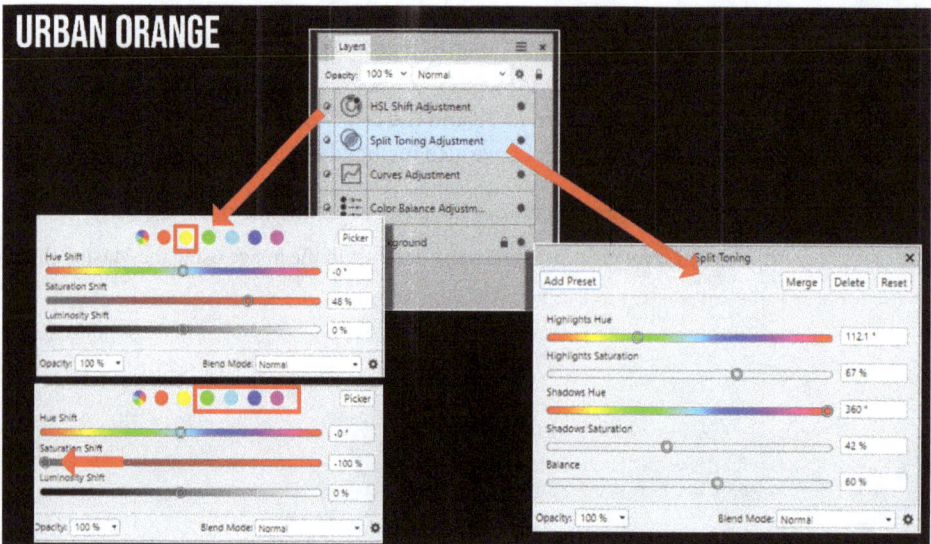

Figure 13.19 – Split toning and HSL adjustment settings in the stack

## Color grading practical edit 3 – teal and pink neon

This is a popular look made popular by Brandon Woelfel, who used twinkle lights and some incredible techniques to make a signature style. We will recreate this with a few tweaks and some embellishments. I have included the base image in the downloads, as well as the finished file to help you follow along.

Figure 13.20 – Teal and pink look

Because this lesson is about color grading, it is recommended that you use the working file to follow along, as we have added adjustments such as curves, shadows, and highlights, as well as the Bokeh brushes we used previously to create some lighting effects.

Only the color grading aspects of the edit will be discussed in this section, as I assume you know how to do the other techniques as they are covered earlier in the book. See *Figure 13.21* for a visualization of the following steps:

1. We add a **White Balance Adjustment** layer, effectively skewing the white colors toward the blue and magenta.

2. We then add a **Split Toning Adjustment** layer, adjusting the hue, saturation, and balance of the image.

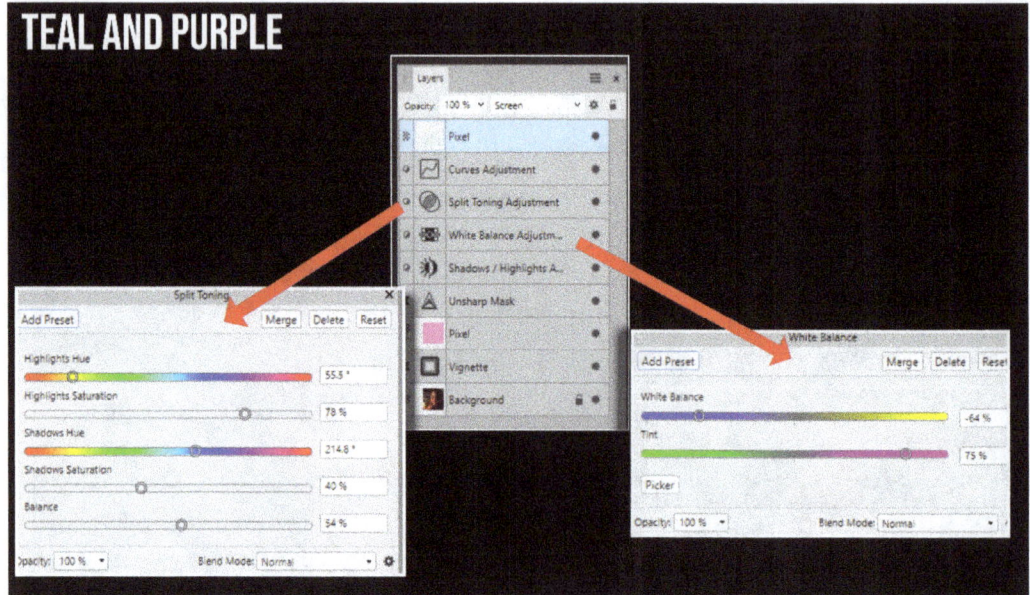

Figure 13.21 – The white balance and split toning adjustments for the teal and pink effect

Notice in this example that the layers and their interaction make a difference; if we had flattened the image with the bokeh embedded, we would not have gotten the same result. This is why color grading is technical. It is not just slapping a filter on but rather knowing how all the layers come together.

# Professional tips, tricks, and important points

Let's go over some important points now:

- I tend to color grade last, but the adjustments you make to tone, sharpness, and so on will have to be adjusted as part of color grading, so expect to make these. This is another reason why a non-destructive workflow is an essential skill to develop early.

- Remember that color sets the mood of the image, so plan accordingly. Also, when working with the style of people you appreciate, don't forget to make their style your own and take inspiration, but not a straight-up copy. Utilizing blend modes on your adjustment layers is a professional move. The colors will interact differently, giving you a unique style.

- I commonly use masks to apply grading only in one part of the image or another. In this chapter, we did not use them, but masking out parts of the image allows you to color smarter and create various feelings.

- The use of color theory on complementary, split complementary, and triadic color chords and harmonies is a good study if color grading is something you want to make a significant part of your editing style.

One of the best to ever teach this is my good friend, Lindsay Marsh, who runs a YouTube channel for designers and has some amazing content. I have included a link to one of my favorite videos on her channel: `https://www.youtube.com/watch?v=bVVtM24x2pY&list=PLP6D1OsIYLkMfz8voiIq5hB5gGZeql9cK&index=18`.

## Summary

We did a lot in this chapter, and this is one of the most artistic chapters of the book. You learned about color spaces, applied five tools for basic color grading, and learned about three popular stacks (teal and orange, neon, and urban orange). So, now you know the isolated tools and you have physical proof that you can combine them into something amazing.

The world of color grading is an evolving one. We have not covered CMYK and LAB aside from mentioning them, and certainly, every great editor out there has their own secret sauce for achieving their signature look. I encourage you to play with other styles, emulate the ones you like, and find yours. I will tell you, after a few years, you can pick my images out of a lineup given the color grading that is done. Now, it is time to turn the page and talk about the dirty side of editing…the black hat side…the destructive side.

Up until this point, we have been exclusively focused on non-destructive editing workflows and techniques, which allow you the flexibility to change your mind, but there is a subset of tools and filters that are destructive in nature and, quite frankly, are leftovers (in my opinion) from the early days of photo editing programs. They are, however, useful in some aspects of editing, and so I would like to spend a bit of time showing you these dinosaurs and pointing out their limitations, as well as their uses. This is what we will cover in the next chapter.

# 14

# Destructive Filters and Tools in Affinity Photo

In the previous chapters, we primarily discussed non-destructive workflows (except for frequency separation during photo retouching). The ideal state is for an editor to practice non-destructive techniques as often as possible, but due to limitations on code and development, sometimes, this is not always possible. These limitations can be seen in antiquated, destructive filters that, once applied, cannot be changed and alter the image permanently. These limitations are accompanied by a set of what I call **dirty editing** tools, such as blur and sharpen brushes. These antiquated tools have non-destructive alternatives (for example, a **Blur Live Adjustment** layer with a mask) but still exist for people that want to do quick and dirty mock-ups or destructive edits.

It is by design I saved this for the end of this book, as I wanted to teach the more professional ways we work before showing you all the ugly parts of programming. However, the time has come to pull back the curtain and expose you to these tools and filters. So, in this chapter, we will start by covering the groups of tools that exist, and then, within each group, remind you of how the same thing is accomplished in a non-destructive fashion.

Following that discussion, we will explore a group of destructive filters that I use most frequently in my edits. We will not be covering them all, because quite frankly, there are many I have never even opened and see no viable use for in my workflow. Hence, this should be looked at as a chapter that starts your journey but will require you to push the boundaries after and ask, *why does this exist?*

In this chapter, we will be covering the following topics:

- Common destructive tools used in editing
- Destructive filters used in editing
- Practical destructive edit
- Professional tips, tricks, and important points

## Technical requirements

The CiA video for this chapter can be found at `https://packt.link/6jxxE`.

You will also need the `Chapter 14` project to follow along.

## Common destructive tools used in editing

In this section, we will be going over some of the tools you will see in the tools area of Affinity Photo. Your question at this point in this book might be, "*Why haven't we explored most of the tools? I mean, we have only covered some, such as basic brushes, cropping, and the move tool.*" The short answer, in my opinion, is that the tools we traditionally had in earlier versions of photo editing are being replaced with non-destructive options. In this section, I have highlighted four common tools (or tool groups such as blur and sharpen that you can see in the default tools menu). I will show you how to use them, and then we will finish the discussion by reminding you of the non-destructive option that currently exists – 99% of adjustments can be made non-destructively these days.

### The Blur and Sharpen brushes

The **Blur** and **Sharpen** brushes allow you to make parts of the image either more defined or less defined, respectively. They come in a matched set of tools in the interface. They can be located in the tools area, as noted in *Figure 14.1* (**1**):

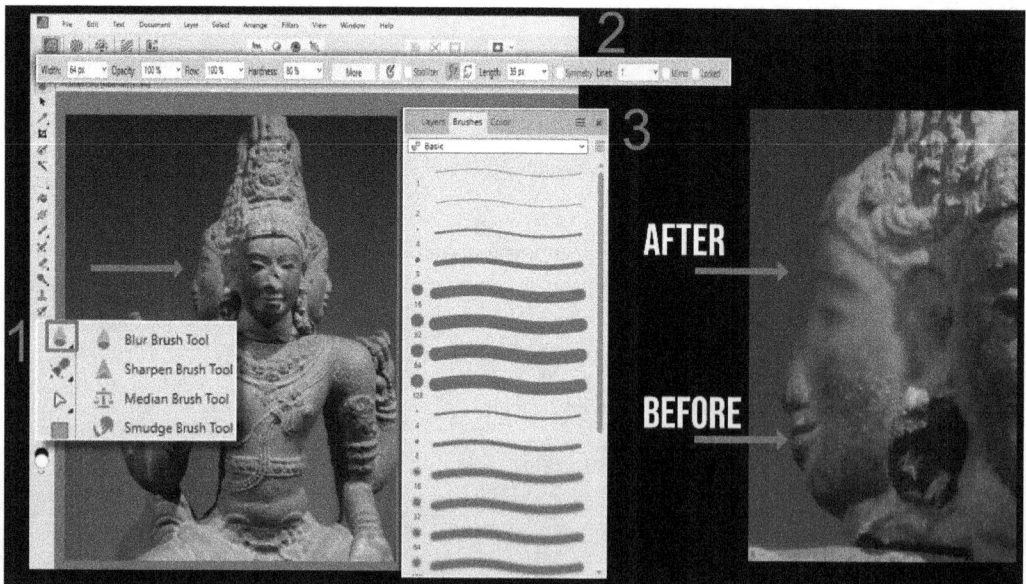

Figure 14.1 – The Blur and Sharpen brushes in the tools area

## Using the Blur and Sharpen brushes

These brushes are utilized just like the normal basic brush with the same adjustments (refer to *Chapter 9* on brush basics for a more detailed description of the individual settings). These brushes have **Opacity**, **Flow**, and **Hardness**, as well as **Stabilizer** options, just like a traditional brush (see *Figure 14.1* (**2**)).

It is worth noting that you can choose different brushes that are used for this function by selecting a new brush from the brushes studio tab. So, 99.9% of the time, I use the basic round soft brush, but if I want to blur or sharpen a texture, I can also choose a separate brush (see *Figure 14.1* (**3**)).

Each instance of the effect is engaged each time you click. This is an important designation because if you just click and hold and start making multiple passes, you will not see the effect add up. This is because you just click each time, so you click and apply, then release, and then click again. This is not a brush where you click and drag over for the application – you must release the button and click again for another instance.

Note that the application is a *slow* process, and by slow I mean it takes several applications to see a change in your image. Even at 100% flow and opacity, it takes several passes to get a significant amount of blur or sharpening.

In *Figure 14.1*, I have included a before and after example of my image, which took 15 to 20 applications.

Also, notice the **Layers** panel as you work – no adjustment layer or garbage layer will be applied; you are just destroying the layer that you are on.

Now that we know the destructive version of the blur and sharpen brushes, let's look at how Affinity contains the more modern alternative.

## What is the non-destructive version?

The non-destructive version of this technique can be found in the adjustment layers we have already explored. To replicate this, simply use a **Live Filter** layer for blurring or sharpening and then use the basic brush to either conceal or reveal the layer where you want it to apply.

In *Figure 14.2*, notice I used the same image, applied a **Live Filter** layer, and then used black to conceal the layer except for the area where I want the face to be blurred (I intentionally overdid the blur so you could see it properly):

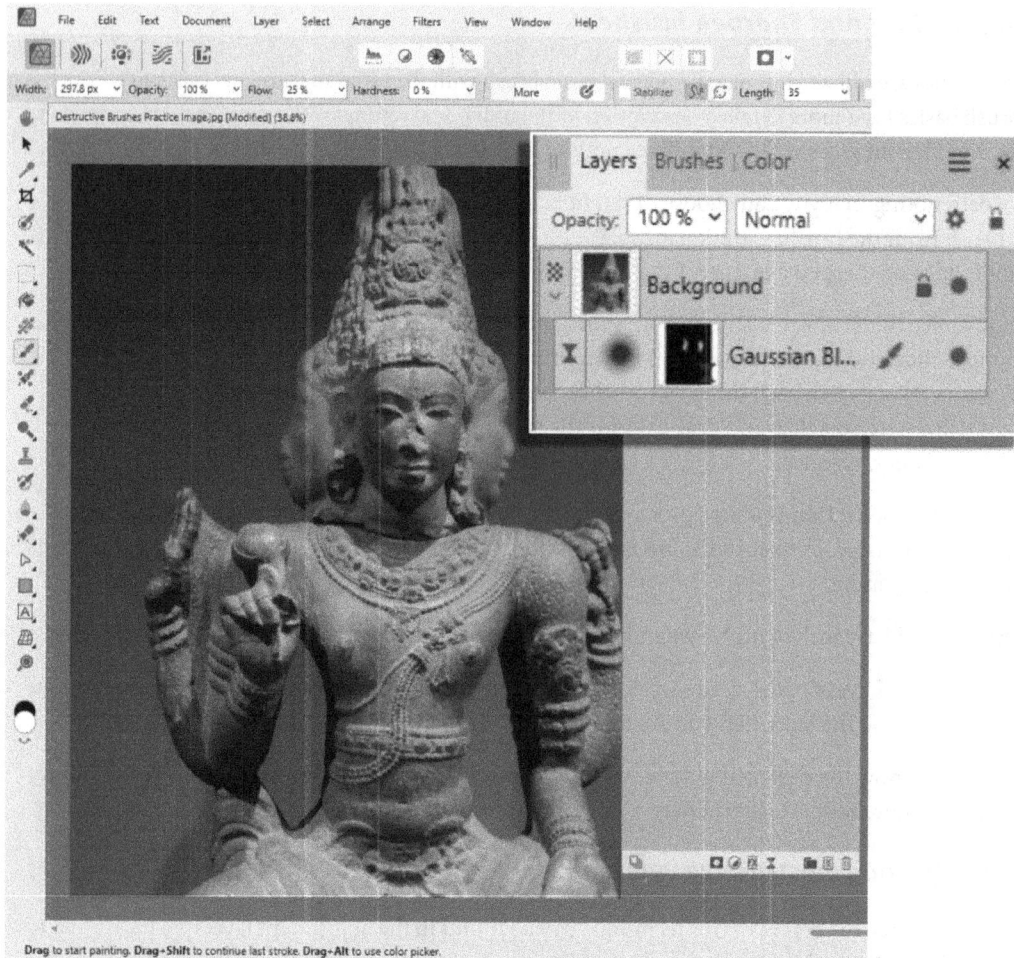

Figure 14.2 – Example of a blur Live Filter with a mask applied

Let's move on to the next set of brushes.

## The Dodge and Burn brushes

The **Dodge** and **Burn** brushes come from the earliest photo editing programs. The fun fact about the terms dodge and burn is that these come from the age when photos were exposed to chemicals. If you expose areas of the photo to the chemical for too long, it will make it darker, and thus *burn*, while if you reduce the exposure and mask out certain areas, you will *dodge*.

We have included the base image for this destructive adjustment in the downloads if you would like to follow along (see `Destructive image base file #2`).

## Using the dodge and burn brushes

The following figure shows the location and the Context toolbar for the **Dodge** brush and the **Burn** brush. They work the same as other brushes:

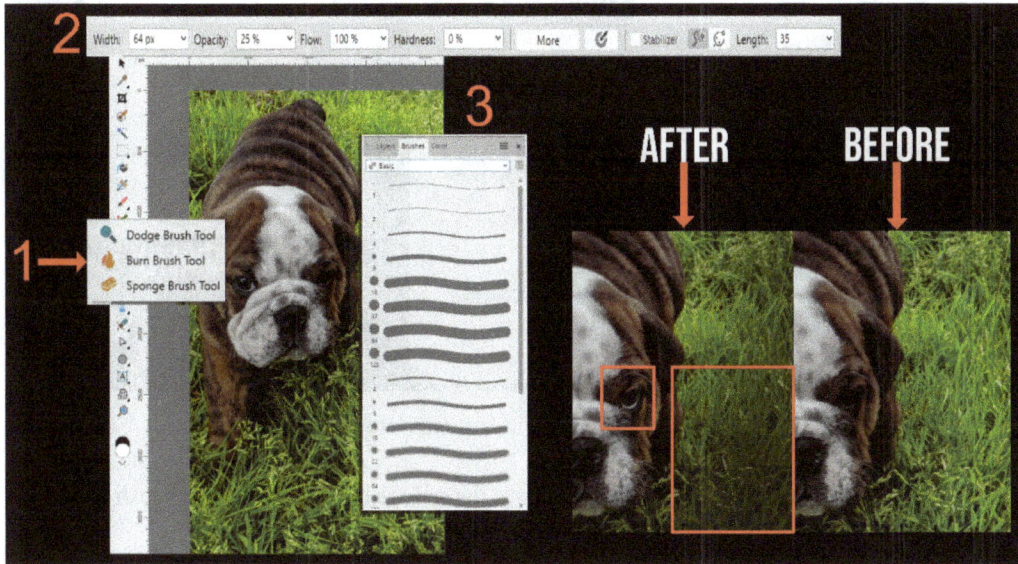

Figure 14.3 – Location and application of the Dodge and Burn brushes

I have kept the flow down so that I don't over-adjust. I click instead of drag on these brushes (personal preference) since a little bit of adjustment goes a long way.

*Figure 14.3* also shows a before and after of the grass (where I used the **Burn** brush) and a highlight in the eye from the original where I used the **Dodge** brush.

## What is the non-destructive version?

The non-destructive version of this technique was shared in the previous section and involves using two curve layers (one turned down to mimic burning and one elevated to mimic dodge adjustment). We then apply a fill bucket tool using black. This effectively masks the layer, concealing it.

Lastly, using a simple basic soft brush, we can paint white on the curve layer, thus revealing the adjustment only where we want it.

The process can be seen in *Figure 14.4* and is carried out in the CiA video (see the *Technical requirements* section for the link), which shows the technique since I know it is difficult to explain through just the written word:

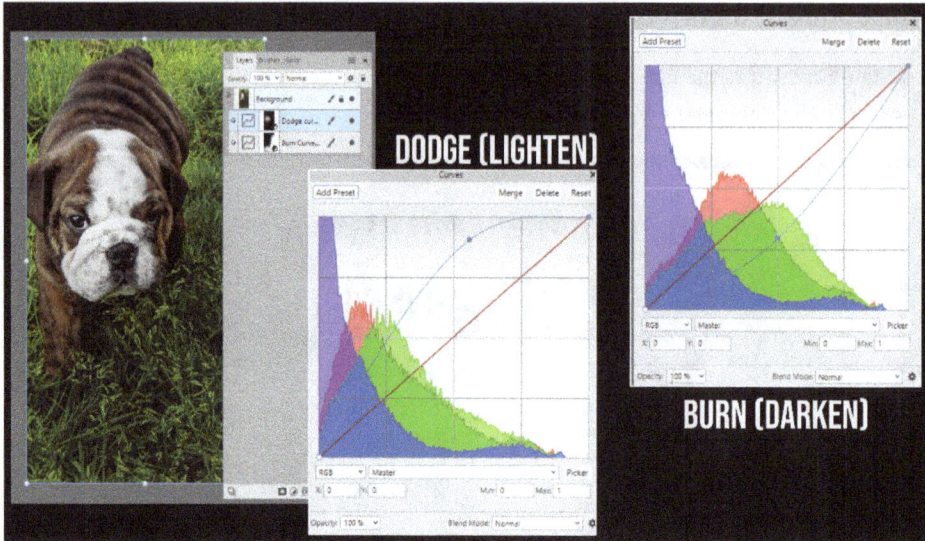

Figure 14.4 – Curve-based adjustments that replicate dodge and burn

One of the underappreciated tools in this set of tools is the sponge brush. The Sponge brush can add or subtract saturation from the image. So, let's look at this unsung hero.

## The Sponge brush

The **Sponge** brush can be found alongside the **Dodge** and **Burn** brushes, and it is used to **desaturate** or **saturate** the image (see *Figure 14.5*):

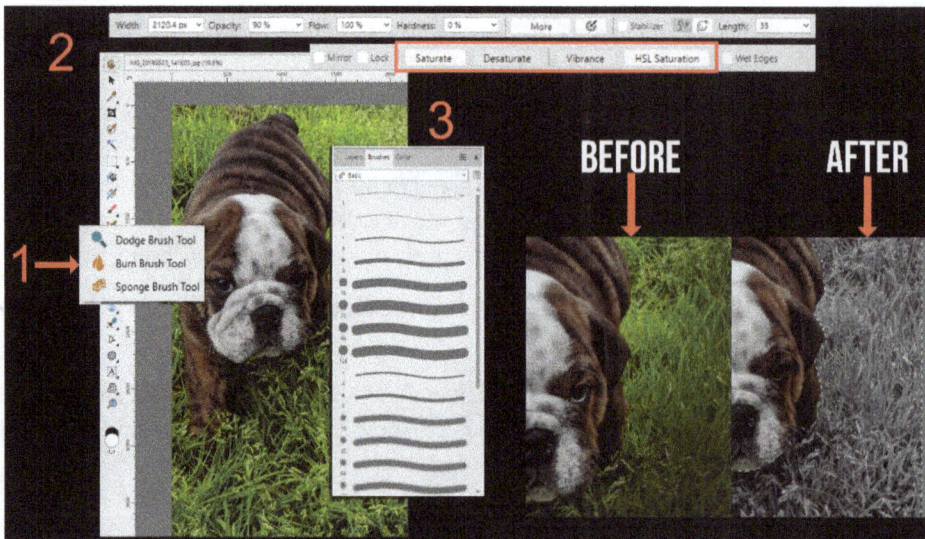

Figure 14.5 – Sponge brush application

The brush can both **saturate** and **desaturate**, so make sure you know which mode you are utilizing in the context toolbar.

### Using the Sponge brush

Just like every other brush, the same adjustments are prevalent (see **2** in *Figure 14.5*). There is also an option to use this brush while targeting vibrance or the HSL of the image.

Choosing **Vibrance** allows a more gradual desaturation, while **HSL Saturation** is a far more aggressive option.

### What is the non-destructive version?

The non-destructive options are the **HSL** and **Vibrance** adjustment layers. They have been covered in earlier edits and chapters, so we will not spend too much time on them. Simply reduce the saturation in an HSL layer or reduce the vibrance in the adjustment layer.

Masking the adjustment layer can be the only way to apply the effect to parts of the image.

## The Mesh Warp tool

Before V2 of the software, this destructive method was the only way to morph a pixel-based image. It can be found in the toolbar, as shown in *Figure 14.6*:

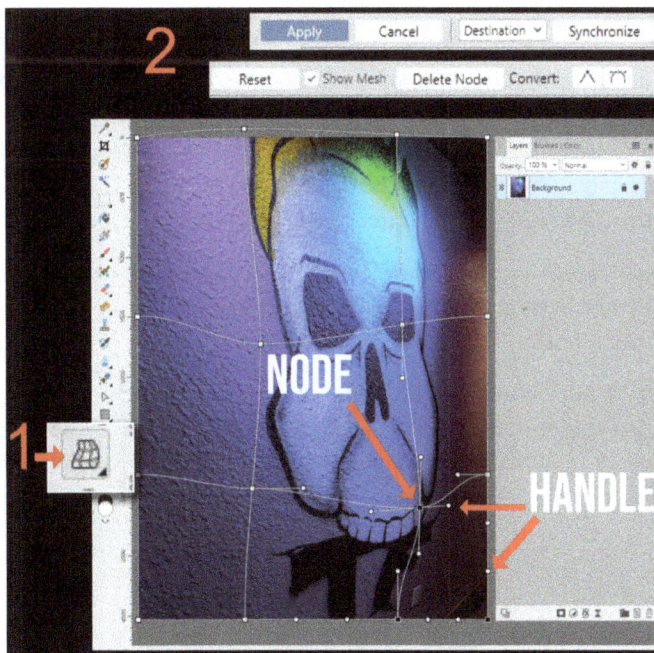

Figure 14.6 – Application of the traditional Mesh Warp destructive method

This tool allows you to create a mesh grid and distort the image along that grid.

### Using the mesh warp tool

To use the Mesh Warp tool, simply apply the tool to the layer you want to work on. Make sure you copy your original image, as this is destructive. Once you've done this and the tool has been applied, you will see four corner nodes that make up a square.

Clicking on the four nodes reveals the handles (shown in *Figure 14.6*).

Here's how to manage the nodes:

1.  Add additional nodes by double-clicking inside the mesh. By default, this will create a round node.
2.  Adjust the node type to square by using the context toolbar and choosing the **Convert** function.
3.  You can delete a node by selecting it and hitting **Delete Node** in the context toolbar.
4.  You can select multiple nodes by clicking and dragging them, turning them black (meaning they are selected), and moving them as a group.

When you use the Mesh Warp tool, there are two options in the context toolbar:

*   **Source**: I have never in my career used this option. It is used with **Synchronization**, which, again, I have never used. A quick review of the literature out there reveals that no one else has successfully found a use for this either… so, stick with **Destination**.
*   **Destination**: This option moves the image in real time. I use this option exclusively.

### What is the non-destructive version?

With the advent of V2, we now have a non-destructive version of the tool, and it can be found in the **Live Adjustment** layers under **Distort | Mesh Warp**. Notice that this creates a Live Filter layer nested into the image. From there, you just work the same as with the destructive tool concerning the nodes.

With that, we have explored the tools that have destructive components and their non-destructive alternatives. There really should be no reason to use destructive techniques in modern software related to tools… but what about filters?

There is an entire menu of destructive filters in Affinity, but as I mentioned previously, there are many that have non-destructive versions. An example of this is Gaussian blur, which we have explored since the beginning. So, in the next section, we will cover some of the more obscure ones that we have not heard of yet but that I find valuable for beginners.

# Destructive filters used in editing

Destructive filters can have a place in photo editing, as many filters have not been updated yet with a non-destructive alternative. So, I want to cover them here to introduce you to them, and wherever possible, I will give you the non-destructive option, as I did with the tools.

## Motion Blur

**Motion Blur** is a cool filter that allows you to take an image and, as its name implies, apply a certain amount of blur to make the image appear to be in motion. We have included a working file for you to follow along with (see the `Motion blur practice file` inside the `V1 Project Files` or `V2 Project Files` folders depending on the Affinity version you are using.).

In this edit, we will be making the car appear to be traveling by using **Motion Blur** on another instance of the car. In the working file, we separated the car as a selection and then created a layer.

Move the layer slightly behind the car, but make sure you keep it on top of the background layer (as shown in area **1** in *Figure 14.7*):

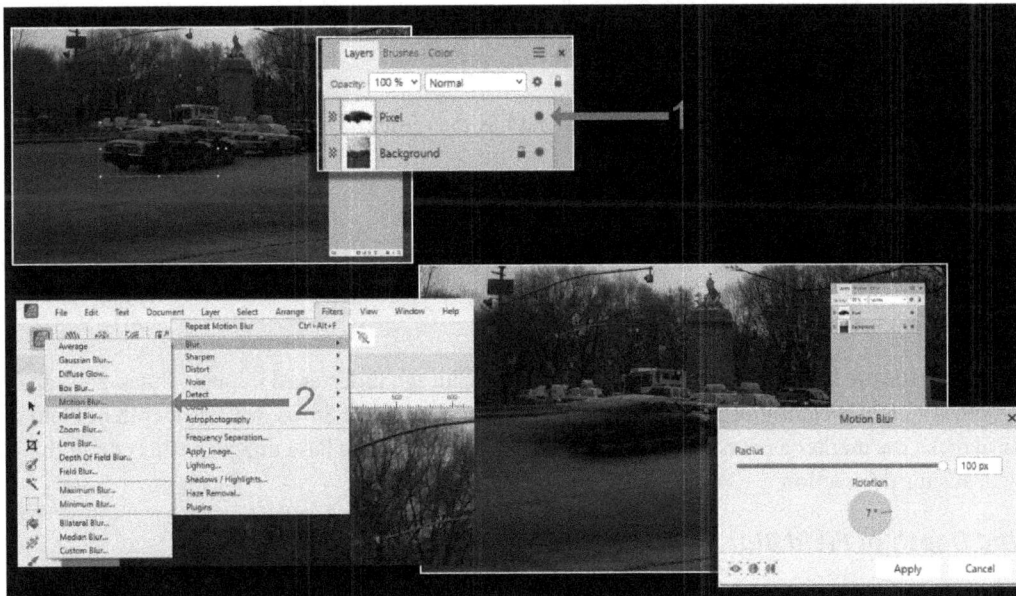

Figure 14.7 – The Motion Blur filter

Now, we are ready to apply the filter.

### *Using Motion Blur*

Follow these steps to apply the **Motion Blur** filter on the image:

1.  Select the filter from the menu, under **Filters | Blur | Blur-Motion**, as shown in area **2** in *Figure 14.7*.

2.  Now. click on the layer titled **Pixel** and drag it to the right and up slightly. The length of the drag controls the amount, and the direction of the drag controls the direction. I have included my settings in *Figure 14.7*.

3.  As an alternative, you can directly enter the number for both the pixel drag amount and the angle in the dialog box for the filter simply by clicking and typing.

### *Uses of Motion Blur*

I use this filter a lot, and I typically pair it with mask layers and then apply a gradient to get some cool transparent motion blur that makes my effect pop.

Another way to use it is to create multiple instances of the layer (in this case, the car) and then apply various amounts of blur to get the car to appear more detailed as it gets closer to the end of the blur.

### *Non-destructive option*

There is a **Motion Blur Live Adjustment** layer that has the same functionality under **Layers | Live Filter Layer | Blur | Motion Blur**.

## Depth of Field Blur

**Depth of Field Blur** is one of my favorites, but the effect can be accomplished by masking the **Live Filter** layer for Gaussian blur.

As a practice file, we have the grave of famous American Chicago mobster Al Capone. But as you can see, there is a lot of background pulling the attention from his marker. So, we will use **Depth of Field Blur** to make the marker a primary focal point. We have included the base image for this edit in the downloads for this section.

### *Using Depth of Field Blur*

Open the practice file from the download. The name of the file is Depth of Field practice. Once you have opened the image, go to **Filters | Blur | Depth of Field Blur...** (see *Figure 14.8*, area **1**). This will bring up three circles on the image (shown in area **2** in *Figure 14.8*) and a dialog box (shown in *Figure 14.8* (area **3**)). I have included my settings and my after photo in *Figure 14.8*:

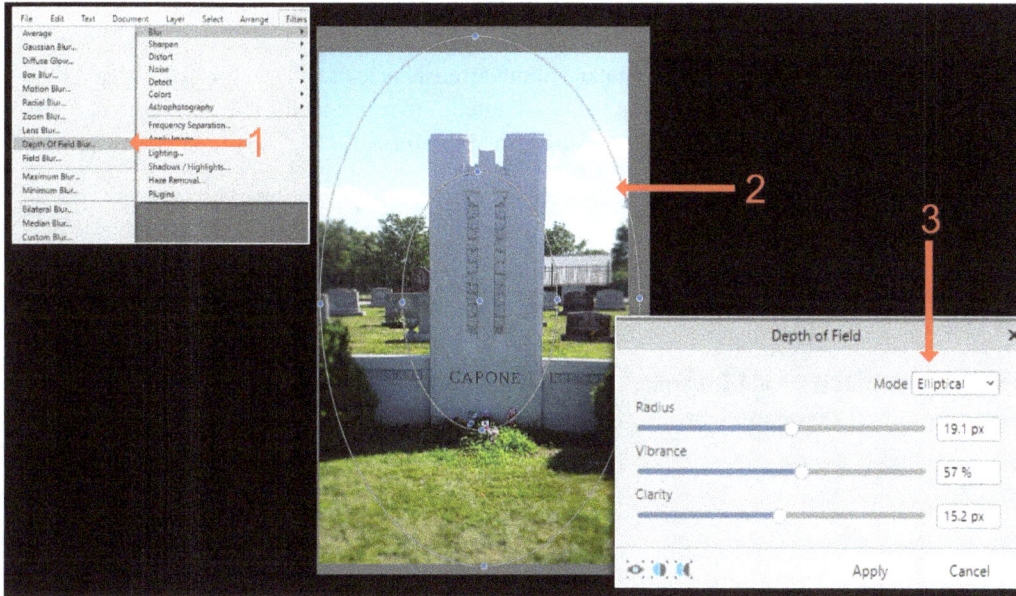

Figure 14.8 – Using the circles of the Depth of Field Blur filter

The circles affect the position and shape of the blur, not its strength. So, first, you must set the position:

- The outermost circle controls the limits of the effect. If you want more of the image included in the blur, adjust the inner and outer circles; for less, contract the outermost circle. You can also change the shape by adjusting the nodes (blue dots) to get a more oval shape.

- The middle circle controls the ease of how the effect is applied between the center and the outer circle. For a harsher transition, push the inner circle closer to the outer circle or inner dot. For the smoothest transition, keep it at the center. Lastly, you can change the shape of this circle as well – just drag the nodes.

- The center dot controls the focal point of the blur. If you want to change the focus point to the left or right, drag the circle.

In the next section, we will understand how to use the **Depth of Field Blur** dialog box.

## Using the Depth of Field Blur dialog box

There are three adjustments you can make based on the position and shape of the circles. This section controls how intense the blur, vibrance, and clarity are. Let's go over the different options:

- **Radius**: Refers to the radius of the blur outside the focal point

- **Vibrance**: Makes the areas close to the focal point more vibrant

- **Clarity**: Adds more detail close to the focal point

### Uses of Depth of Field Blur

Remember earlier in this book when we talked about atmospheric perspective, where I stated that things in the foreground are more in focus, more saturated, and more detailed? Well, **Depth of Field Blur** takes care of these things. All that is missing is the brightness.

I use this adjustment when I have a single subject that I want to bring attention to because you cannot specify multiple points of application. This is not something I use when I have several subjects in a foreground spaced out.

### Non-destructive option

There is a **Motion Blur Live Adjustment** layer that has the same functionality under **Layers | Live Filter Layer | Blur | Depth of Field**.

## Ripple

Ripple is one of the filters I use a lot in my work. Sometimes, it is to add disturbance to water, while other times, I use it in a shockwave if I am doing a superhero-style composition. The effect has multiple uses, so don't let the single example we give in this book dissuade you from trying it for yourself.

We have included the practice file in the downloads for this example so that you can follow along (see `Ripple practice image`).

### Using the Ripple filter

I have recorded a follow-along CiA video for this effect (see the *Technical requirements* section for link), so if you get lost, feel free to have a look and see how this is done. Also, for this example, we will be using the free hand selection tool because, in the image, we want to add some noise and disturbance to the still water in the lower left-hand corner. To use the freehand selection tool, follow these steps:

1.  Select the freehand selection tool from the menu (shown in *Figure 14.9*). This might be in a group with the marquee selection tools. If you do not have it, simply add it from the tool menu.

2.  Draw a selection around the still water. This will allow us to apply the filter only to the selected area, as the rest of the water is very choppy.

3.  Select the **Ripple** effect by navigating to **Filters | Distort | Ripple**.

4.  In the lower left-hand corner of the interface, we can see there are instructions, such as drag to set center. I drag to create the center (in this case, we drag down), and I set the intensity using the dialog box.

5.  Once you have the look you want, deselect using the top menu using select and deselect.

I have included my settings and selection in *Figure 14.9*, which shows the following:

- The center I chose (red circle)

- The selection area I used with the freehand selection tool (red line)

- A point showing the effect (red arrow)

- The intensity setting I used for the **Ripple** effect:

Figure 14.9 – Applying Ripple to an image

## Non-destructive option

There is a **Motion Blur Live Adjustment** layer that has the same functionality under **Layers | Live Filter Layer | Distort | Ripple**.

Now that we know the relationship between destructive and non-destructive filters (there is generally a non-destructive option for each filter), let's practice doing a destructive edit.

# Practical destructive edit

This edit follows part of the same workflow before, but with a couple of extra steps due to the destructive nature of the edit. We have included both V1 and V2 versions of the finished files for this edit in the downloads if you get lost. In the working files for this edit, it is best to make all the layers invisible and begin at the original layer and then click each one visible at each step. This will allow you to see where I applied edits, and how much I did. Typically, you do not have to do this process in a non-destructive workflow because the adjustment layers are all there and we can click them off and on. However, we can't do this in a destructive edit. Let's get started with the steps:

1.   Duplicate the **Background** layer and then lock it. Make the background invisible, and rename the new layer `Editable 1` (see *Figure 14.10*):

Figure 14.10 – The layer's structure and the two channels created by the selection

2.  Create spare channels for the background and foreground (see *Figure 14.10* for the marching ants around the foreground to see where I drew the line). We covered this in *Chapter 5*, so we will not rehash it here:

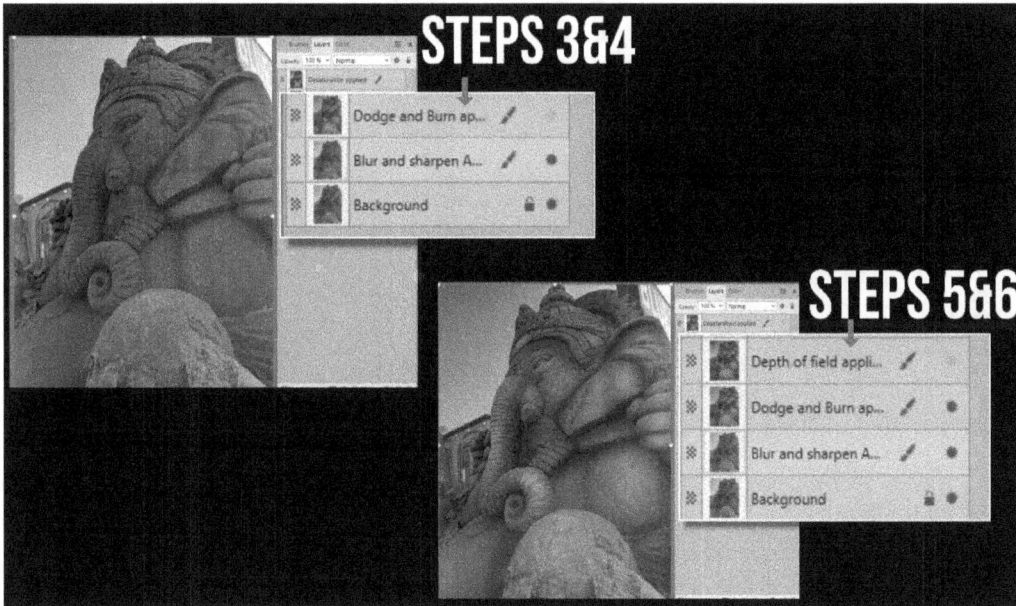

Figure 14.11 – Application of the various filters on the layers

3.  Using the **Blur** brush, apply the blur to the background and, using the **Sharpen** brush, apply some sharpening to the foreground, focusing on the ball in the lower part of the image and the eye area of the statue.

4.  On the editable layer, change the name to `Blur and sharpen applied`, and then duplicate the layer. Change the name of this new layer to `Dodge and Burn applied`.

5.  Utilize the **Dodge** and **Burn** brushes to sharpen and lighten the image in the areas you want to darken or lighten. Remember to build it up slowly as a little goes a long way.

6.  Duplicate the layer you just finished and rename the new layer `Depth of field applied`:

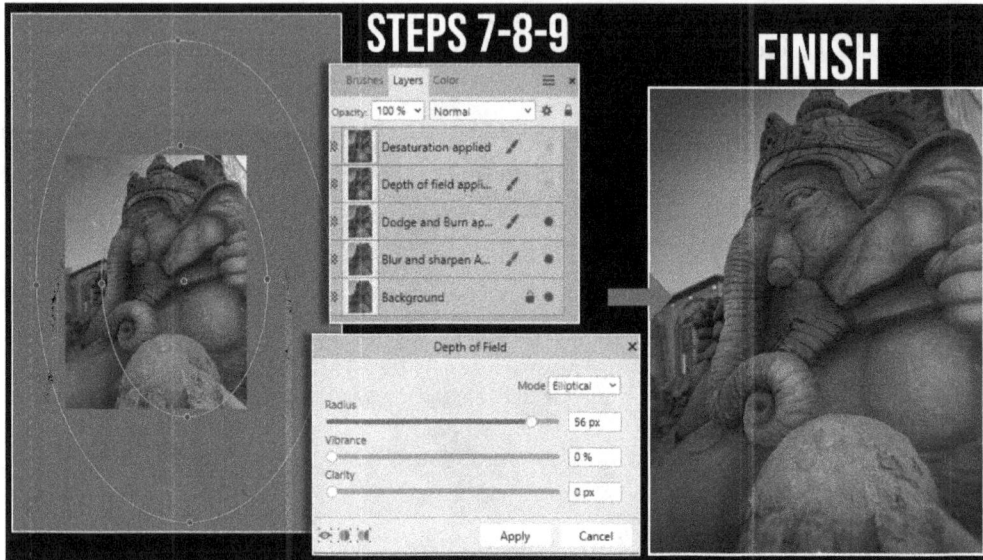

Figure 14.12 – Application of Depth of Field and a vignette

7.  Remove all the selections from the image and apply the **Depth of Field** filter (covered earlier in this chapter). Apply the center point on the statue and adjust it until you have something you like. I have included my setting for where I applied it, as well as its strength, in *Figure 14.12*.

8.  Using a destructive filter, go to **Filter | Color | Vignette** and apply a subtle vignette (I know there is a non-destructive version, but we are working destructively).

9.  Lastly, duplicate the layer and rename it `Desaturation applied`. Then, use the **Sponge** tool on the **Background** area to **Desaturate** it (remember, we saved a channel just for that) and then switch to **Saturate** and apply it to the **Foreground** area.

This is a simple way to work destructively. Now that we've seen its limitations, I urge you to always work non-destructively when possible.

## Professional tips, tricks, and important points

Here are some takeaway points from this chapter:

*   When working destructively, always make sure you save a copy to work on instead of your original image. So, the first steps should always be to duplicate the background, lock it, and make the other layer invisible.

*   After each large step, I right-click in the **Layers** area and choose **Merge Visible**, and then relabel my layer with the operation I just performed (for example, **dodge and burn completed**). This step allows me to go back in the event I want to work on an image in a different path.

- Utilize selections to apply the destructive effects only to part of the image (as we did with the **Ripple** effect). Most destructive filters do not need to be applied to the entire image. Use the channel selections we did previously to make sure you do not ruin your image.

- 90% of all destructive operations can be completed using non-destructive flows, so try and find a non-destructive way to work if possible.

## Summary

Destructive editing is a leftover from the early years of photo editing programs, and as technology gets better and better, the option to work non-destructively gets greater and greater. While I do not like working in this way, there are many filters (especially distortion and edge detection-related filters) that are still in this antiquated system. In this chapter, we learned what the typical destructive tools in photo editing are, and the viable alternatives from a non-destructive workflow perspective.

Some of the typical destructive filters exist in the various menus of the application, while the non-destructive alternatives are available in the software.

You now know how to use these destructive methods, giving you a starting point for your exploration of workflows. So, if you are going to work destructively, you know where to start.

This was an important lesson because, in the next chapter, we are going to enter a section that requires the use of some of these styles of filters. We will be going into an area near and dear to my heart as a composition artist, and that is the area of special effects, including layer effects. We will apply the filters we just explored, along with the layer effects and all the selection techniques we addressed earlier, which makes the next chapter my personal favorite. I hope it stirs up something inside of you as well and you see all the wonderful fanciful things Affinity Photo can help you create. So, let's get it crackin'.

# 15
# Creative Effects and Specialty Brushes in Affinity Photo

Of all the chapters in this book, this one is my favorite. Due to the nature of the work that I do, specialty brushes, special effects, and superhero fantasy is the area of photo I have chosen to specialize in with my editing. This chapter is the result of not only my effort in developing some of these tools and techniques but also a way for you to create things that are fanciful and fearsome, and really take your edits to places that subtle retouching never could. This chapter also demands every skill gained from previous chapters because, to create effects like these, we need to understand selection, masking, blend modes gradients, and pretty much everything we have previously covered.

In this chapter, there will be more CiA videos than in previous chapters, as the technical directions could consume more pages than this book has. So while I will break them down, consult the videos to assist in your learning for this chapter

We will start with specialty brushes and build off the basic brush chapters previously in the book to create flames, fog, and particle brushes. Following that, we will utilize the layer effects portion of the **Layers** panel to create styles, and finally, we will finish by putting it all together to create a rain overlay layer utilizing basic filters.

In this chapter, we will cover the following topics:

- Advanced brush concepts
- Working with layer effects
- Specialty brush creation
- Creating an atmospheric overlay – the rain overlay
- Professional tips, tricks, and important points

# Technical requirements

The CiA video of this chapter can be found at `https://packt.link/BJoFp`.

You will also need the `Chapter 15` project to follow along with me.

# Advanced brush concepts

While the basics of brushes were covered in a previous chapter (see *Chapter 9*), we will take the basic knowledge and expand it in this section, teaching you how to make specialty brushes using different nozzles and techniques primarily to create special effects. So let's get started.

## Different types of brushes

So far, I have shown you how to create categories, rename categories, and import and export brushes into different brush categories. However, in the **Brushes** tab, there is one area that we still have yet to explore. Here is where you can create different types of brushes. These options can be seen highlighted in *Figure 15.1*:

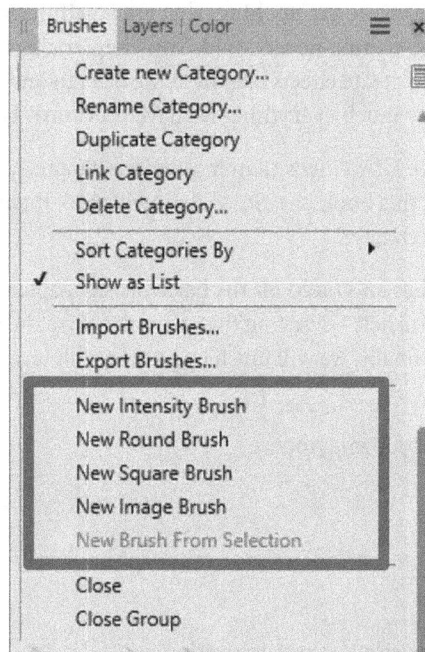

Figure 15.1 – The different brushes you can make in Affinity

Let's go over the options highlighted in *Figure 15.1*:

- **New Intensity Brush**: This brush style uses values of black and white to generate the image. This is the primary type of brush we will use for this section's brushes.

- **New Image Brush**: This allows you to use a PNG image as a brush. We will cover this in its own section later in this chapter.

- **New Round Brush**: This creates a basic round brush that can adjust softness, size, and so on (not the subject of this chapter).

- **New Square Brush**: Never in my career have I used this. All this does is create a simple square brush in various sizes.

Now that we have focused on the intensity brush, it is time to discuss it in more detail.

## Breaking down the intensity brushes

As previously mentioned, intensity brushes utilize values of black and white to denote what is revealed and what is concealed (observe how this builds on the concept of masking presented earlier in the book, it is the exact same thing). So, in short, if you understand masking, you understand intensity brushes. However, it is flipped on brushes—black is actually what is revealed, and white conceals (I know, but I cannot explain why they did this; just accept it is flipped).

In this section, we will start with the absolute basics of making your own intensity brush and end by creating a fully functioning Pixey dust-style brush that you can use in your work.

### Reading the intensity brushes

In the downloads for this section, I have included the flat .png images we will export so that at the end of the process, we are all working on the same .png files, but feel free to follow along and create your own; after all, this is the artistic part.

These square flat PNG images are the basis for a brush. So all brushes are created from a flat square file, which forms the "STAMP" style of the brush, so what is on this square tells Affinity what *the brush shape is like*. I have included the bullets for creating a brush in the following points—you will see all of these present in *Figure 15.2*:

- Black will show up in the brush, but white will not.

- You must have a white background to get the "intensity" between black and white.

- Shades of gray will create some transparency, so if you want the brush to apply fully opaquely, make sure it is fully black.

- When you export it, it *must* be in a .png format, as .jpg is not acceptable:

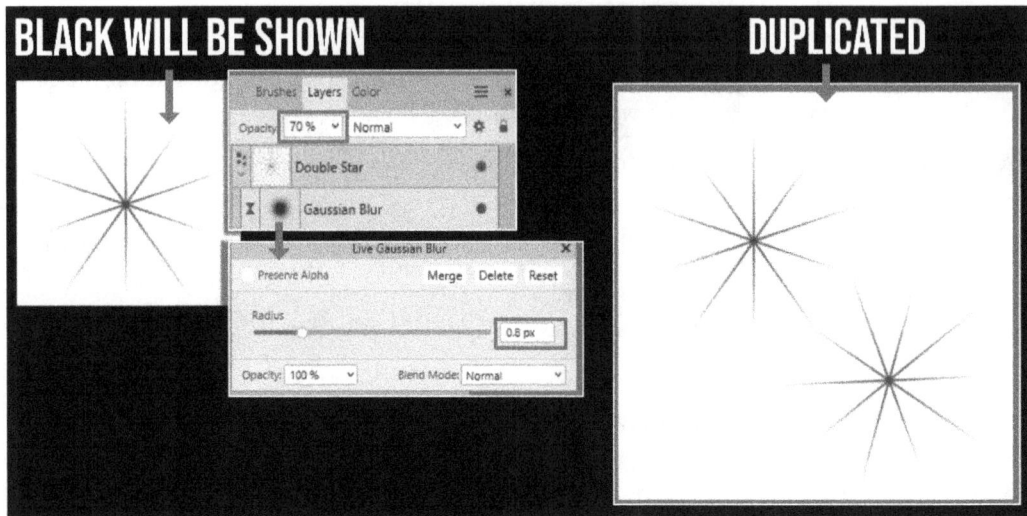

Figure 15.2 – The effect of opacity on the brush behavior

### Creating a simple intensity brush

All brushes begin as square projects. In *Figure 15.2*, the size of the canvas is 1000 x 1000 at 72 DPI. I personally use a 1000-px canvas because it minimizes the pixelation while working and is a good size to start. It is not required that you use 1000 px, but if it is too big or too small, it may pixelate a bit in the brush. I have experimented and found this to be the optimal size.

Follow these steps to create an intensity brush:

1. Create a perfectly white background.
2. Paint your pattern, shape, or texture on top of the background (in this case, I used a double tool shape, increased the point count, and reduced the radius to get the shape).
3. Save the image for later (best practice).
4. Export the image as a .png file to a folder; this will be the basis for your intensity brush.

See the CiA video (see the *Technical requirements* section for the link) for an example of the brush creation *steps 1-4* in the process.

> **Pro tip**
>
> I reduced the **Opacity** value of the layer to **70%** and added a blur so that when we make this a lighted effect, it does not have a hard edge, and it is never truly opaque. If the edges are too sharp and the shape too opaque, it does not look right. Think about how infrequently we use a hard round brush. Most of our editing has been done with a soft round brush. This is because, in the real world, there are very few sharp lines. So, I keep a bit of blur on my brushes, and when doing a light-style brush, I keep it at a lower opacity.

## Moving the images into being an actual brush

Once the flat PNG file has been exported, it is time to make it a brush, so to do this, we need to follow these steps:

1.  Go to the **Brushes** studio panel and locate (or create) the category you want to add the brush to. I am going to create a new category called PACKT-Special Effects for the brushes we make in this chapter.

2.  Select **New Intensity Brush** and load the PNG image from the dialog box. You will now see your brush with a brush preview and a stroke preview.

Now we can adjust the basics and dynamics of the brush the same way we did in the bokeh lessons in brush basics (*Chapter 9*). I have included my settings for both the **General** and **Dynamics** tabs in *Figure 15.3*:

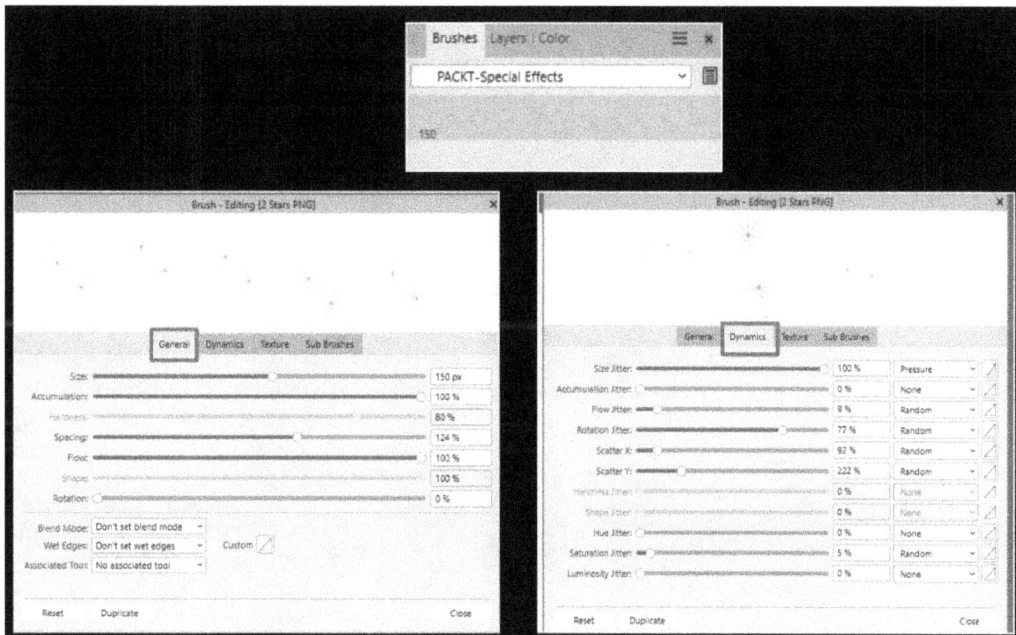

Figure 15.3 – The General and Dynamics tabs adjustment for the PACKT-Special Effects brush

## Testing the brushes

To test the brushes, you want to create a test space; any size will do, but you want to create an environment that duplicates the application. In this case, we are making a light brush, so I will use a black background. Plus, I am using a pressure pen and the size jitter, so I want to practice with them to ensure the pressure variations work.

For a light brush, we will have to add layer effects (covered in the next section), so all I am interested in is, "Does it work mechanically?" It may not look good because we have not finished it yet, but we have the base.

I have included my test in a CiA video (see the *Technical requirements* section for the link). We will make the glow effects in the *Working with layer effects* section. I just wanted to show you what is possible moving forward, so do not sweat it, we will create them shortly.

Now that we have explored intensity brushes, let's create an image brush and show you how to make a repeating random pattern brush from natural elements.

## Breaking down image brushes

Image brushes are pretty simple. They allow you to take an image and make a brush out of it. I have used them for creating destructive scenes where rocks fly, adding leaves to an autumn portrait, or trees added into a scene that didn't have trees.

In this project, we will use a series of images I shot just walking around my yard; it is autumn here, and so the leaves have fallen. In the downloads of this chapter, you will find a folder named `Isolated Image Brush PNG files` I created. I isolated them simply with the selection tool, so there is no magic there.

### Creating image brushes

First, begin with isolating the image you want to use as the brush image. Each individual image should be saved as a separate PNG file. I typically create a 1000 x 1000 pixel workspace just to size my isolated images against. This is because if the image is 4,000 px, which will make a bad brush, so I always isolate the image and then size it against a known size to make sure it is controllable (see *Figure 15.4*):

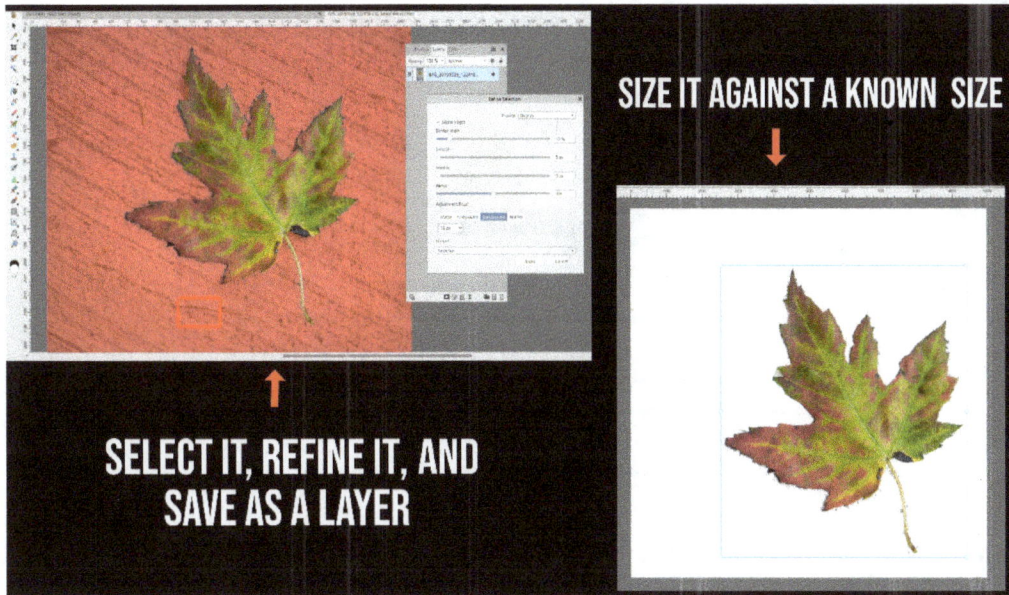

Figure 15.4 – Selecting the leaf for the brush

Notice my selections on the PNG field are rough; often, individuals stress over selection. In the brush, these subtle amounts of black will not be an issue. However, to be perfect in the selection, utilize a pen-based clipping mask to get a crisp edge.

### Moving the PNG image into an actual brush

Once the PNG images are created, select **New Image Brush** from the menu on the **Brushes** tab. Remember to make sure you are in the right category.

Select the PNG image, and then the brush will show in the category.

Set your **General** and **Dynamics** properties (I have included mine in *Figure 15.5*) as well as in the brush file so you can follow along:

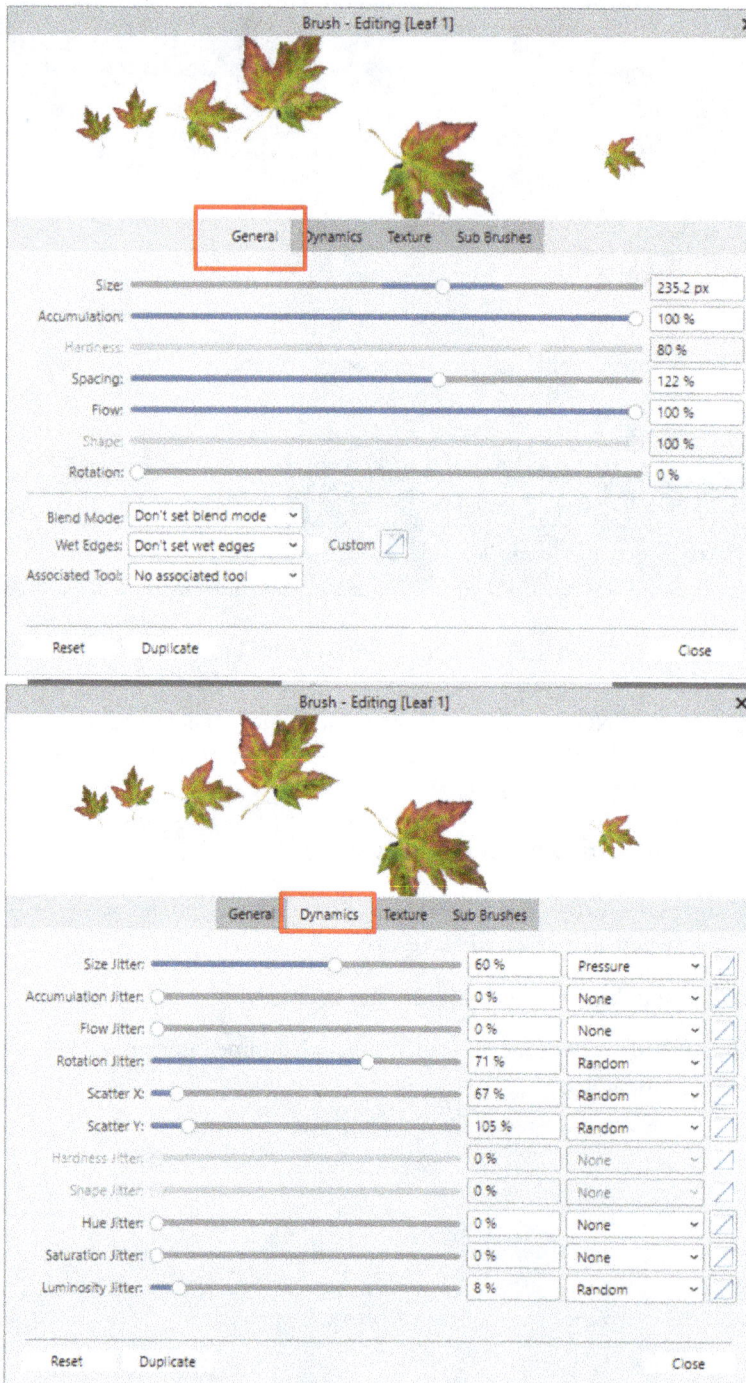

Figure 15.5 – Tab settings for the image brush

## Adding additional images to the brush

In this sort of brush, we want to alternate the types of leaves we show, so there is an advanced section in the brush dialog box called **Texture**:

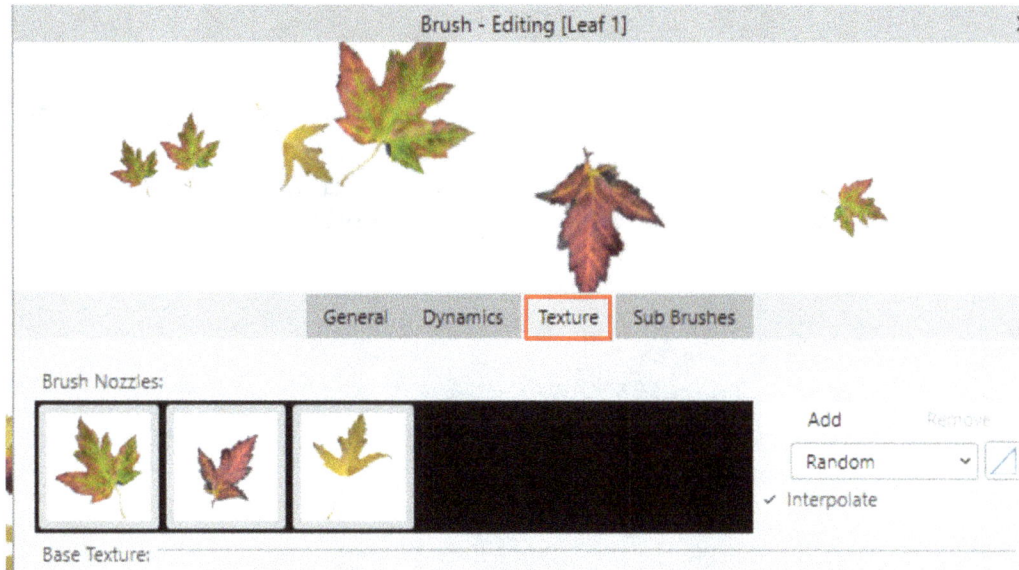

Figure 15.6 – The Texture tab

What the **Texture** area does for image-based brushes is allow you to add additional images, thus random leaves. So to accomplish this in the brush, let's add **Leaf 2** and **Leaf 3** from the images. This is shown in *Figure 15.5*.

To add an image, select **Add** from the **Brush Nozzles** section and then upload your images.

Now we can see in the test we have an alternating leaf pattern each time. This helps pull off the effect.

### Testing the brush

Once the brush is made, we can test and adjust it. I have included a nice fall image for us to work with. Plus, in the PACKT-Special Effects Brushes complete set brush file, there is a bonus brush where I used the image brush technique to make a background style leaf brush to assist.

Notice that when you apply it, the brush will not be perfect; you will always have to adjust the images and edit them to match the scene; it is 100% compositing with an image brush. So have a play with them and see what you can create. I have included my trial in the CiA video (link mentioned in the *Technical requirements* section), and even the leaves look out of place and will need adjustment and compositing to look right. Primarily what I am concerned with at the testing stage of the brush is, "Does it work mechanically?"

Now that we have made a couple of simple brushes, we need to explore the relationship that layer effects have with brushes. A good solid brush without a functioning layer effect will create the desired effect for things such as lightning and energy. So let's look at the second half of this dynamic pairing in this section.

# Working with layer effects

Off and on through previous chapters, we have touched on various layers such as **Pixel** layers, **Mask** layers, **Adjustment** layers, and so on. However, there is another aspect of layers we need to talk about if we are discussing special effects and fancy brushes, and this is the idea of **layer effects**. Layer effects are found at the bottom of the **Layers** panel, displayed by small **FX** symbol (see *Figure 15.7*):

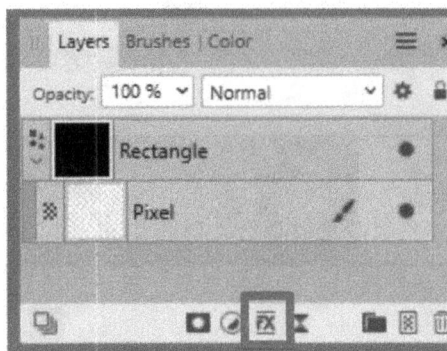

Figure 15.7 – The FX tab location

Layer effects are more of a factor in Affinity Designer, as we apply them primarily to vector-based objects. In Affinity Photo, there are live filter layers for several layer effects (such as blurring an image). So, you may not use them as much as an illustrator would, but it is essential when discussing specialty brushes. While we will *not* go through all of them, I will give you the most frequently used by me in my work relating to advanced brushes and photo editing.

## Applying layer effects to images

As the name implies, layer effects are applied to layers, so make sure you are on the layer and then click on the **FX** tab. A dialog box will appear (see *Figure 15.8*), and then you can select the layer effects you would like to apply to the layer. You can use more than one:

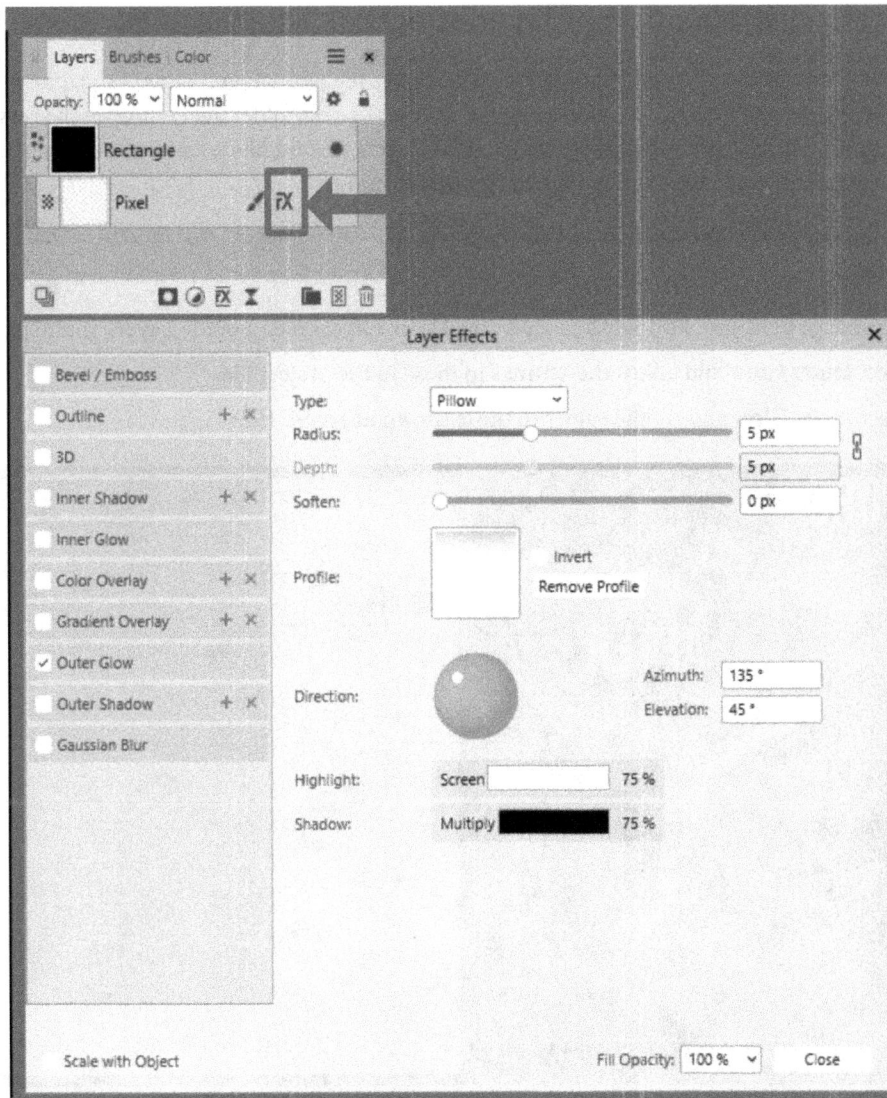

Figure 15.8 – Dialog box for Layer Effects

Each effect has its own options. There are options to the right in a dialog box. You can use the combination of sliders, checkboxes, and drop-down fields to select your special combination to get the look you are after.

Once you apply the effect, you will notice the **FX** symbol appear next to the visibility point in the layer. Clicking on this will allow you to open the dialog box and change the effect; thus this qualifies as a non-destructive operation.

## The practical application of layer effects

In your downloads for this chapter, I have included a practice file (see `Layer styles practice file` in the `V1 practice files` or `V2 practice files` folder depending on the version of Affinity you are using). In the file, there are two **Pixel** layers above a black rectangle, one is labeled **Large lightning bolt**, and one is labeled **Background lightning**.

The following steps will show you how to take the black-and-white images into something that looks like lightning:

1.  With **Large lightning bolt** selected, open the **Layer Effects** menu in the **Layers** studio panel.

2.  Select **Outer Glow** and adjust the settings to those in the project file.

3.  Select **Inner Glow** and set the values to those shown in *Figure 15.9*:

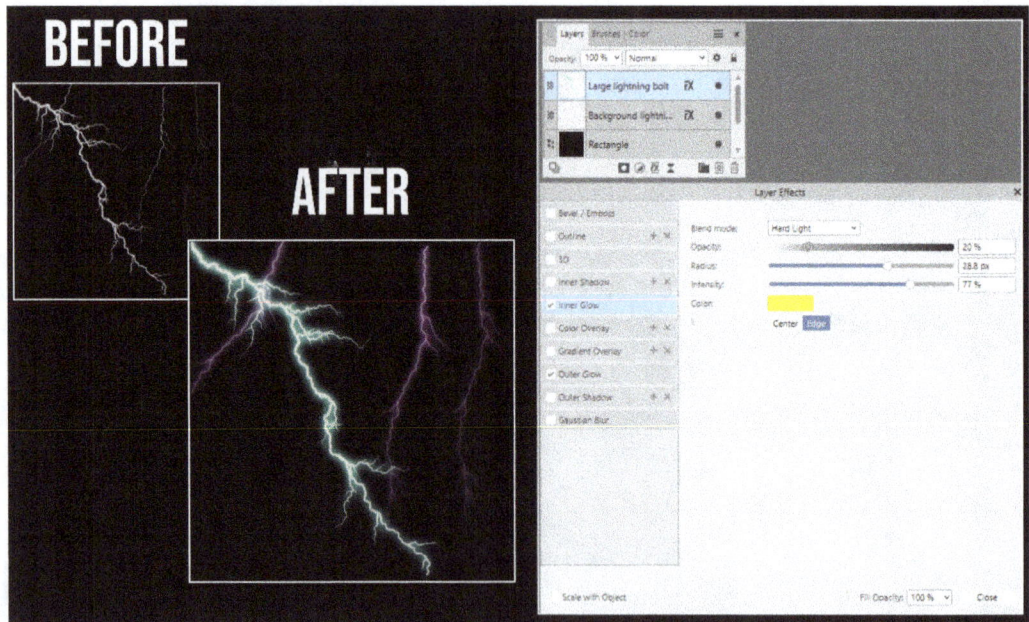

Figure 15.9 – Before and after for the lightning

Do something similar on **Background lightning** and adjust it to your liking. I have included the project files with specific values as well as a CiA video (see the *Technical requirements* section for the link) where I adjust both layers.

This is how the layer effect changes the mundane nature of a brush (in this case, simple black and white lightning images into really cool pieces of art).

Now that you have this style, it would be great if we could save it for later. This is where styles come into play. Let's explore how to make styles for later use in Affinity.

## Saving layer effects as styles

Once you have a set of layer effects you like (such as the lightning adjustments you just made), you now want to save them so that you can apply them to other layers. The mechanism for this in Affinity Photo is called **Creating a Style**. The **Styles** studio panel is found in the studio section, shown in *Figure 15.10*:

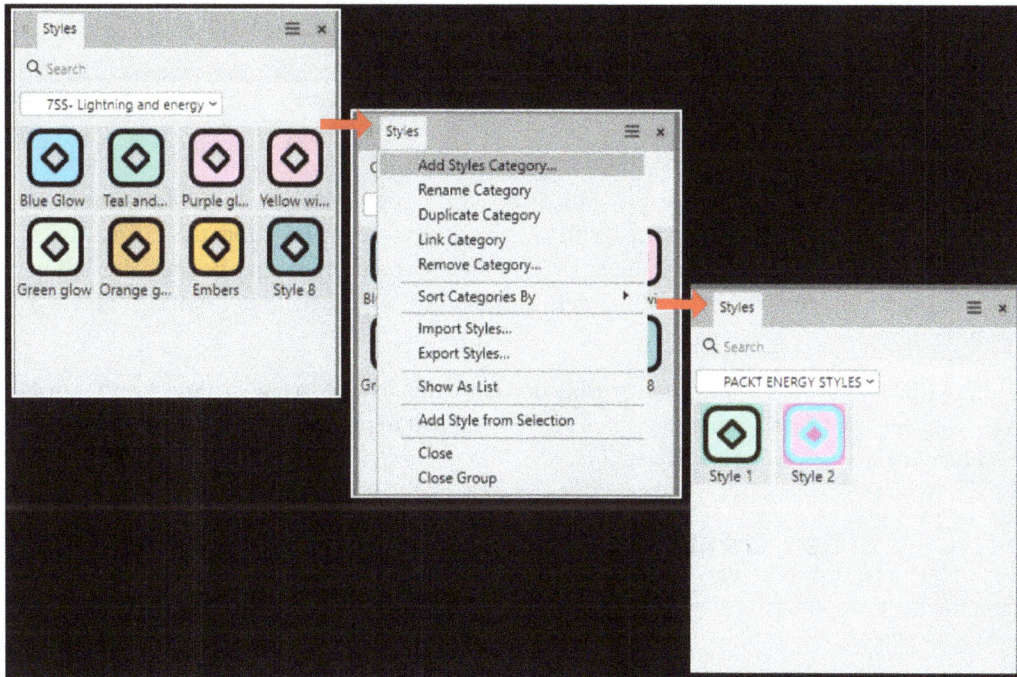

Figure 15.10 – Styles management graphic in Affinity Photo

To add your adjustments to the style, you need to first create a new category for the style:

1. To create a category, click on the menu icon on the studio tab (the lines in the upper-right corner) and choose **Add styles category**.

2. Create a name for your category (I am going to use PACKT ENERGY STYLES).

3. Select the layer you want to take the style from. I am going to use both **Large lightning bolt** and **Background lightning**, so I will make two styles. Go to the menu (the four lines on the studio panel) and select **Add Style from Selection**.

A new style will appear in the panel; now this travels with Affinity Photo and can be exported to other Affinity users. You can also import styles from other users and websites. I have included the styles file I made here in the downloads (see PACKT ENERGY STYLES).

Back to the cooking analogy, the layer effects are the spices you add to the dish. Structurally, you have created a fine meal, the composition is right, and the values are well balanced, but it is missing something, the layer effects. I personally use some form of specialty layer effect in all my images. Even if it is *not* a special effects composition, I may add some subtle light effect with some outer glow to highlight something I want the viewer to see.

So bottom line is, do not be afraid to utilize layer effects with your images; they will make a markable difference to them.

## Applying styles to new layers

If you add a new layer you want to add a style to, simply create your layer, go to the **Styles** menu, and click on the style. The layer will be updated with the style, and you can move from there. Even styles will need adjustment in *each* composition, as the style in one composition may be adequate, but in another, it may be too powerful or need a slightly different hue. So, use them as a starting point and adjust from there.

Now that we have the basics of styles and brushes down, we will dive into a few specialty brushes that are very common for photo editing, and some are very difficult to produce well, so let's up our game now with a very special type of brush creation.

## Specialty brush creation

In this section, we are going to cover two types of special effects brushes very commonly sought after by artists; smoke and fog brushes. Both are complicated, and capturing a good smoke image that we use is a very specific skill set in photography, but not to worry, as I have captured some images we can use. Let's get started with specialty brush creation.

### Creating a smoke/flame brush

Smoke brushes and fire brushes work fundamentally the same as base brushes. We will create the images for these two brushes the same way as we did in the previous section. The only thing that will change is the blend mode and certain things such as the softness. In your downloads is a base image that has been adjusted with a black-and-white layer, and the contrast has been increased to make the black and white more pronounced (see the Smoke brush base image file in the Smoke brush files folder in the downloads of this chapter).

## Initial selection

To select the smoke, go to **File | Select | Select Sampled Color**. It seems counterintuitive because there is no color, it being a black-and-white image, but values of gray have Hex codes and **hue, saturation, and lightness** (HSL) values, so there is, in fact, something the program can use to sort by.

To select by color, select the initial color you want to use for the selection, the location of my selection is shown in *Figure 15.11* (the red dot). The location matters because it is from this value we will add **Tolerance** with the slider. So the color you select needs to be somewhere in the middle of what you want (i.e., do not select pure white because then it will include some of the darks in **Tolerance**). I have also included my **Tolerance** value in *Figure 15.11*:

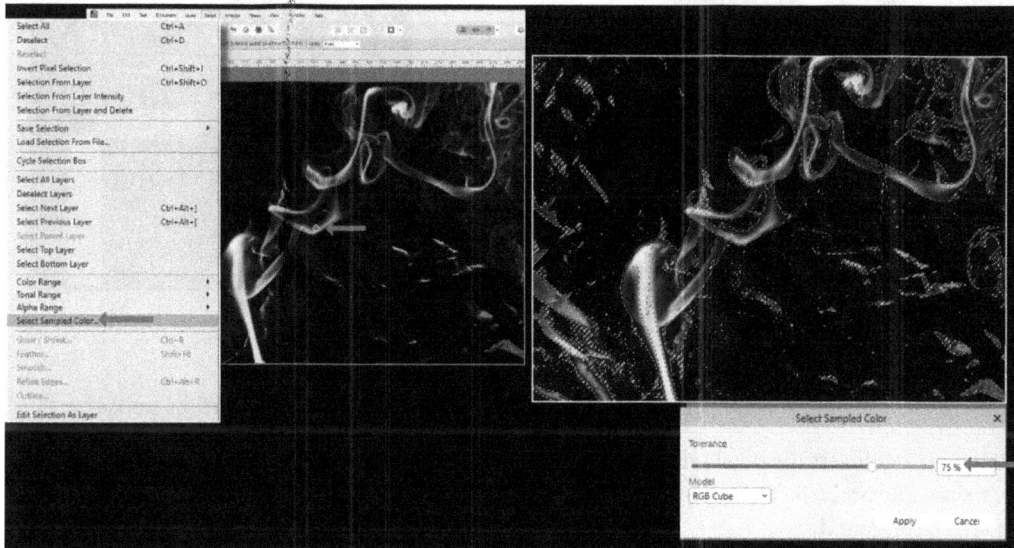

Figure 15.11 – Where to select the smoke (target shown by the red dot)

Use the selection brush to remove anything outside of the smoke (this includes the obvious areas outside of the smoke you do not want).

I have included a CiA video showing this process, the link to which is mentioned in the *Technical requirements* section.

## Refinement of the selection

Now select **Refine** in the context toolbar—the default overlay matte of red will come up. Switch **Matte** to **Black and White** (shown in *Figure 15.12*). It is clear that we need to refine it.

Using a large brush size in **Refine Selection**, paint over the large undetailed areas of the image; this will force Affinity to do a better job. I have shown my brush size and the *single clicks* I made in *Figure 15.12*:

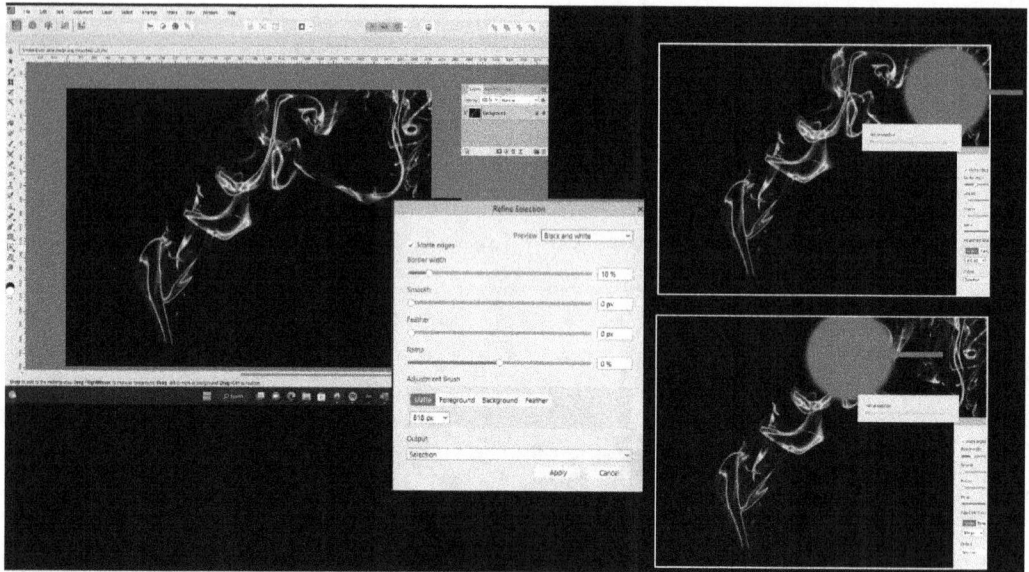

Figure 15.12 – Refining the smoke selection

After the refinement output as a selection, you should now have a good rough version of the smoke brush.

Lastly, go up and invert the pixel selection from the **Selection** menu; this will select everything that is *not smoke*, and press the *Delete* key. You will be left with the smoke shape in gray and a transparent checkered background.

If you get lost in these steps, I have included the *raw* intermediate selection to this point as a PNG file in your downloads; we will have to refine this in the next steps, but I wanted you to have something to follow along with as this brush is a pretty advanced move (see `Intermediate smoke image` in downloads).

I have also included these steps as a CiA video, so check out that too.

### Finishing and exporting the brush base

Create a 1000 x 1000 square workspace with a black background in Affinity Photo, and then cut the smoke from the background and paste it onto this plate; this assures the square brush background and the white background.

Now we need to use the eraser brush to apply some artistic thought to where we want to erase the parts that still exist. I will remove the solid stem from the smoke and some of the hard outlines along the edge.

Once you are happy with the brush, export the entire thing as a PNG file, and you now have your brush base file. Now, it is time to make a brush with it.

In the **Brushes** tab, go to the category you want to apply this brush to and select **New Intensity Brush** from the menu.

Import the PNG file you created, and this will appear in the brushes for the selected category (see *Figure 15.13*):

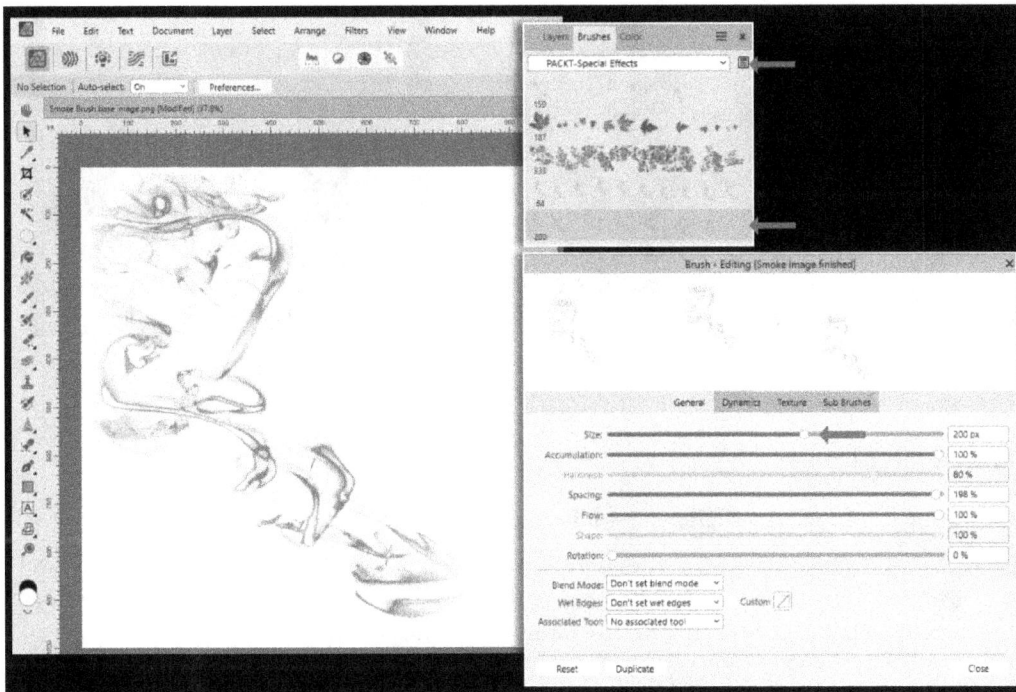

Figure 15.13 – Adding the smoke image to a brush

I do not like unpredictability in my smoke brushes, so I set the size (200 px) and then the spacing. (See *Figure 15.13* for my settings.) These are the only adjustments, and I always set the blend modes in each image I use them in, as I never know the underlying values.

I have included a CiA video for you to follow along with, showing the finishing steps.

If you get lost, I have included the final PNG image for you to follow along with in the downloads.

## Using the brushes

*Figure 15.13* shows the layer structure and the process for the application with the steps outlined after the image. I have included a base image where you can use the smoke brushes (see the Smoke brush application file in the V1 practice files or V2 practice files folders depending on your version of Affinity):

Figure 15.14 – Example of smoke brush application in a project

Let's look at the steps now:

1. For the brush, select a slightly off-white color (smoke brushes do not work well with very dark colors).

2. Use the brush stamp (ONE Instance of the brush) on a new **Pixel** layer of the image, then adjust the blend mode to taste (I typically use **Overlay**, **Screen**, or **Multiply**).

3. You can add some color tint to them as you want.

4. Apply a **Gaussian Blur** adjustment layer and then mask the area of the smoke you do not want to see.

You will notice I duplicated the layer and rotated the smoke to break up the pattern.

I have included a CiA video to show how I use them in the image (the link to the CiA video is mentioned in the *Technical requirements* section), but it's totally up to your preferences.

Next, let's understand how to create fog brushes.

## Creating fog brushes

Fog brushes utilize the destructive **Perlin Noise** filter, so we will create a flat layer, and apply **Perlin Noise** to it, then we will remove the area we want to make the smoke brush while exporting. This **Perlin Noise** effect is also the basis for the particle brush later in the chapter.

I have included a CiA video for this process (see the *Technical requirements* section for the link) so you can follow along.

### Setting up and isolating the image

The following steps show you how to set up your working file for a successful fog brush:

1.  Open a new project at 1500 x 1500 px.

2.  Apply a **Pixel** layer to the image and flood it with 50% black (see *Figure 15.15* for how to do this in the color wheel).

3.  Apply the **Perlin Noise** filter by navigating to **Filter | Noise | Perlin Noise** at the top of the screen. Adjust the settings until you get something you like (I have included my settings in *Figure 15.15*), and then apply the filter.

4.  Use a soft round brush with the eraser tool and drop the flow to 10%; erase around the edges as you do not want anything touching them (see *Figure 15.15*):

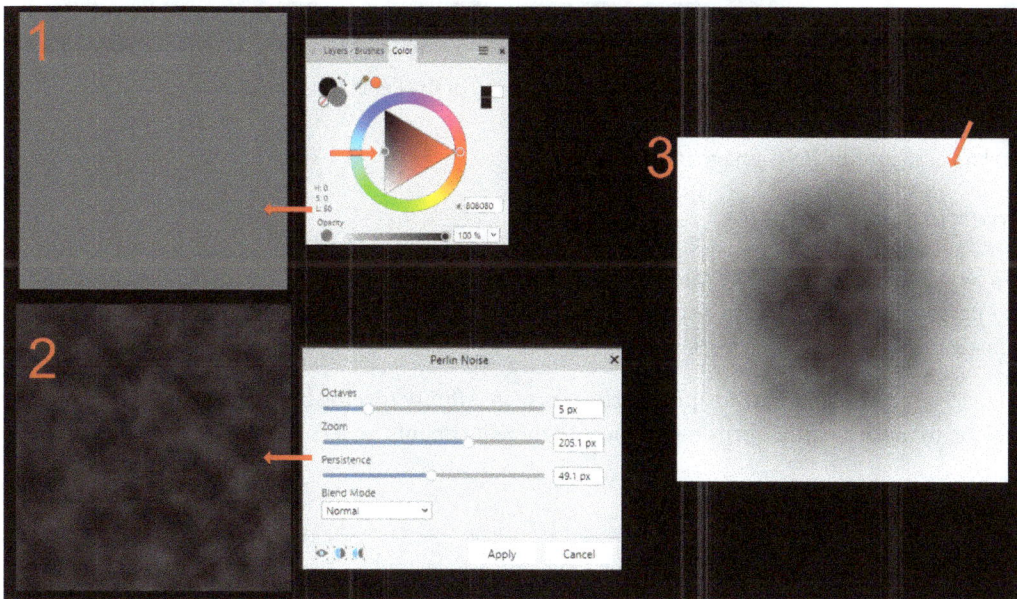

Figure 15.15 – The Perlin Noise settings

5.  Use the Move Tool to select the **Pixel** layer and shrink it down, so you can verify you have removed all the corners and dark spots around the border.

6.  Apply an **Invert Adjustment** layer to invert the image; this will become the PNG image for your intensity brush:

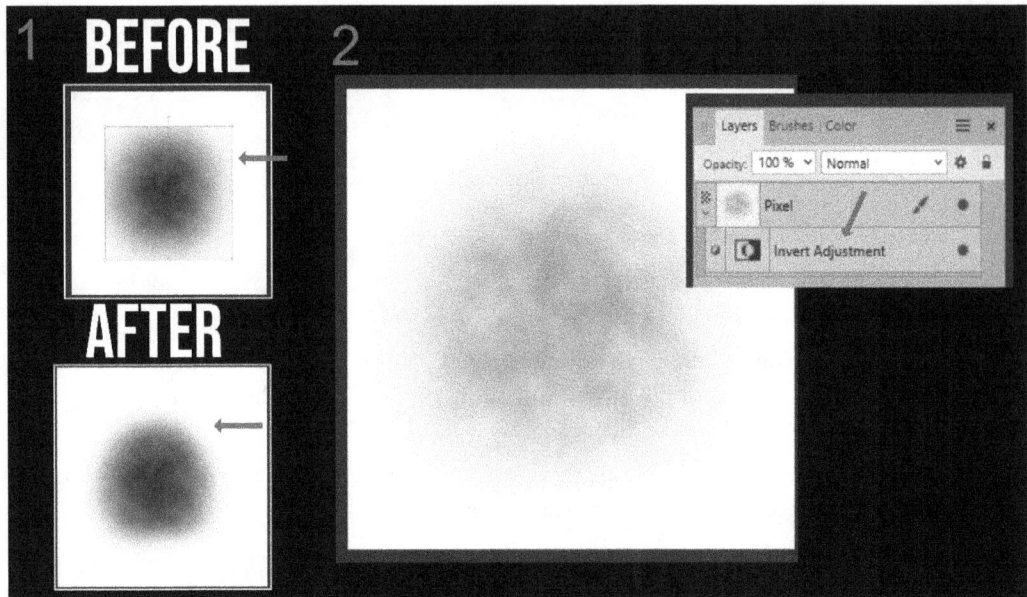

Figure 15.16 – Inversion and centering of the Perlin Noise layer

Next, let's convert this PNG image into a brush.

### Converting into a brush

Add a new intensity brush in the category you want to use, and then adjust your general spacing and size. I personally prefer a larger brush size for fog brushes and enough spacing so I can apply only one instance.

Now that we understand how to apply Perlin Noise in Affinity to a fog brush, we can use the same technique to make a particle brush with some slight adjustments, so let's get into it.

## Creating particle brushes

The particle brush is simply a variation of the same process we used for the fog brush—we just need to increase the contrast using **Levels Adjustment** and **Invert Adjustment**. However, the idea of *creating the layer*, adding noise, and then erasing around the image is still the same. I personally use particle brushes for tons of projects, from texture to embers in flame. Particle brushes are an indispensable part of my tool kit.

I have included a completed CiA video for this brush (the link to the CiA video of this chapter is mentioned in the *Technical requirements* section).

## Setting up and isolating the image

Follow these steps to set up and isolate the image:

1.  Create a 1000 x 1000 px workspace.

2.  Add a **Pixel** layer.

3.  Change the color to black (this is important because the **Perlin Noise** filter we add needs color to show up on the **Pixel** layer, so choose black before you add the **Perlin Noise** filter to the **Pixel** layer or it will not show up).

4.  Add **Perlin Noise** to the layer, set **Octaves** to **0**, and zoom far out; the zoom determines the size of the particles. My settings are shown in *Figure 15.17*:

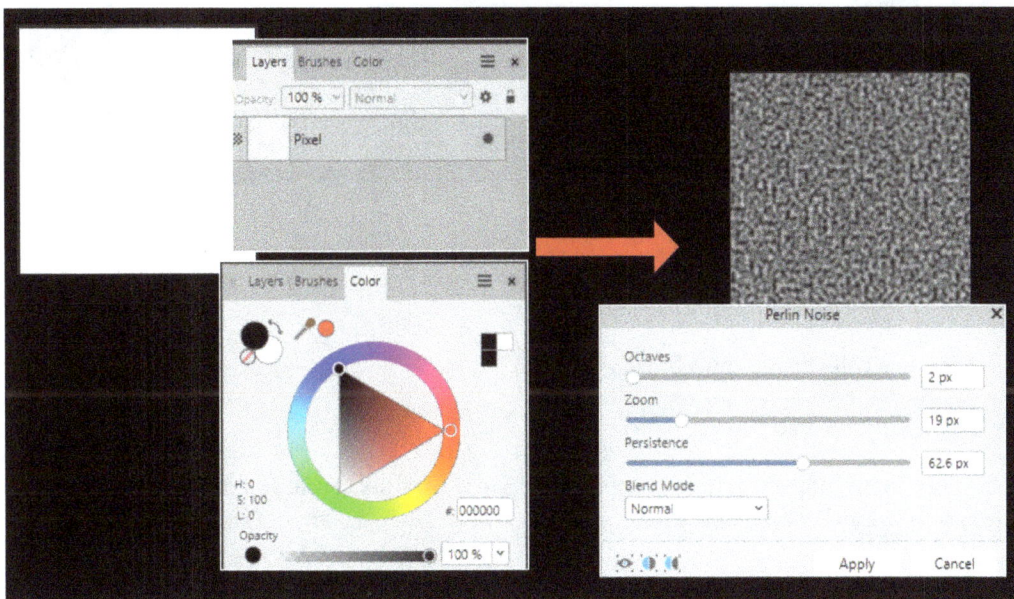

Figure 15.17 – The Perlin Noise settings for the particle brush

5.  Now add **Levels Adjustment** above the layer and move the **Black Level** slider to the left until *most* of the noise disappears, leaving the particles. Move the **White Level** slider to the right. This increases the contrast and creates really clear particles. The white areas will represent what the particles will look like (my settings and result are shown in *Figure 15.18*).

6.  Now add **Invert Adjustment** to make the white black, as an intensity brush shows the black parts.

7.  Once this is done, use an eraser tool on the **Pixel** layer to erase the particles near the edge of the brush and make it round:

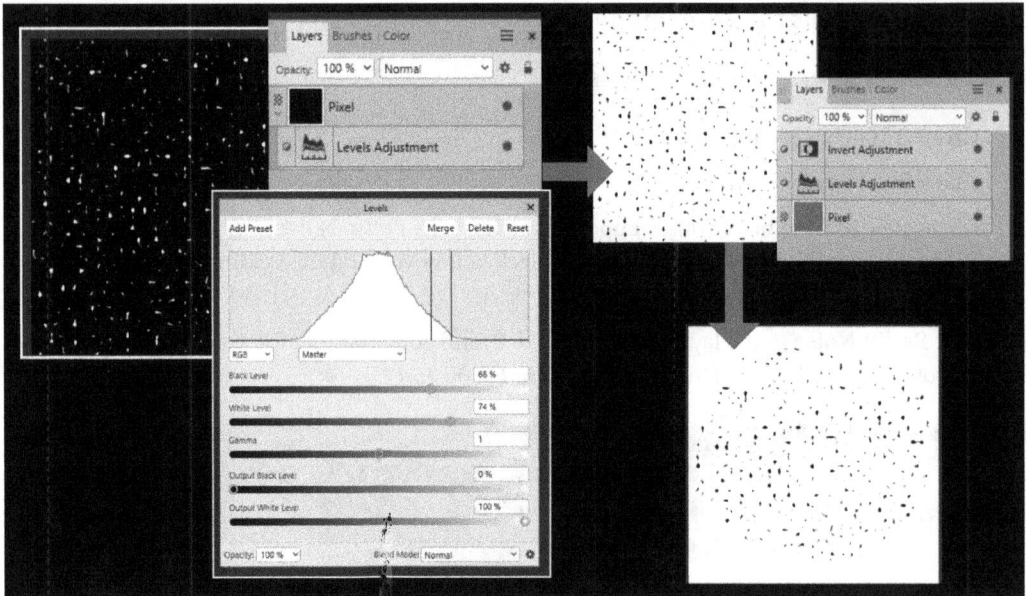

Figure 15.18 – Adding levels and inverting

8. Export the PNG image for brush creation.

Let's convert this PNG image into a brush.

### Converting into a brush

This has been shown numerous times: simply create a new intensity brush, add the PNG image, and then adjust to size and spacing. I use these as stamp-style brushes usually, not stroke-style brushes.

### Applying the brush

I have included a baseline image (see `Particle brush practice image` in the downloads); this was taken at Mount Rushmore in the winter, but I added some snow to enhance it.

Figure 15.19 – Applying the particle brushes

Notice in *Figure 15.18* that there are two layers, one where we made the snow larger and added a **Gaussian Blur** layer and a **Fine Snow** layer where I applied the brush at a smaller diameter but then used the same particle brush to erase some of the snow. Both layers had their opacity lowered to make sure the effect was not too much.

Now that we know how to make various particles using **Perlin Noise**, in the next section, we will add **Motion Blur** to make a functioning rain layer in Affinity Photo.

# Creating atmospheric layers – the rain overlay

Now we can apply the same filters and techniques we have discussed with brushes to apply them to full-blown atmospheric effects. In this case, we will be making a rain layer effect that you can apply to your images.

I have included a CiA video of the entire process (see the *Technical requirements* section for the link).

## Setting up the effect

My layer stack is shown on the right-hand side of *Figure 15.20*; we have discussed all of the adjustments and live filter layers previously, so you can see how I applied them:

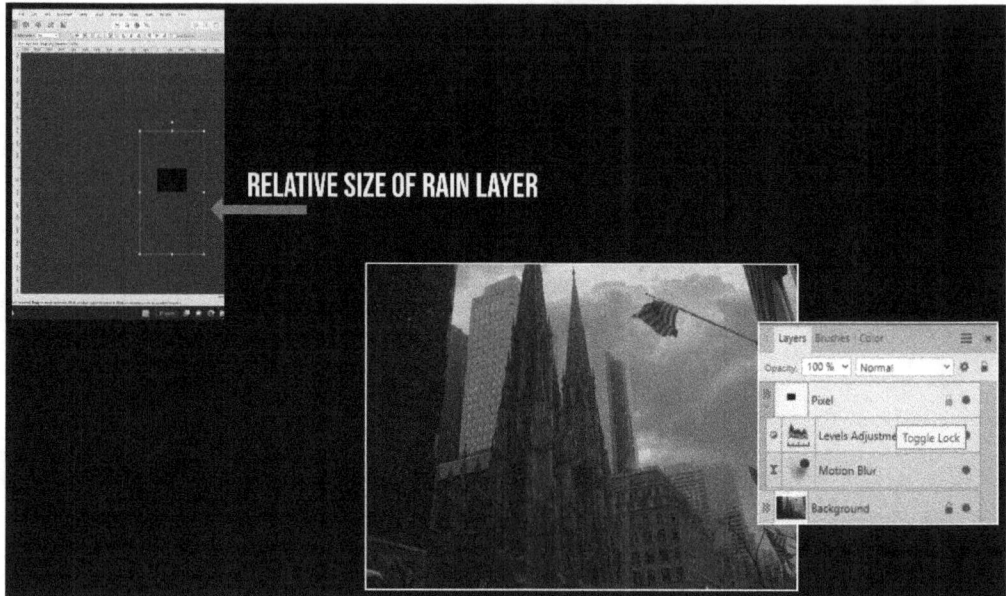

Figure 15.20 – Setting up for a rain layer

Let's go over the steps now:

1. Open a layer on top of the image you want to apply it to (in this case, I have given you a practice image in the downloads – see `Rain layer base image`).

2. Fill the new pixel image with black using the fill bucket.

3. Then, go to **Filter | Noise | Add Noise**.

4. Max out the noise and then select **Gaussian** and **Monochromatic** as the options (**Monochromatic** is essential for the effect).

5. Once you hit **Apply**, it will look as if the noise disappeared, but it is still there. You now have to zoom out and make the layer much bigger than the document, as in, 4–6 times bigger. This makes the noise big enough to see (this is shown on the left side of *Figure 15.19*; it shows how large I made my layer relative to the image).

6. Change the blend mode of the rain layer to **Screen**. This will only allow the light to show through.

7. Add **Motion Blur** to make the rain move (zoom in to see the effect).

8. Finish with **Levels Adjustment** on the **Rain** layer – this makes it more pronounced by kicking up the contrast.

# Professional tips, tricks, and important points

When creating special effects and brushes, you will *never* create the perfect brush for *every* occasion, and they will *all* require some adjustment on an image-by-image basis. For these tools, I want you to think of them as "baselines." Even for the tinker bell brush we made and created a style for, you will have to adjust based on the colors of your image.

Most specialty style brushes come down to the base image (the PNG they are pulled from, so spending time on getting a good dynamic selection and thus a dynamic image will be more important than becoming a wizard of dynamic settings). I can do more with a single stamp-style brush than most people can do with fancier brushes.

I avoid the large "1,000-brush" packs for artists; you will find in your career you really only need four or five of each style to achieve 99% of the effects.

I personally do not set the blend modes for my brushes, tending to instead set the layer blend mode when I use them.

# Summary

Take a look at the following artworks:

Figure 15.21 – Artwork 1

Figure 15.22 – Artwork 2

Figure 15.23 – Artwork 3

Wrapping this chapter up, I wanted to point out that these types of images are why I got into photo editing. I never liked landscapes, architecture, or portraits. I loved the idea that you could take an image and make it more, or combine a mundane image with another and create something no one has ever seen; editing allows me to do that. So far, I have not included many of my artworks but as we are getting close to the end of the book, I wanted to give you a word of encouragement. You have almost every technique I use to create my own style of art, so I added a few of my images to show you how far you can climb with just this little bit of knowledge. Throughout this book, I have kept saying, "You do not need to know everything and a handful of techniques will get the job done," and I absolutely believe that.

In this chapter, we took the idea of a brush and made several brushes using a very small handful of techniques, including:

- Muti-nozzle brushes
- Repeating images brushes
- Smoke and fog brushes
- Particle brushes
- Special effects rain layers

So far in this book, we have focused exclusively on how to create great images, fantastical compositions, and special effects; now it is time to add text to images and this involves shapes to create things such as photo frames, text on paths, and so on. So we are stepping solely out of the area of photo editing and we will go briefly (for one chapter) into the area of design using photos and shapes to augment your photos.

# Part 4:
# Finishing Your Edit and
# Building Your Own
# Artistic Palette

As we finish the book, you now have undeniable physical evidence of your ability to create work in Affinity Photo, but now it is time to get you from a simple user to working like an actual digital artist. In this section, we will cover the various personas that professional artists use, as well as begin to build our asset library for reusable pieces in future projects.

This part comprises the following chapters:

- *Chapter 16, Working with Text and Shapes in Affinity Photo*
- *Chapter 17, Editing in Other Personas in Affinity*
- *Chapter 18, Exporting and Artist Efficiency Tips*

# 16
# Working with Text and Shapes in Affinity Photo

So far in this course, we have dealt with the idea of image editing and largely ignored the role of design in the editing space, but now is the time when we will stretch outside of what we think of as traditional editing and dip a toe into the pool of design. When the discussion turns to design, two elements of Affinity Photo come into play that have yet to be discussed in depth: the elements of shapes and text. Shapes and text are used in conjunction with photos to create things such as magazine layouts, print pieces for advertising, and the ever-popular cultural phenomenon that is the meme.

In this chapter, we will look at shapes and text as they relate to the artistic application of the elements, not the design of elements, so what is the difference? For this book, we will be looking at how to create elements, concerning ourselves with the artistic aspects of those elements, such as how to create text to use in a meme or utilize text as a clipping mask for an image, not whether we have included enough space in the piece according to universally accepted design rules, or whether we followed the text size hierarchy mandated for multiple rows of text in design theory.

So, to begin, we will discuss the idea of text layers and how they differ from other layers, including the text box and the popular text-on-curve features. I will recreate some popular text looks to teach the idea of alignment, justification, and so on to teach you basics of typography (notice I said basics, not complete) that I use 99% of the time. Following that, we will discuss the idea of shapes and how they differ from traditional curves and why you would want to use them in creative photo applications.

So, as we leave behind the realm of pure photo editing, we now cross the hybrid bridge between editing and design to give your art that added punch that the written word adds to your images.

In this chapter, we will cover the following topics:

- Basics of text in Affinity Photo
- Shapes in Affinity Photo
- Setting up the project

- Making a personal gallery
- Professional tips, tricks, and important points

# Technical requirements

The CiA video for this chapter can be found at `https://packt.link/TIPbO`.

You will also need the `Chapter 16` project to follow along with me.

# Basics of text in Affinity Photo

When it comes to text in Affinity Photo, you will find a host of traditional settings, visualizations, and adjustments that you are used to in other popular software; however, as you may know, typography is an area of design that some people specialize in. In this book, I do not seek to make you a typography master but rather show you some of my most frequent techniques used to create text-based images.

Things such as leading tracking, kerning, leading, and baseline will still need to be set and adjusted on a case-by-case and font-by-font basis, so it is important you go through some of the exercises where we adjust the text to make it work.

When it comes to text, there are two distinct tools in Affinity Photo:

- Artistic Text Tool
- Frame Text Tool

Let's go over these tools in more detail.

## The Artistic Text Tool option

The **Artistic Text Tool** option is the most frequently used in my work, as text is typically used to augment an image. Utilization of the **Artistic Text Tool** option is fairly straightforward, so in the next few subsections, I have outlined the steps to find it, use, and adjust it; then later, we will build on that to do things such as convert it, bend it, use it as a clipping mask, and so on.

### Creating artistic text

To locate **Artistic Text Tool**, move to the tools section on the right-hand side and navigate down to the **A** icon (see **1** in *Figure 16.1*). Notice the icon has a white triangle in the corner, if you do not see the **Artistic Text Tool** icon, you may have the **Frame Text Tool** option visible. If it is not in your tools area, add it using the **View** menu:

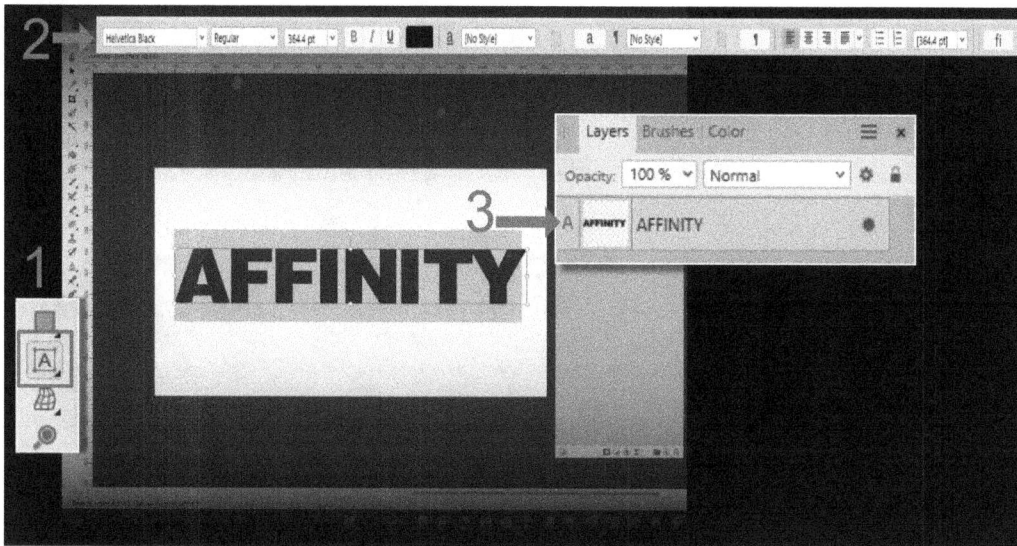

Figure 16.1 – Location of Artistic Text Tool

Once you click on this tool, the mouse cursor will change to the letter **A** with a + sign. Simply click and drag the text to the size you want to use (don't worry, you can change it later) and then type your text.

You can adjust the text size by clicking in the corner and dragging the box, resizing all the text, or selecting the individual characters and using the context toolbar (see **2** in *Figure 16.1* that the text shown in the image is **364.4 pt**.

A layer will also be created in the **Layers** panel (see **3** in *Figure 16.1*).

### Adjusting the artistic text

There are two ways to adjust the artistic text in Affinity Photo. We can do this in one of two areas:

- The context toolbar
- The **Character** studio panel

Both panels will be required for the text to look correct, and all the text needs to be adjusted in every piece (nothing comes perfect out of the box). What I have included in the following subsections are the most common adjustments in both these areas that new editors need to make their vision a reality.

## The context toolbar

In *Figure 16.2*, I have broken down some of the most frequently used adjustments in my work relating to text to help you understand how to read some of the terms present in the context toolbar:

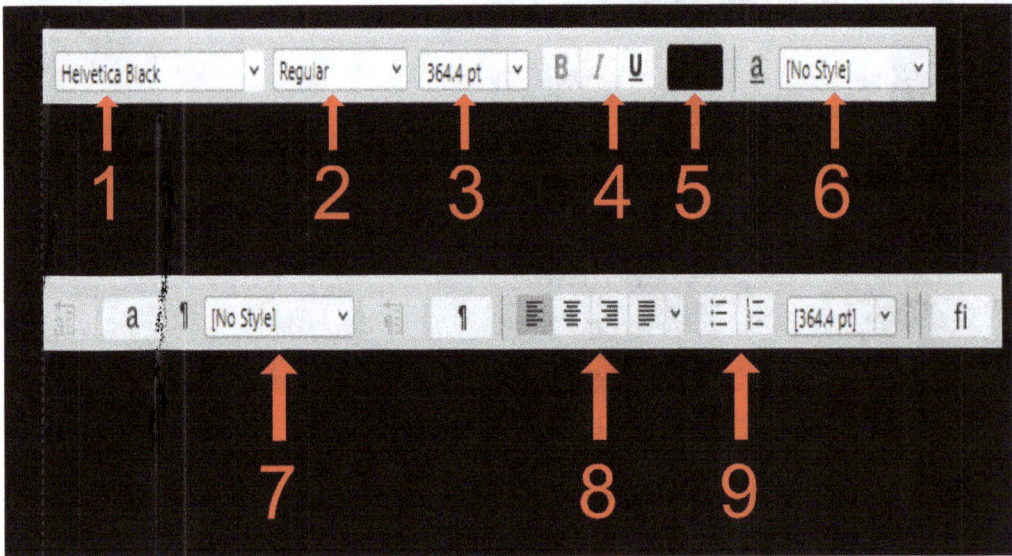

Figure 16.2 – Breakdown of the context toolbar for the Artistic Text Tool option

If I do not cover one and you are curious, simply hover over the item in the context toolbar, and it will give you the name of the option. Let's go over the options seen in *Figure 16.2*:

- Font name (**1**): By default, Affinity picks up both **TrueType font** (**TTF**) and **OpenType font** (**OTF**) currently installed on your machine, so there is nothing new or special to do there.

- (**2**): If there are various versions included in your font file, then these versions will show up here; for example, sometimes there are the **Bold**, **Italic**, and **Regular** options for some fonts. If there are multiple options, then this is where you select them. Not all fonts have multiple options; refer to your font file to see whether there are multiple.

- (**3**): This shows the current size of the text; to be specific, simply click in the box and tell Affinity the size you require.

- (**4**): The common **Bold**, **Italic**, and **Underline** options commonly available in any word processing program.

- (**5**): This displays the color of the font; click to select and adjust the color.

- (**6**): I personally do not use styles ever (styles of text are presets for the text you make for common adjustments you use), but this displays styles of the text; this is different from item 7, as styles can be saved, like presets for text.

- (7): This displays the style of the paragraph. Again, in my work, I have never set this option; this is more something for those that do layouts and work in publishing.

- (8): This is the alignment of the text (align left, middle, and right).

- (9): These are bullets and numbers. Simply select the text and apply the bullet option.

These options will be largely familiar to most people who have worked in a word processing program, so this portion should be rather intuitive but can be made more complex in the event you want to do higher-level typography work.

## The Character panel

To locate the **Character** panel, go to **Window** (in the menu at the top of the screen), and select **Text**. From there you will see the various panels used in text adjustment, but we will focus on the **Character** panel. The **Character** panel is shown in *Figure 16.3*:

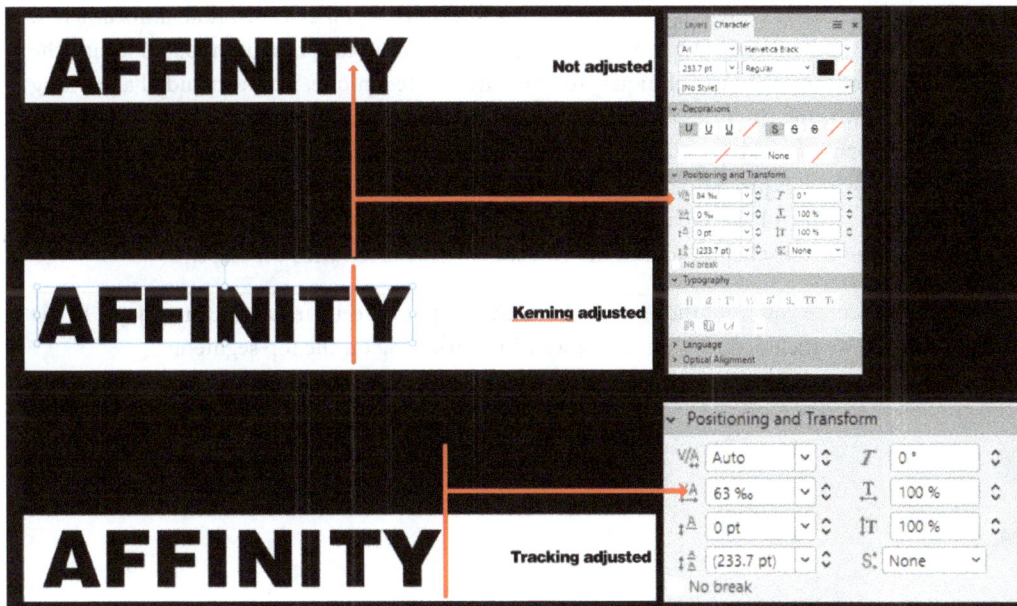

Figure 16.3 – The Character panel breakdown showing adjustment

What we have created in *Figure 16.3* is a simple piece of text, **AFFINITY**. At the top is the **Not adjusted** version that comes natively from the text tool, but this needs adjustment. For this example, we will adjust both the kerning and the tracking. Kerning refers to the space between letters; notice the lack of space between the **T** and **Y** characters in the **Not adjusted** version of the text. To adjust kerning, the adjustment is located in the **Positioning and Transform** section of the **Character** panel.

To adjust the kerning, simply click once between the characters; this will place the cursor in between the two characters you want to space. Following that, click the up and down arrows to adjust it to a spot that looks good to you. In our case, for *Figure 16.3*, we increased the space between characters by **84%**.

If kerning refers to spacing one letter from another, tracking refers to increasing the spacing of the entire selection. To adjust the tracking, simply click and drag selecting the text you want to track, and then adjust the tracking adjustment in the **Positioning and Transform** area. Notice in the **Tracking adjusted** portion of *Figure 16.3* that the word is much longer; this is because the spacing between letters has been adjusted by **63%**.

These types of adjustments are necessary because when the program makes a guess, we still need to trust our eyes.

### Creating text on a curve

One of the most frequent uses of text is placing it on a curve, such as in logo design or following a drawn path. While not quite used as often in the photo editing world this is still a desired skill; unfortunately, it is a weakness of Affinity Photo. Other programs make this process easier, so be prepared to ride the struggle bus a bit in learning this technique. To illustrate this technique, I have included a working file for you to follow along (in both V1 and V2 folders. These folders can be found inside the Text on curve practice files folder.). In this example, we will be using a pre-drawn curve – called a segment – which is just a drawn-out shape. See the Text on curve practice file in the downloads . Lastly, I have included a video showing this technique, so refer to the *Technical requirements* section of this chapter to find the video. Let's get started with the steps:

1.  Identify the curve you want the text to follow. This can be a freely drawn line or a shape. In the practice file, we included two segments; we will be working on the top segment:

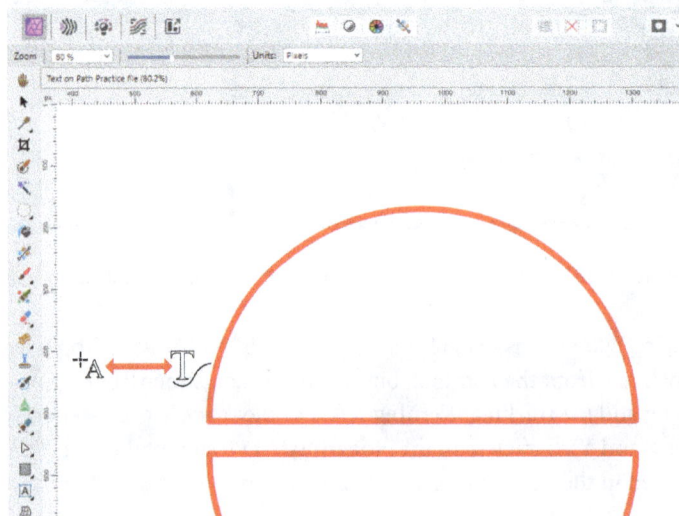

Figure 16.4 – Identifying the text on curve function

2. With **Artistic Text Tool** selected, move the cursor to the part of the curve you want to use. The cursor will change from an **A** character with a + symbol next to it to a **T** character with a line, indicating that Affinity recognized a curve, and we are now in a text on curve mode (see *Figure 16.4*):

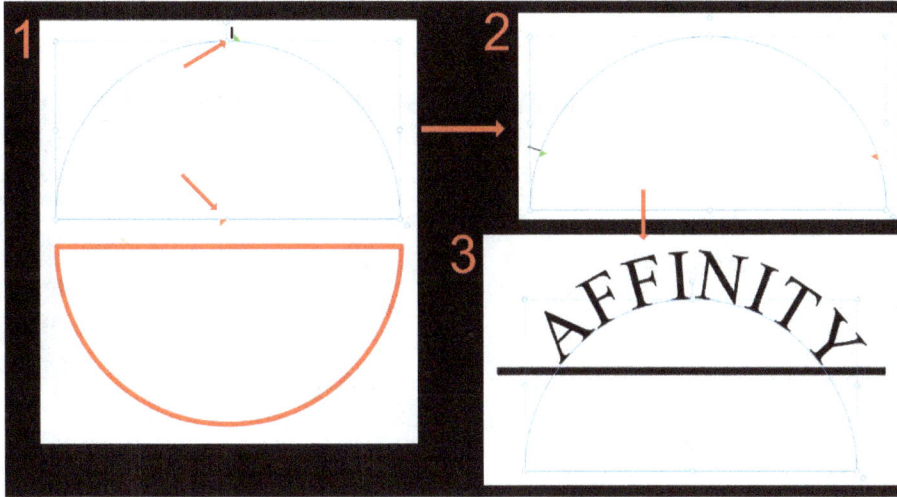

Figure 16.5 – Positioning text on curve

3. I set my text size to **30** in the context toolbar for this example. Click on the curve where you want to start the text. Make sure you click on the outside of the circle; this will place the text on the outside of the path (Affinity is not good at recognizing this, so it may take a few tries).

   Notice that the creation of text on the path is destructive; there is no longer a segment and three things will happen (see **1** in *Figure 16.5*):

   • A green triangle will appear; this will be where the text will initiate

   • A red triangle will appear; this is where the text will terminate

   • A black line will begin to blink; this is where you will write your text

4. Reposition the beginning and end points to the top of the segment (see **2** in *Figure 16.5*), and then type in the text.

5. Adjust the text to size and position it to be straight (you will see a black rectangle in the image. This is the technique I use to make sure I am even) (see **3** in *Figure 16.5*).

6. From here, we can adjust the baseline, direction, and so on from the context toolbar.

While we have discussed text from a pure character point, we can also use text as a vector-style shape to create a clipping mask for images. So, let's look at this now in the next section.

## Converting artistic text to a clipping mask

This is a very common request from editors and a very popular technique by graphic artists, and it is simply a play on using the text as a clipping mask. So, I figured it was applicable and useful to include here as a frequently used technique. To accomplish this, simply follow the steps that we will cover next. I have included a CiA video (see CiA 16.2) to follow along (the link can be found in the *Technical requirements* section of this chapter). Lastly, I have laid out the numbers in *Figure 16.6* to follow along:

Figure 16.6 – Clipping mask breakdown showing adjustments

Let's get started with the steps:

1.  Create a rectangle project that is 5" by 7" at 300 DPI and create a black rectangle as your background.

2.  Create a text layer using an **Artistic Text Tool** option that matches the feel you want to use (use a thicker font so you can see the image). Notice in the **Layers** panel (see item **2** in *Figure 16.6*) that there is a layered effect applied to the outline. This separates the text from the background.

3.  Now find an image using the **Stock** tab – I used an island image – and nest it into the text layer (see item **3** in *Figure 16.6*). Make sure it is nested, as this pulls off the entire effect.

4.  Add another text layer somewhere just to practice.

5.  Duplicate the image and then enlarge it. This creates visual interest, placing it outside of the text layer and behind the text layer (see item **5** in *Figure 16.6*).

6. Add an **HSL Shift Adjustment** layer to this duplicated layer (item **5** in *Figure 16.6*) and move the saturation to **0**.

7. Add an adjustment layer to this image and choose the lens filter; I chose a blue one.

8. Change the blend mode of the image layer to **Average**.

Believe it or not, this is an FAQ we get most often in our classes because people are familiar with clipping masks being curves, but do not consider text as usable as a clipping mask.

This concludes what I would consider the most important functions and everyday uses for the **Artistic Text Tool** option, let's turn our attention now to the **Frame Text Tool** option.

## The Frame Text Tool option

The second type of tool in the text capabilities of Affinity Photo is the existence of **Frame Text Tool**. Now typically, I use **Frame Text Tool** when doing print work, such as a magazine or a project that will have actual paragraphs. I rarely use this as **Artistic Text Tool** provides me with 99% of what I need for the work I do; however, since we are covering all aspects of text tools here, I have included it.

To utilize **Frame Text Tool**, simply go down to the area of the tools that includes the **Artistic Text Tool** option, and click on the little black corner of the icon (shown in *Figure 16.7*). This will bring up the option for **Frame Text Tool** (also shown in *Figure 16.7*):

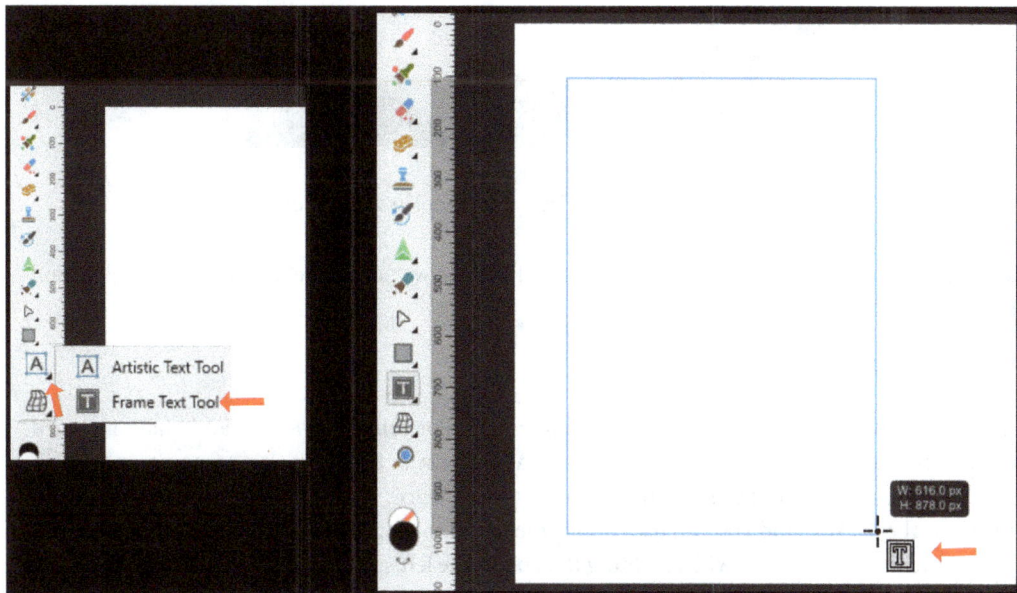

Figure 16.7 – Location, icon, and application of Frame Text Tool

Click and drag to create a text box the size you want. This is the primary purpose of **Frame Text Tool**, creating a predefined area of text to assist in the layout of documents. This is different from the **Artistic Text Tool** option. **Artistic Text Tool** primarily creates the text and allows you to adjust your size to application needs. Notice in *Figure 16.8* we have some example text typed into the box; this is what **Frame Text Tool** does:

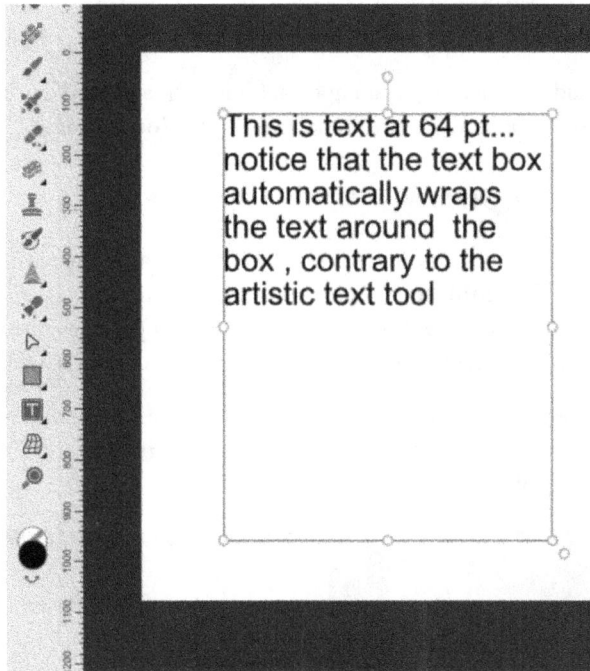

Figure 16.8 – Example of Frame Text Tool in application

Now that we know what the tool does and how to make a box, we need to define a practical application of the tool in the context of editing.

### Using Frame Text Tool

In this example, we will use a famous poem by Edgar Allen Poe called *The Raven*, and I have included two files in the downloads for this project (see `Frame text tool base image` and `Text for the raven project` inside the `Frame text tool project` folder). I have created an image for the background you can use (notice the use of the fog brushes we made in the previous chapter in the base image). We will be adding the text to the left side of the image (see *Figure 16.9*).

I have included the CiA video of this project in the *Technical requirements* section for you to see how it was put together (see CiA 16.3)

To do this, we will use **Frame Text Tool** and draw a square. Then, we will insert the text inside the square by using the copy and paste approach to get the text from the Word document into the Affinity file (see *Figure 16.9*):

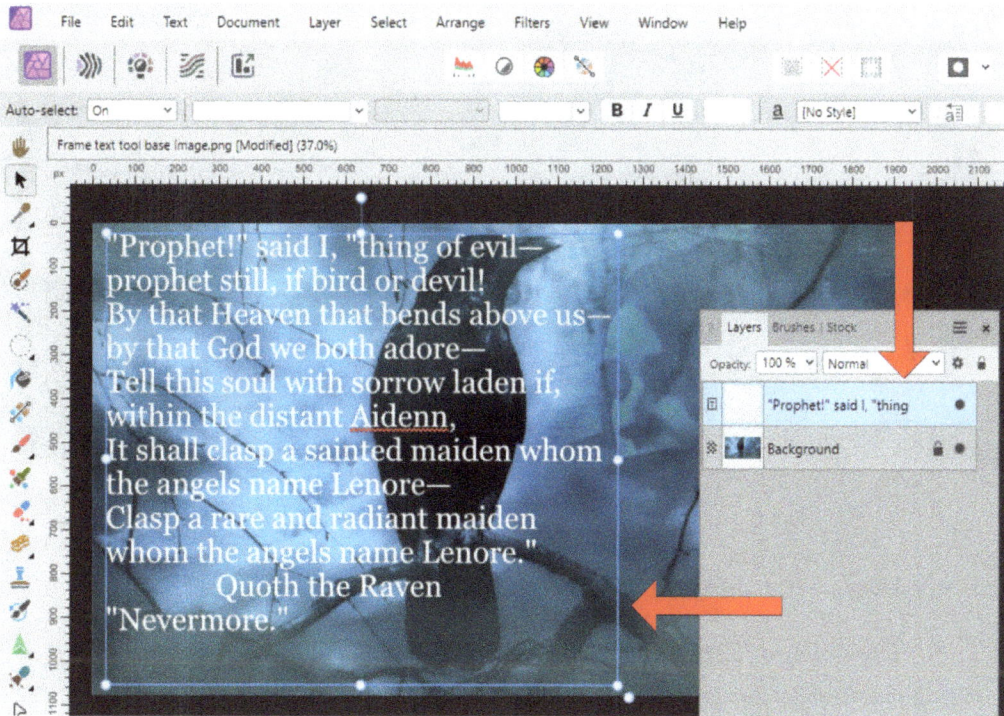

Figure 16.9 – The location and presence of the framed text on the image

Now adjust the text to your liking. In my image, I have adjusted the color, font, and font size.

Notice in *Figure 16.9* that there are two dots at the lower right-hand corner of the box. The outermost will adjust the entire box size, the text size included. The inner corner point will readjust the box size, not the text size.

Lastly, for this image, I am adjusting the blend mode and adding a gradient mask. It is not too important for the readers to be able to read the entire text; I want it to be identifiable for people familiar with the poem. In this edit, I am making it part of the image and not communicating an idea as a magazine would (see the finished image with a breakdown of the layer structure in *Figure 16.10*):

Figure 16.10 – Finished frame text image with breakdown

Notice the following items in the layer structure of *Figure 16.10*:

- The use of the gradient mask running top to bottom on the text layer
- The highlighting of the term "Nevermore," a famous line from the poem, makes it identifiable without having to read the entire verse
- The use of the **Soft Light** blend mode

This is the typical application of the tool for me in my work; I know some of you out there may be doing layout work as part of your editing, but now at least, you know how to read the tool to do what you need it to do

Now that we know about text, it is time to move into the world of shapes (to make objects such as the segment we used during that text on path activity).

## Shapes in Affinity Photo

In the first part of the book, we alluded to the use of shapes in some of our various activities and projects; however, in this section, we will explore the subject formally to give you an idea of how these things are actually used in photo editing and discuss how to edit them to get you the desired shape.

## Creating shapes

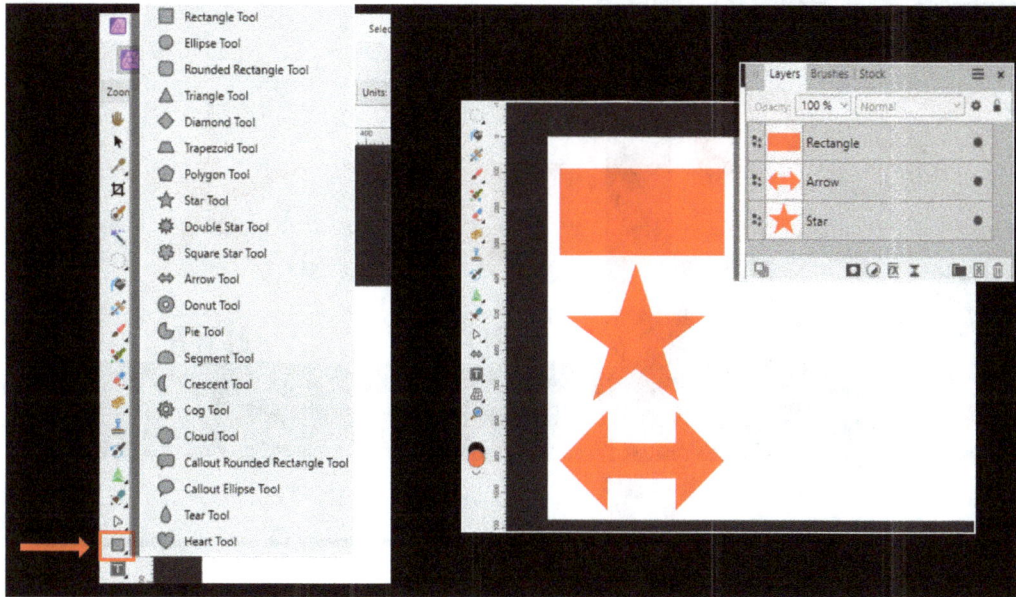

Figure 16.11 – Location of the shape tool and three shapes we will be using

There is a small triangle in the lower corner that has all the shapes in it. Notice that there are the normal shapes (such as rectangles, ellipses, and so on) but also some of the more abstract shapes (such as callout clouds, polygons, arrows, and so on).

To select a shape, simply choose the tool and click on the workspace, dragging the shape to the size you desire. See *Figure 16.11*, where I have used a rectangle, star, and arrow shape. If you would like to follow along, I will explore the topic of shapes using this example. I did not make a template to follow along, as at this point, we are more than capable of creating three shapes on a workspace.

To pull a perfectly symmetrical shape (an oval that is actually a circle or a rectangle that is actually a square), simply hold the *Shift* key during your drag to drag out of the shape.

## Adjusting shapes

The adjustment of shapes uses two distinct points located on the shape and then the context toolbar for the shape. All shapes have their own context toolbar options. In this example, we will only be looking at the three shapes we identified, so if you are using another shape, please pay attention to the options for that particular shape:

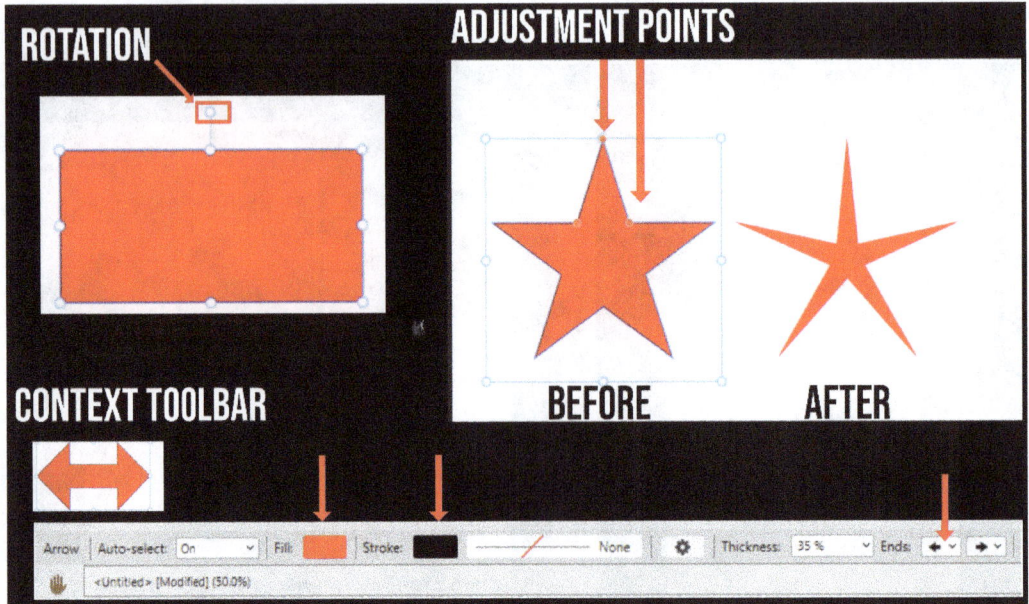

Figure 16.12 – Adjustment points on shapes

Referring to *Figure 16.12*, we will go over the various adjustment points on the shapes:

- **Rotation handle**: Some objects need to be rotated, and some do not (a circle does not need to be rotated, as it is perfect in all orientations). However, if the object needs to be rotated, the rotation handle is present, as identified in *Figure 16.12*.

> **Important note**
> Holding shift while rotating will rotate at 15-degree increments.

- **Adjustment points**: If the shape has adjustment points, they will be noted with orange points that can be adjusted. In the example of the star, there are adjustment points for the inner diameter of the star, and the movement of these points creates different effects.

> **Important note**
> You have to have the shape tool selected to see the adjustment points, they will not show with the Move Tool.

- **The context toolbar**: The context toolbar has specialty options for each shape, so it will be on a shape-by-shape basis that you make your adjustments, so unless we want this chapter to be 120 pages, I will leave it largely up to you to have a play and try some things. However, in *Figure 16.12*, I have included the context toolbar for the double-ended arrow we were working with, and I have highlighted some of the important points for this specific tool:

  - **Fill**: This is how you fill a shape

  - **Stroke**: This is where you can apply stroke to a shape

  - **Ends**: If you want a single-ended arrow, make your adjustments here

Now that we know what shapes are and how to adjust them, we need to know how to use them in an actual edit. For this, we will be making a simple YouTube placard for a video using only shapes, text, and an image.

# Practical application of shapes – making a YouTube thumbnail

In this project, we will use the following techniques:

- The generation and adjustment of shapes
- The generation and adjustment of the artistic text
- Using shapes as a clipping mask

I have also included the following in the downloads for this section (see the YouTube thumbnail project folder for both the V1 and V2 versions):

- The image we will be using
- A template with the elements already placed
- The completed file so you can reverse-engineer it

## Setting up the project

Follow these steps to set up the project:

1. Create a 1920x1080 project at 72 DPI.
2. Generate a rectangle and stretch it to cover the entire surface – make the fill color black.
3. Generate another rectangle and rotate it (see *Figure 16.13* and rotate it changing the fill to a **Linear** gradient using the **A27C00** hex code and then fading to black).

4.  Create a circle in the upper right-hand corner. This will be where we include the content creator's picture.

5.  Add some artistic text to the area of the gold rectangle. This will be the title of the video.

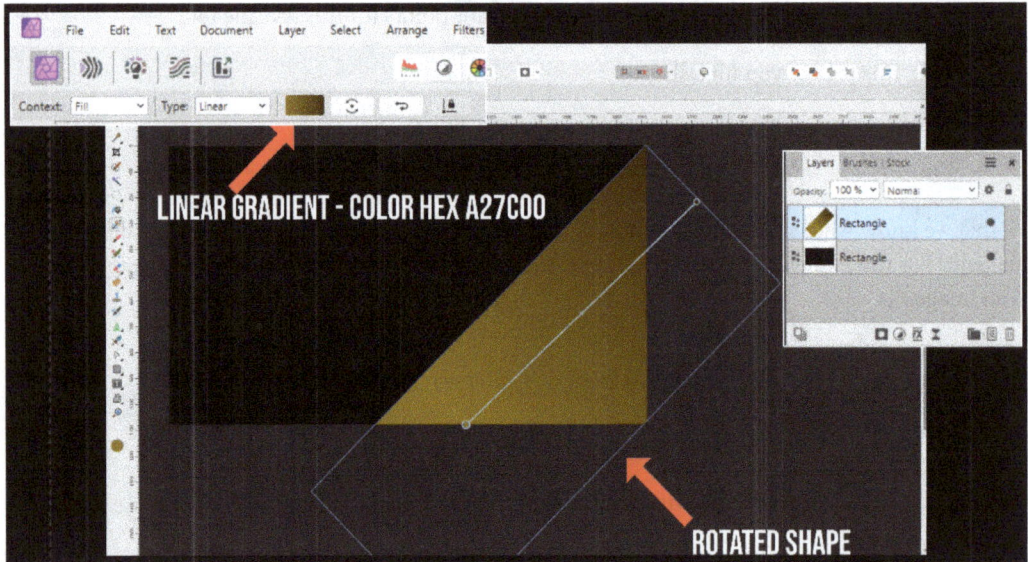

Figure 16.13 – Laying out the shapes – rotation and gradient image

I have included the template for both V1 and V2 in the downloads, so if you got lost following along, you know how it was done.

Now that we have set up the project, let's populate it with text in the next section.

## Populating the project

For this section, I have included the reference images we need (see two solo images in the YouTube thumbnail project folder), and for this fictitious project, we will be making a double exposure image using Affinity Photo.

Follow these steps to work through the process of populating the thumbnail you are working on:

1.  Use the artist's image, place the image above the circle (shown in *Figure 16.14*), and then nest the image inside the circle shape. This effectively creates a clipping mask with the circle. Adjust the artist's image to wherever you think it looks good.

2.  Place the Image of work.jpg file in the thumbnail behind the rotated rectangle.

3.  Change the text to whatever you would like:

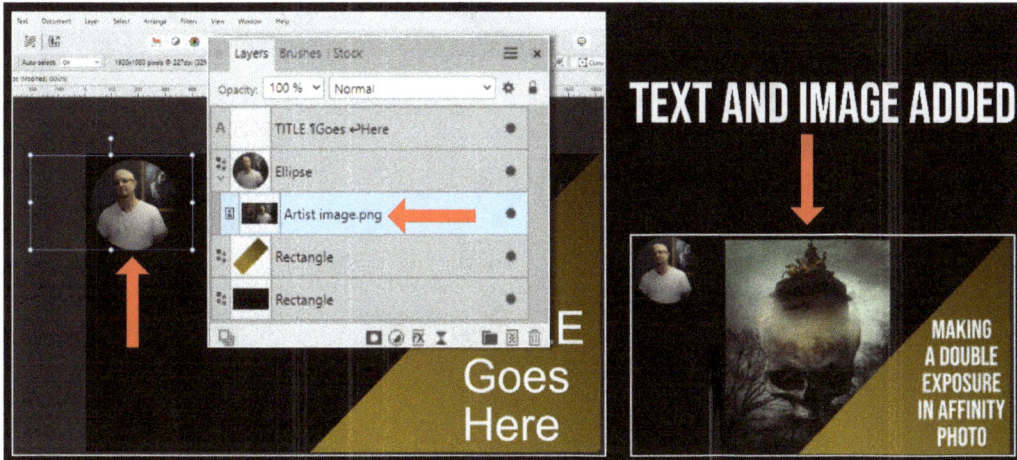

Figure 16.14 – Placement of the images

Now you have your basic image and text elements in place, the last step in the workflow will be to get a polished look to it.

## Finishing the image

The last part is filling some of the empty space and unifying the image. We will duplicate the `Image of work.jpg` file (in this case the skull double exposure), and then enlarge the image to make it very large behind to give unity to the piece:

1. Duplicate the `Image of work.jpg` file (the skull double composition image), and on the lower image expand it to be much larger than the thumbnail area (see *Figure 16.15* for my sizing).

2. Change the blend mode to **Average**.

3. Reduce opacity to **50%**.

4. Export it:

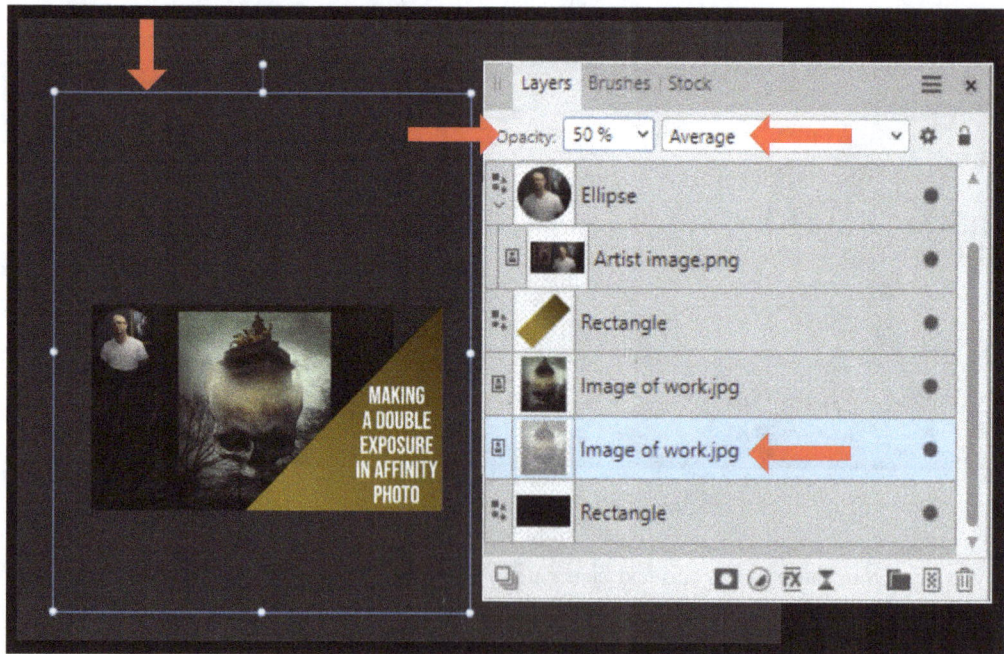

Figure 16.15 – Finishing the image

While typically, projects like these are completed by hiring a designer, if you understand the simplicity of shapes, you can use Affinity Photo to create very similar art for your personal projects. Along that line, let's do our last project and make a gallery for your art in the next section.

## Making a personal gallery

This is a project I use to display all the art that I make from time to time, and it is a great way to display your art and realize how far you have come. So, for this project, I will be using art from my personal collection, but at this stage in the book, feel free to use the art you have created in completing this book. This project works on the following key areas:

- We will be faking 3D using warp of the floor
- We will be using textures over gradient rectangles to create wall textures
- We will be using multiple lighting layers to create mood and color
- We will be adding reflection from glass to mimic reflection

I have included the images to set up the gallery (see the `Gallery project` folder found in the downloads for this chapter), but not fill it; if you do not have enough photos to fill the gallery right now, you can always use stock. Let's get started.

## Setting up the gallery

Follow these steps to create your own gallery space:

1.  Create a 1920x1080 workspace at 300 DPI. This will be the size of the image.

2.  Find the image labeled Cement floor and place it in the workspace, covering about three-quarters of the way up the page.

3.  Go to **Layers | Live Filter Layer | Distort | Perspective** and apply a perspective warp. Notice in *Figure 16.16* where I placed the points. This gives the cement an illusion of depth and perspective. When we add lighting, this will also give the floor some texture.

4.  Now create a medium gray rectangle and place it under the floor. This will form the base wall color.

5.  Apply the Steel wall image on top of the gray rectangle and set the blend mode to **Overlay** and drop the **Opacity** value to **75%**. This adds texture against the tone of the wall and is a trick from the 3D art world. Once the lighting is added, this layer will provide textural information:

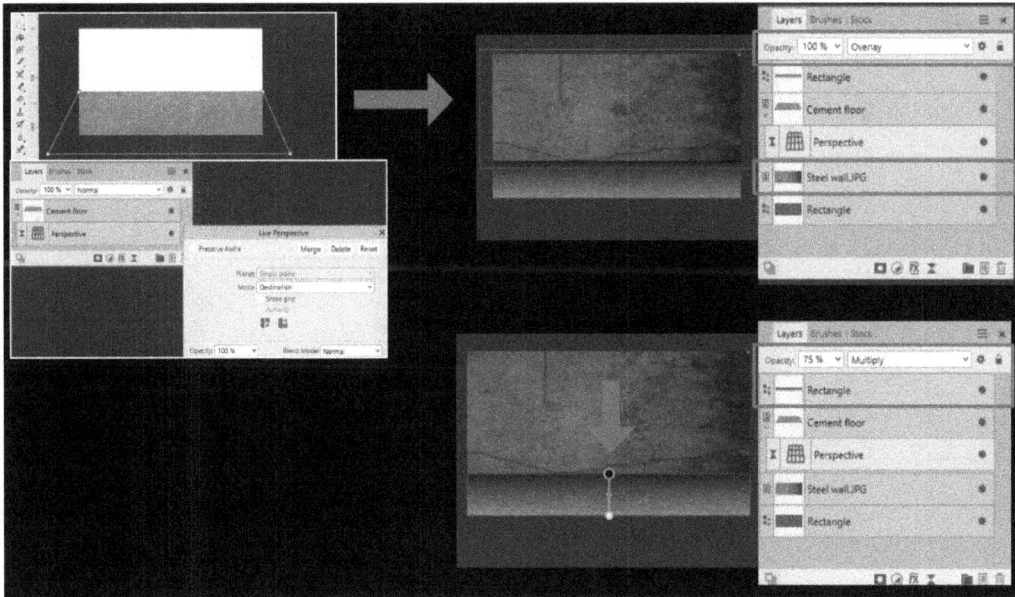

Figure 16.16 – Creating the gallery space

Now, we have a floor and a wall, the space is coming along well. Now, it is time to hang all the amazing art you have made along the journey in this book.

## Adding the frames

Now we have to add the frames, and the size and style of the frames are up to you. I will use simple squares with an outline and a bevel (layer effects), and then I will simply duplicate it. However, here are two suggestions if you want to go to the next level:

- Round frames or different shapes

- Find stock images of frames and insert your pictures into them

So, to begin with, we will make one frame and then duplicate it. We will have to adjust each frame to get the lighting right when we add lighting in the next section, but for now, we will focus on the frame and the fill. Follow these steps:

1. Create a square or rectangle, and apply a stroke color close to black, but not pure black, as we will still need shadows.

2. Apply a layer effect of outer shadow to the frame. We will adjust each one as we do the lighting, but I want them on there.

3. Add a **Bevel / Emboss** layer effect. My settings are shown in *Figure 16.17*.

4. Duplicate these frames as often as you want, and alternate the size and shape to match your art:

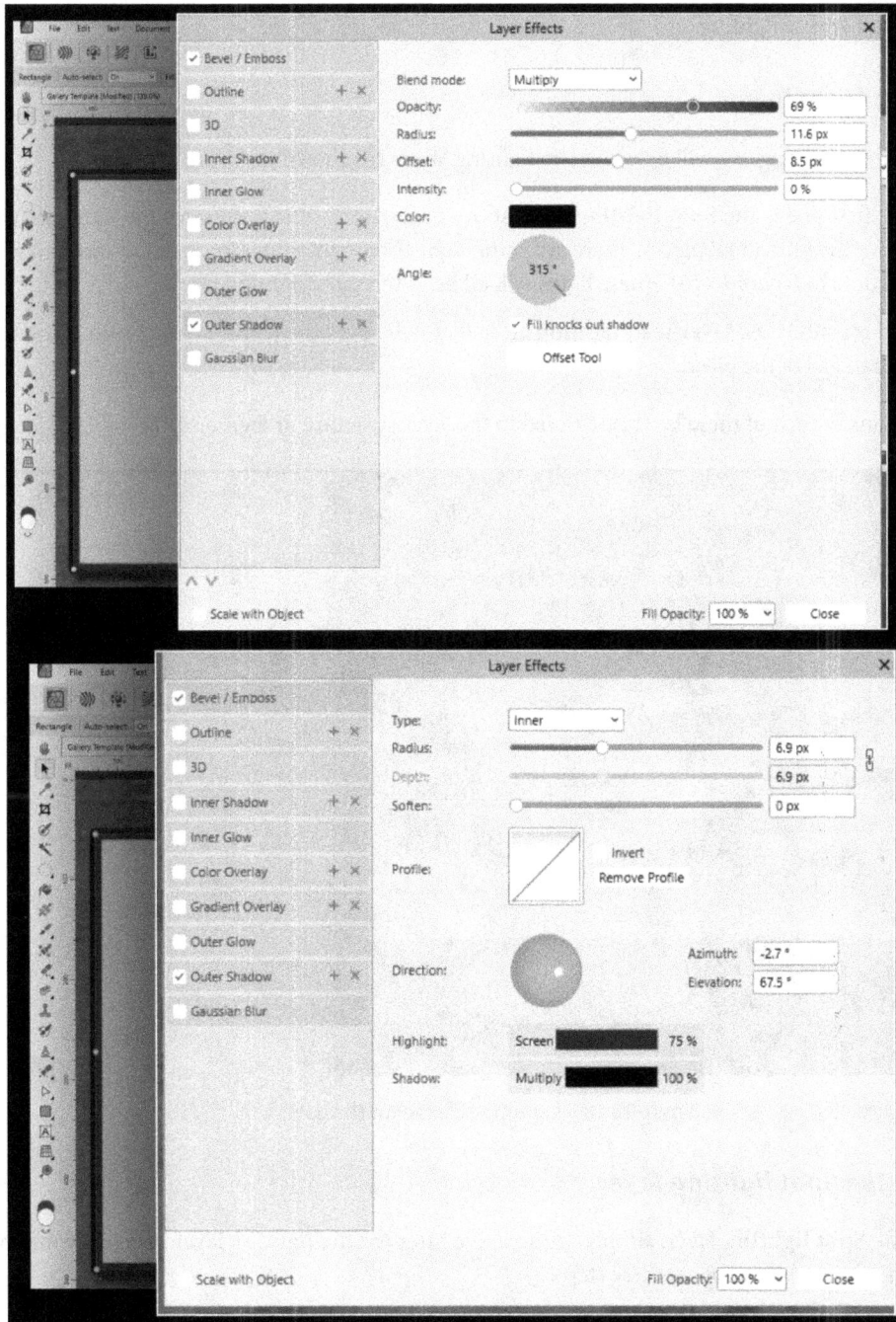

Figure 16.17 – Creating frames and settings for shadow and embossing

Next, let's add some lighting.

## Adding lighting

To pull off the lighting, we will be using two lighting layers (as shown in *Figure 16.18*):

- The first one is the **Spot lighting** layer above everything. This represents the lights that are above the individual pieces. There are going to be three spotlights (the locations are shown as red dots in *Figure 16.18*). These lights will all be in the same lighting layer.

- The second is the **Overhead lighting** layer that will cover the texture of the walls, floor, and overall feel of the piece.

The locations of each of these layers are noted in the layer structure in *Figure 16.18*:

Figure 16.18 – Lighting scheme for the gallery

### Adding the Spot lighting layer

To add the **Spot lighting** layer, simply create a live filter for the lighting layer and place it above everything else and then follow these steps:

1. In the dialog box for the lighting layer (click on the flashlight icon to open), where it by default puts in a spot light, change the **Type** option to **Point**.

2. Position the point light in one of the positions shown in *Figure 16.19*, slightly outside the frame.

3. Using the dialog box, adjust the color and settings to the settings present in *Figure 16.16* to *Figure 16.19*. You can adjust as per your need, as your image may be different from mine.

4. In the dialog box, click on **Add**. This will create another light. Change the **Type** option to **Point** and position it over the center. Adjust it to your liking.

5. Lastly, create one more light by hitting **Add** and then moving and adjusting it:

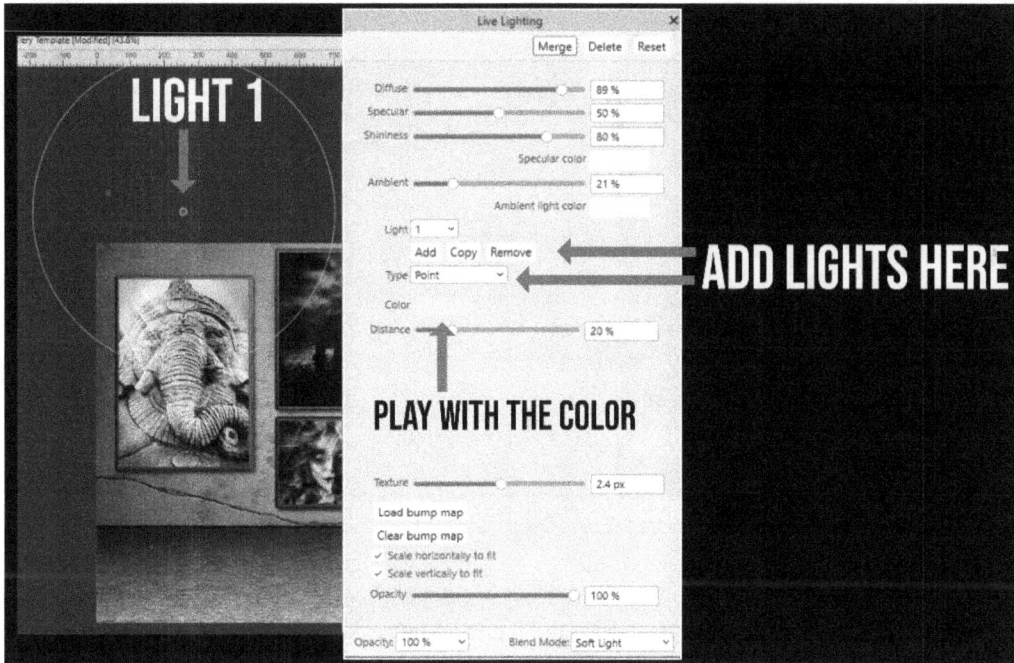

Figure 16.18 – The Spot lighting layer

Notice we created all three point lights in the same layer. This is important as we did not have to create three lighting layers to do this.

### Adding the overhead light

Now we have to add the overhead light, and the process is largely the same.

Follow these steps:

1. Choose a lighting layer by going to **Layers | Live Filter Layers | Lighting**.

2. Place the layer down in the stack below the frames, as you do not want the layer interacting with the images.

3.  Adjust the overhead light above the image (the position is shown in *Figure 16.19*) and adjust the settings to your liking.

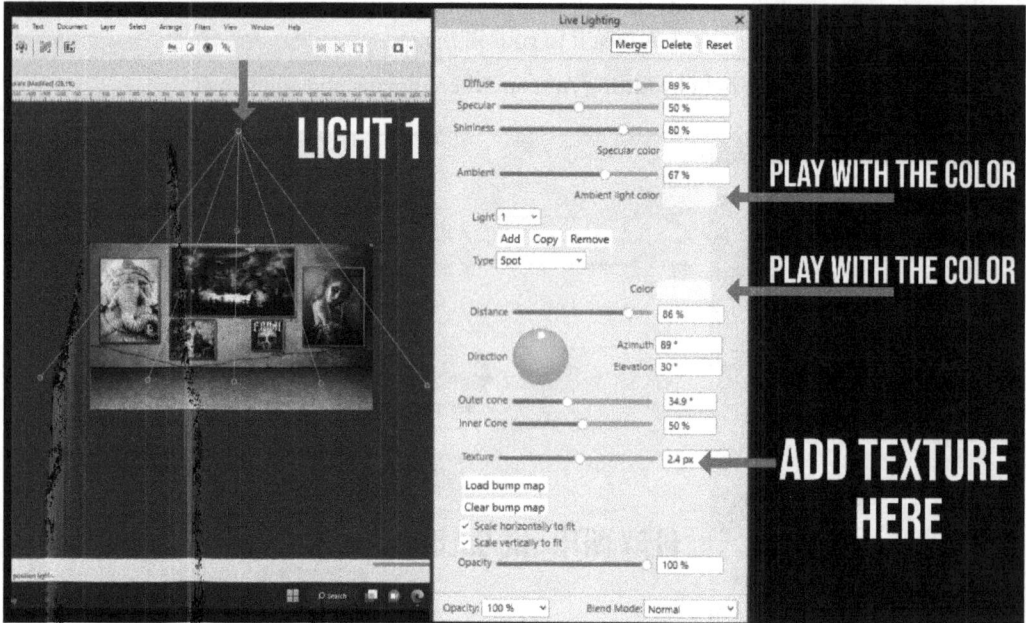

Figure 16.19 – Overhead lighting

Notice the **Texture** slider is adjusted on the overhead. I personally like it to add texture to my scenes; it is quite strong, so please play with the effect, and if it is not for you, simply keep it at zero.

This gallery can be used again and again with *infinite* possibilities, add images, subtract images, change the feel, put your logo out against the wall, or add a watermark; it really is limitless, so we have included the gallery templates for you in the `Gallery Project` folder found in the downloadable of this chapter.

## Professional tips, tricks, and important points

When it comes to text and editing, the biggest mistake I see editors make is picking fonts that do not *help* but actually *harm* the image because they do not match. Either they do not match the feel, or they don't match the theme, or they have not adjusted the text for saturation, opacity, and so on, and so the text overpowers the piece.

Pixel-based editors typically do not spend much time learning vectors, but the shapes in Affinity Photo can be converted to curves to create some amazing clipping mask options, so while you may have no interest in being a vector artist, please learn how to use the Pen Tool, as it will help substantially in your career.

Remember, your text has to tell a story, so do not forget to pay equal attention to your text to make it as amazing as your picture. The following image was created in Affinity Photo, and the chrome text was equally as challenging as the skull. Notice the amount of effort put into the text to *support* the image:

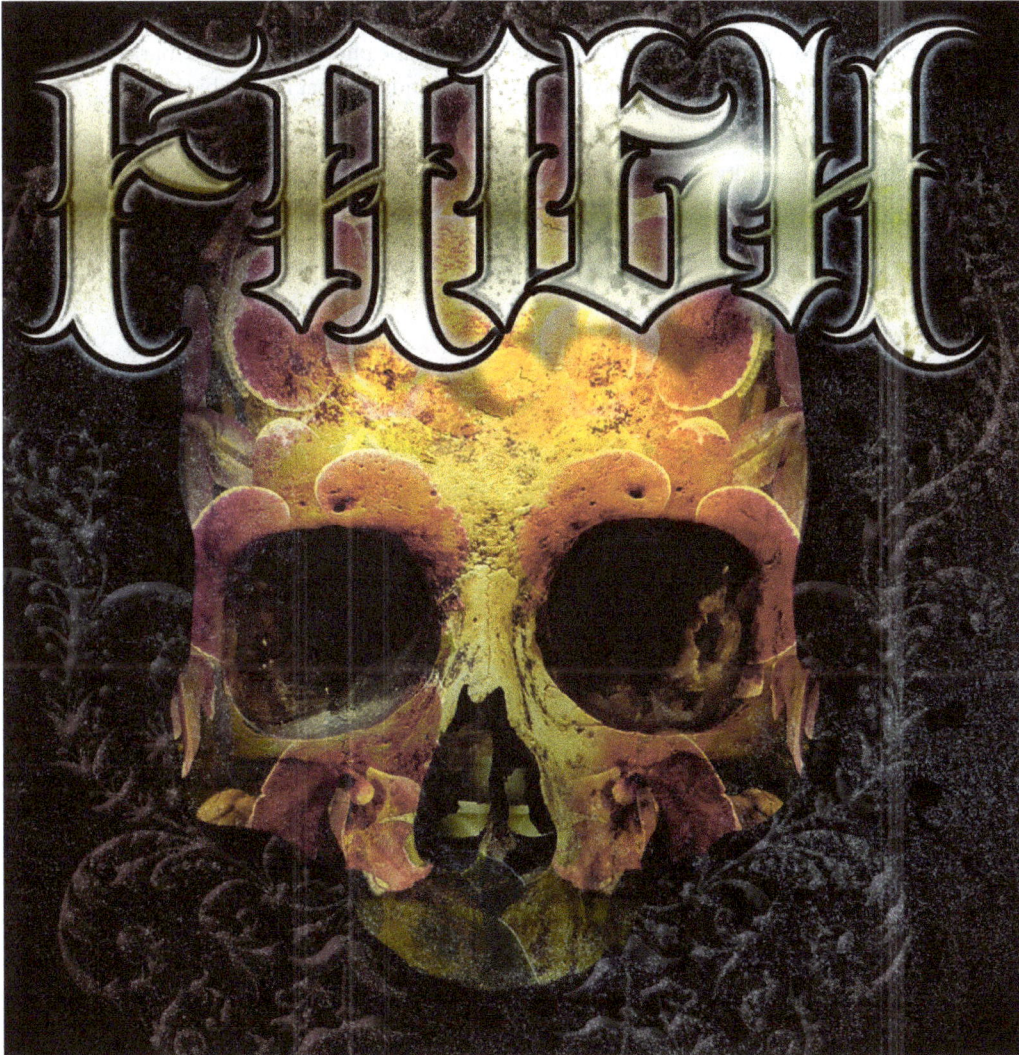

Figure 16.20 – Text with skull image

## Summary

Wrapping this chapter up, we covered two main points; working with text in photo editing and utilizing shapes. I taught you text on curve; text is a clipping mask and makes not only consumer goods such as postcards but also social media assets such as YouTube thumbnails. This is what makes Affinity Photo so much fun; it doesn't matter whether you are a landscape photographer or a marketer, there is something for everybody here; we just have to open our minds to what is possible.

We will now turn the page into the other personas in Affinity and look at the various other ways more advanced editors can use Affinity; things such as raw editing, tone mapping, liquifying, and dedicated exporting, will be covered in the next chapter, so let's leave the photo persona behind and deep dive into the alternatives Affinity offers.

# 17

# Editing in Other Personas in Affinity Photo

So far in this book, we have been working completely in what is called the Photo persona, but Affinity has more personas that we can work with to accomplish our photo editing goals. In this chapter, we will explore three additional personas and how we can leverage them to build one last building block on the very solid foundation we have established so far. In this chapter, we will be covering the following topics:

- The Develop persona – also known as editing in RAW

- The Liquify persona – a holdover that was replaced in a later version of V1 by the non-destructive option for a Liquify Distort Live filter

- The Tone Mapping persona – an editing space that allows us to adjust **High Dynamic Range** (**HDR**) images, but also create amazing effects

While we will be covering some of these in a great deal of depth (such as the Develop persona), we will not spend much time on others, such as the Liquify persona, as there are non-destructive options currently available and their presence is obsolete in modern workflows.

These additional personas will not be adequate by themselves to achieve anything significant and require your artistic eye, as an editor, to make them useful based on your firm knowledge of the Photo persona. This is why I have saved them for the end of this book. Since we're coming to the end of this book, we will go back to the cooking analogy – that is, the preparatory and plating steps, which make for a well-planned dish that can be delivered. They are not enough in isolation but are a source of significant value when used with the fundamentals you have learned so far.

## Technical requirements

The CiA video for this chapter can be found at https://packt.link/kd4it.

# The Develop persona

The Develop persona is the persona that photographers use to develop their RAW files and is located in the top left-hand corner of the interface. If you shoot in RAW format, then working in this persona is the first step in the workflow because the act of developing a RAW file into something we can work with within Affinity destroys the image. Only in V2 of the Affinity software did they add the functionality to switch between a RAW image that had been "developed" and allow it to be turned into the raw version to be edited later. So, for me, RAW image development happens at the beginning as a preparatory step before the image can be moved into the normal photo workflow.

On the subject of developing, this is the official term for finalizing an image in RAW and converting it (compressing it) into the Photo persona. So, in the Develop persona, you will see a button called **Develop**; this is what that word means in this context.

> **Professional note on following along with the examples in this chapter**
> Save a couple of backup copies of RAW files as we will use them to explore the various persona adjustments and effects. That way, you can start over from the original if you're not happy with the results. Once you like the results, you can delete the copies to save storage space. I have seen far too many RAW images lost due to destructive workflows with less-than-stellar results.

## What is a RAW image and why shoot in it?

A raw image is a setting in the camera where you shoot the image (you cannot make an image raw if it was not shot this way). Each camera manufacturer has a format, file extension, and so on. In this case, I am going to use .RW2 as I shot my image on a Panasonic Lumix camera. I am mentioning this because if you are a photographer and shooting images, then you have to know how to set your camera up to capture raw images to use this function.

A RAW format contains exponentially more data than a flat image such as a JPG, and if you have a camera that shoots in RAW, it is not uncommon for the file sizes to be more than 10 MB bigger than that of a JPG. In *Figure 17.1*, you can see a .RW2 file shot at 18.7 MB and the corresponding .jpg file at a fraction of the size:

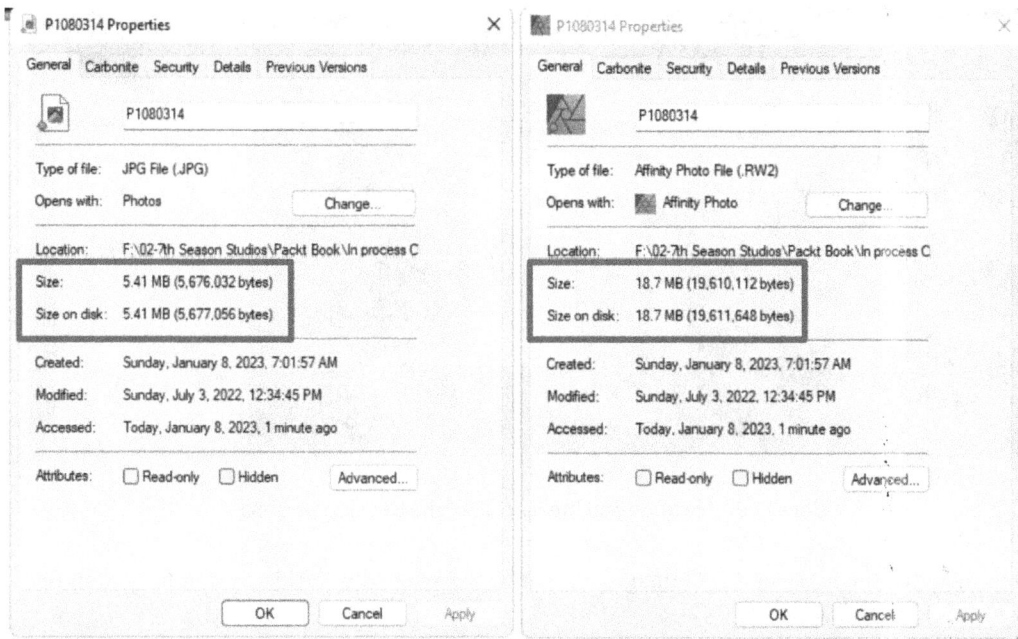

Figure 17.1 – File size differences

Artists shoot in RAW because, with the added data, there is more color information and it can lead to better edits as the blacks can be deeper, the lights can be brighter, and the color information is just overall better. Many photographers shoot exclusively in raw; if your storage situation will handle the excessive file size, I recommend using RAW as a starting point.

## Editing in the Develop persona

When editing in the Develop persona, some key areas of the interface need to be explained, such as some of the strange ways that the persona chooses to *mask* areas. However, it is mostly similar to the Photo persona we have been working with throughout this book. So, in the following subsections, I have tried to highlight the differences that make the Develop persona different.

### Navigating the raw space

As noted in the following figure, the Develop persona can be found in the top left-hand corner of the interface:

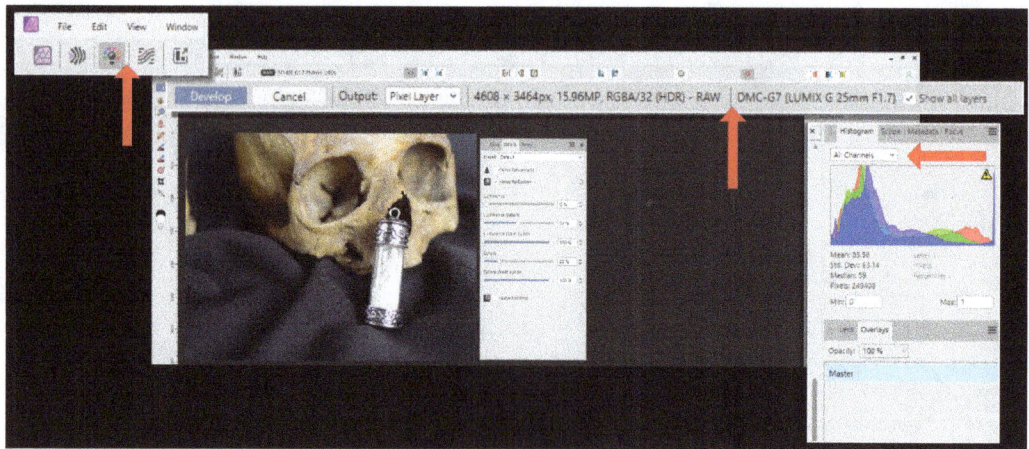

Figure 17.2 – Location and breakdown of the Develop persona

I have included the RAW and JPG images we will be working with in this section for you in the downloads for this chapter.

The image we'll be working with is of a simple skull and vial I shot from my collection. I used a very specific lens to create a shallow depth of field, so the image has *not* been modified at all (see the `.RW2` file in the `Develop persona base images` folder). By default, it will open in the Develop persona if you double-click on it since Affinity Photo recognizes the format as RAW.

With the RAW image open, I would like to call your attention to the areas of the interface present in *Figure 17.2*:

- The location and icon for the Develop persona are shown in the top-left corner of *Figure 17.2*.

- When the **View** tool is selected (the hand), the demographics of the image are shown. The image I have provided was shot with a Lumix G7 and displays the F-stop values and lens type, and clearly shows the file as a RAW file.

- Each persona has a panel that's unique to that persona and can be adjusted in the same way (in *Figure 17.2*, you can see the studio panels that were pulled into the viewport, as we have been doing throughout this book). If there are studio panels you do not see, you can add them using the menu at the top of the screen and select **Window**, just like in the Photo persona.

It is worth noting that you have to reset each studio in each persona, so the panels you had in the Photo persona may not be present in the Develop persona by default. So, you will have to select them again under the **Window** menu.

## Making adjustments to the RAW image

When working with images in RAW, you will see very similar workflows and adjustments from the studio panels to those that are in the Photos persona, so I will not be covering the basics we've already covered (for example, we will not discuss curve adjustment, as that has been discussed previously). Instead, we will focus on the largest difference that we see in the Develop persona – that is, the way it masks adjustments.

In the Photo persona, we utilize masks to dictate where we want effects and adjustments to be shown or hidden; however, in the Develop persona, the term *overlay* is used. You can think of an overlay as a mask, though it is not precise, it is not pixel-perfect, and it works on the idea of painting on masks, which we covered in the masking chapter.

That being said, let's take a look at how it works.

## Working with overlays

In Affinity Photo's Develop persona, masking is referred to as an overlay. Overlays are painted on using a set of tools, as shown in *Figure 17.3*:

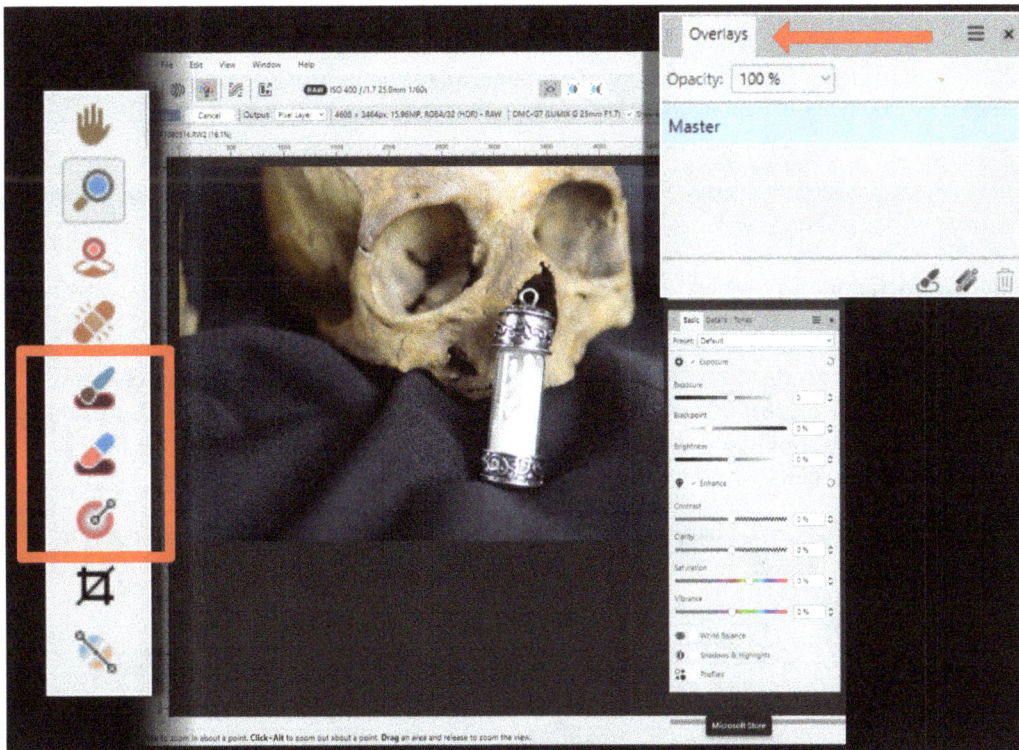

Figure 17.3 – Tools and panels used in the Develop persona

These tools are not in other personas but are simple to grasp.

The following tools are used for overlays:

- **Overlay brush tool**: Used to apply an overlay to the image

- **Overlay erase tool**: Removes an area you previously applied the overlay to

- **Overlay gradient tool**: Applies a gradient-style overlay (working on the same principles as the Gradient tool, which we discussed at length regarding the Photo persona)

So, now that we know what tools are available and where they can be found, let's look at what an overlay tries to do.

### What does an overlay try to do?

An overlay tells Affinity "*I want to apply all of the adjustments I am about to make to "this area".*" To make this work, there is a new panel in the studio for the Develop persona called the **Overlays** panel (shown in *Figure 17.3*). By default, the **Overlays** panel has one overlay in it called **Master**, which means it applies adjustments to the entire image.

> **Important note**
>
> Not all adjustments can apply to all overlays. Some adjustments are only applicable to the **Master** overlay (we will demonstrate this later).

So, before we apply our first overlay, I want you to reaffirm your understanding that, when it comes to overlays, we are making saved selections of the Develop persona.

### Applying a brush to an overlay

In this section, we will be working with the RAW image we have been using so far. I have also recorded a CiA video, the link to which is available in the *Technical requirements* section of this chapter for you to follow along. Note that this will not be a good edit and will simply be a way to illustrate the application of overlays. This means that the colors will be off and the brightness will be off, but hey… this I not meant to be a finished edit.

Using the Overlay Brush tool, I have painted the background of the image, as shown in *Figure 17.4* (see the red area denoted with an arrow):

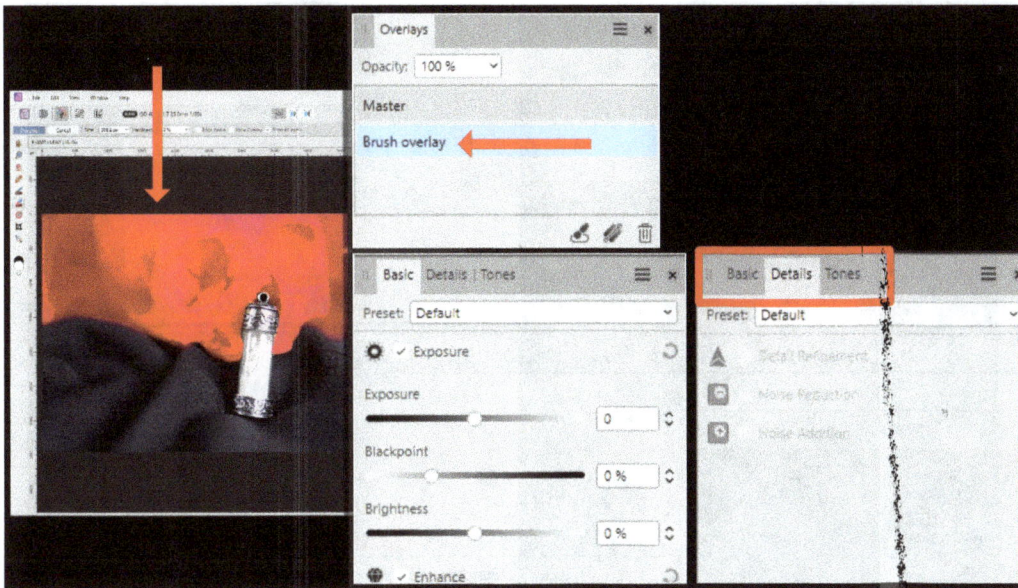

Figure 17.4 – Painting out the background overlay

Doing this will create a new overlay in the **Overlays** tab called **Brush overlay**. In *Figure 17.4*, notice that, below the **Overlays** panel, there are two panels. **Basic** adjustments can be applied to **Brush overlay**, while the **Details** tab cannot be adjusted on any overlay except the **Master** overlay.

I can only assume this is a limitation of Affinity and that it cannot apply adjustments such as detail refinement to just part of the image. It is also worth noting that we cannot save a file at this point for later distribution – we *must* go all the way through the edit and develop it before we can save it.

To fix an over selection, simply use the Overlay Eraser brush to remove areas of red.

In *Figure 17.5*, I have decreased the brightness of the image for the area under the brush overlay I applied to give the foreground more prominence:

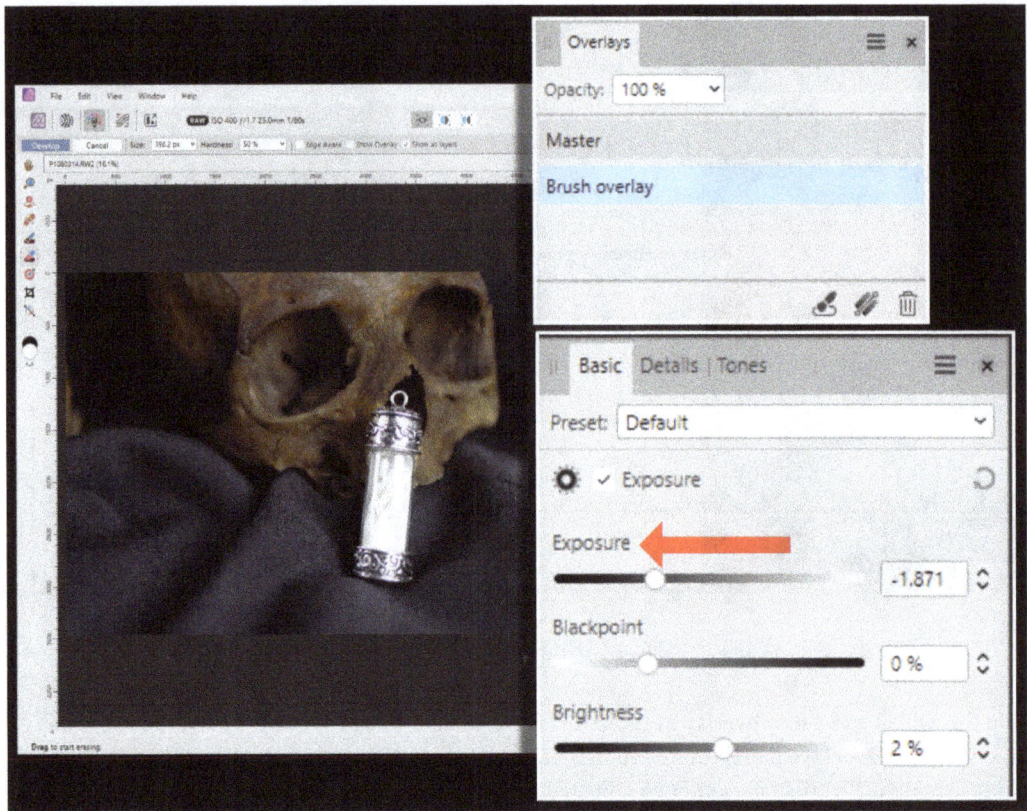

Figure 17.5 – Adjusting the image using reduced brightness in the overlay

This is the basis of overlays – they are thought of as saved selections because you can apply another adjustment to the area we brushed over.

**Applying a gradient overlay**

The process of applying a gradient overlay is the same as using a brush. In *Figure 17.6*, I added a **Gradient overlay** to the image by clicking from the center and dragging outward. Notice the use of an **Elliptical** gradient type:

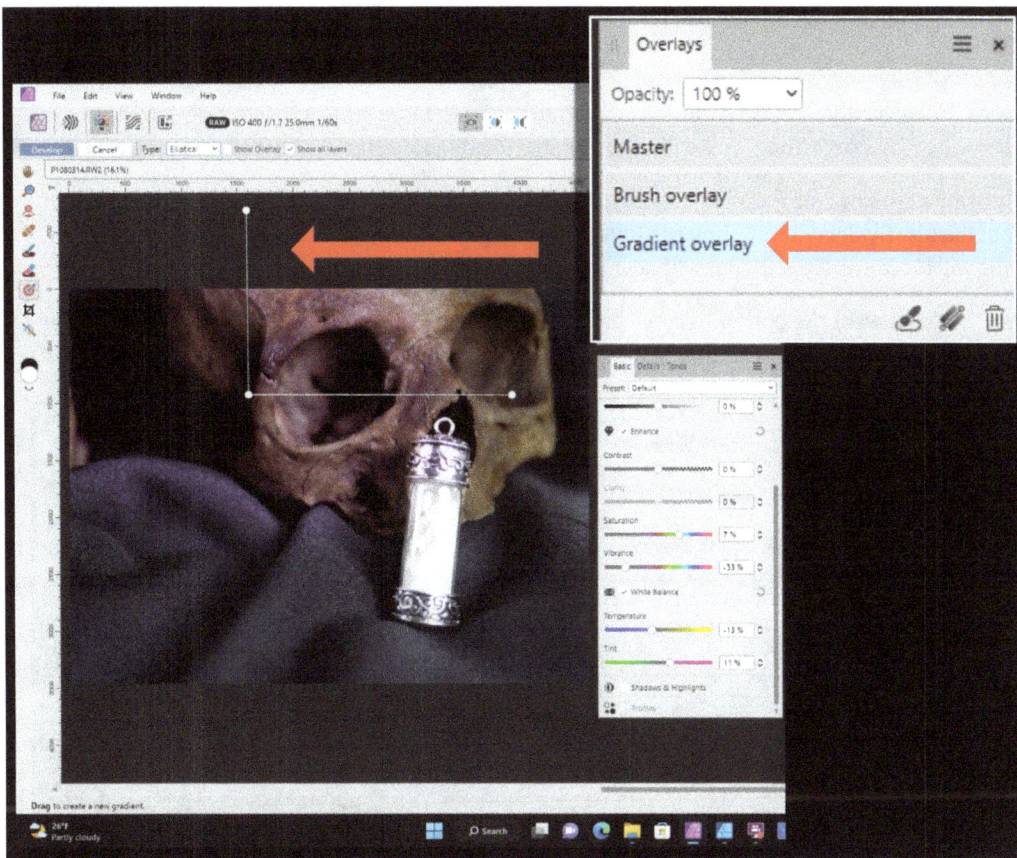

Figure 17.6 – Gradient overlay application

I have also added a CiA video for this, which you can find in the technical resources for this chapter.

I typically use gradient linear overlays to make up for areas of light that may be more prominent on one side when I shot, so I need a stronger edit on the edge and then make it lighter as we go from left to right. Gradient overlays form a majority of my overlay work.

Once you understand overlays and the adjustments preset in the raw persona, the only thing left to do is learn how to switch back and forth between the Develop persona and the Photo persona. This is a common thing and in Affinity V2, this is possible. Note that this is only possible in V2; previous versions do not have this functionality. In V1, once you have developed the image, you can never go back.

## Switching back and forth between raw and developed images

In V2 of Affinity, you can switch back and forth between RAW and developed images in the Photo persona. However, this feature is clunky and requires you to make some choices early in the process. So, if you get this part wrong, you will never be able to go back to the RAW version of your file.

In the context toolbar for the **View** tool (the hand), there is a section that asks how you would like to output the file. By default, the choice is **Pixel Layer**. If you choose **Pixel Layer**, then it's game over – the file will be converted and that will be it.

However, if you want the option to come back and edit again in RAW, there are drop-down options for different outputs, as shown in *Figure 17.7*:

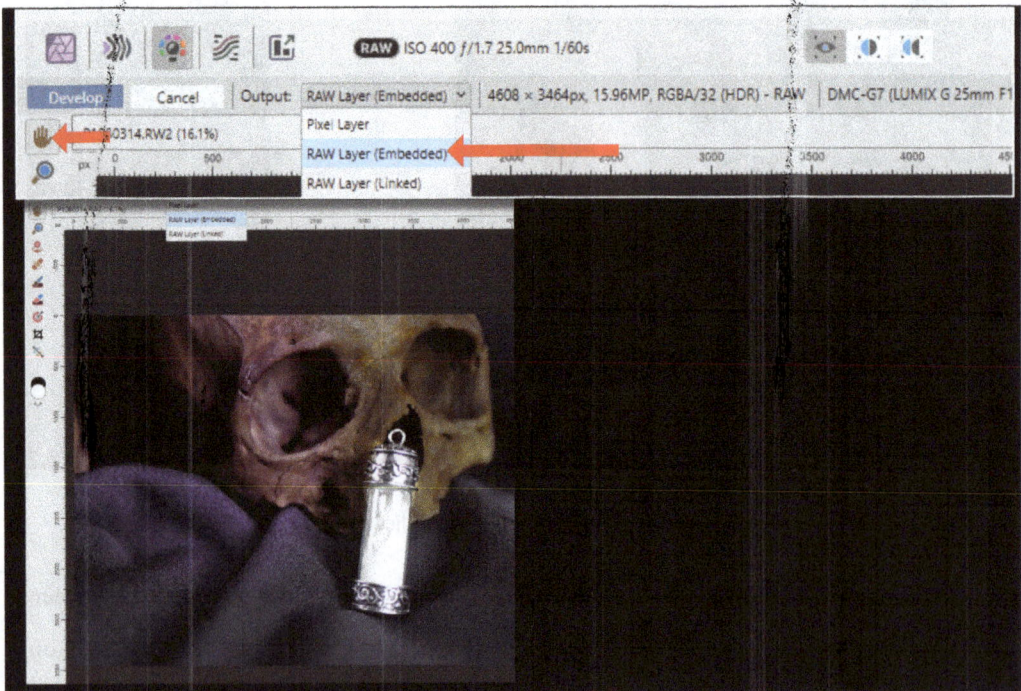

Figure 17.7 – Options for the output of the RAW file

Let's see how these options work:

- **RAW Layer (Embedded)**: This option takes the RAW layer and embeds it in the file so that the RAW layer travels with your developed (Photo persona) file – this is my preferred method. However, this makes the file size quite a bit larger.

- **RAW Layer (Linked)**: This option links the RAW file to a location on your machine. If you change the location of the file after setting this, you will have an issue where Affinity will not be able to find the original file; then, it is game over. This is not my preference, and I can only see this being useful in a situation where you have a very rigid photo storage structure and it is well defined.

In this case, I am going to select **RAW Layer (Embedded)** and then develop.

In *Figure 17.8*, we can see the image in the Photo persona. Here, I added an **Unsharp Mask** and a **Selective Color Adjustment** (this is *not* a good edit; I needed a bold adjustment to show you the existence of the photo edits when I switch back to RAW. I know – it is bold and awful):

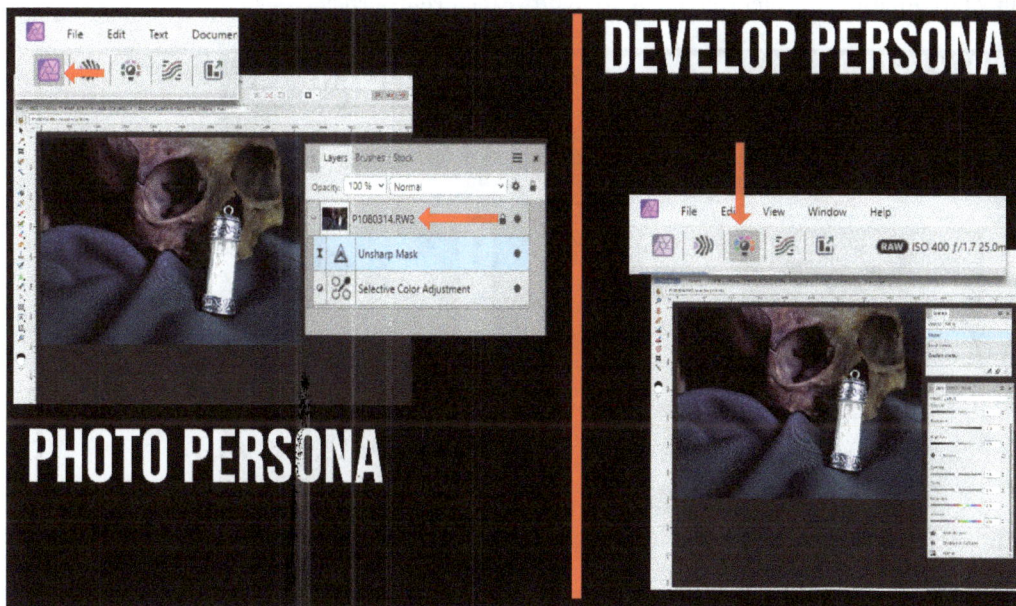

Figure 17.8 – File adjusted in the Photo persona and then reintroduced in RAW format

To switch back to the Develop persona, make sure you are on the image layer, and then click on the Develop persona. The program will carry over your adjustments so that you can work in RAW again.

While for some photographers the ability to work in RAW is essential and is a desired functionality, for me, as an editor working in the style that I do, it has never really been a focus, and outside of adjusting the first image to make up for gross issues, I have little use for it. Adding to that, the idea of RAW development is not Affinity's strength, and there are far more qualified people and better software to handle the world of RAW. So, we will leave it here with the basics and move on to the Liquify persona.

# The Liquify persona

While I am giving the Liquify persona its own section, it will be a short one that will explain the history of the persona. The Liquify persona is a holdover from the older versions of Affinity, before the invention of the live filter layer for Liquify (see *Chapter 11* in this book to see how to use the live filter layer for Liquify).

In the days before this live filter, you would have to take an image from the Photo persona and move it into the Liquify persona, make your adjustments, and then commit them permanently to the image. After the live filter was introduced, there is no longer a reason for this persona to be used. However, for completeness, I have created a side-by-side screenshot of the Liquify persona (on the left) and the Liquify Live Adjustment Filter (on the right) to prove their equivalency:

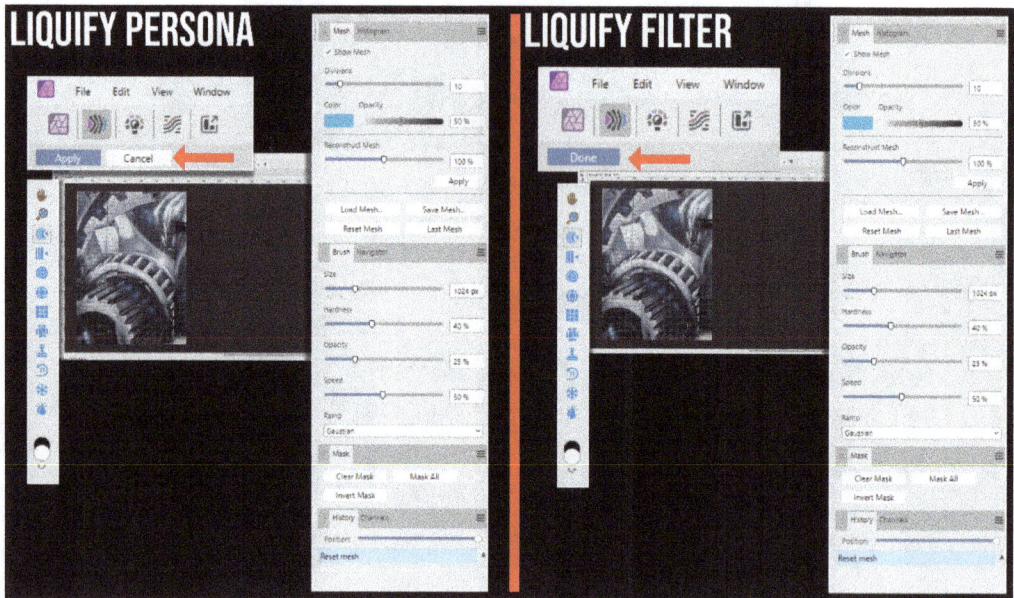

Figure 17.9 – Persona versus Live Filter for Liquify

Notice that, on the left-hand side – that is, the persona side – there's an activation button (notice the arrow showing the **Apply** button on the left-hand side; the right-hand side shows the **Done** button). This designation speaks to the nonpermanent nature of the layer as opposed to the persona.

So, while I have included a section on the Liquify persona, after the invention of the live filter, there is no real reason to use this legacy feature.

Now, let's move on to our next and final persona.

# The Tone Mapping persona

Now, it is time for my all-time favorite persona: the Tone Mapping persona. This is the artistic part, and it is an integral part of my workflow. This persona is the closest thing to being able to fake some of the wild Photoshop filters and effects in Affinity Photo, and I love mixing the images I tone map in with the original image to create some amazing work. So, I will discuss this section from a how-to approach and then show you one of the weird non-traditional ways I use this persona.

In all fairness, I use this persona very differently than some as I do not shoot and specialize in HDR. So, I will show you how it can be used, but I ask you to remain open to its more artistic use in everyday editing.

So, you may be wondering what it does and how to use it. We'll discuss that in this section.

## What does the Tone Mapping persona do?

Previously in this book, we had a deep discussion about color (remember *Chapter 13,* on advanced color). In that discussion, we covered that there were different bit depths of color (8-bit, 16-bit, and 32-bit) and that this bit depth equated to how many colors the image had (the more bit count, the more variations of colors the image had). That chapter summarized that images with greater bit depth create HDR images and that for some photographers, HDR is a specialty and a primary part of their work.

Tone Mapping seeks to take this crazy amount of color and compress it into a viewable digestible image that can be realistically printed with a high degree of quality but with some contrast and good overall output. So, Tone Mapping is an integral part of a photographer's workflow if they exist in this space. You must tone map the image to be printable because your average printer cannot print 12,000 shades of red, even though your camera can capture it.

However, as we are not doing HDR photography, and I wanted this book to apply to those even shooting on a mobile phone, I am going to give you the artistic application of the tool. If you do HDR photography, chances are you will take specialty classes from others in your field, so in this book, I am going to give you the everyday application.

## Working with the Tone Mapping persona

The Tone Mapping persona can be found in the top left-hand corner of the interface (see *Figure 17.10*). I have included the image from Figure 17.10 in the downloads for this so that you can follow along (see `Tone mapping practice image`).

To switch a photo into the Tone Mapping persona, it must be on a single layer. So, for this example, we only have a single layer and we have called it **Background**:

Figure 17.10 – Location of the Tone Mapping persona

The Tone Mapping persona is destructive, so whatever you apply inside this persona will be there when you apply the changes; it is not reversible.

Now that we know what the Tone Mapping persona is, we'll look at how we can work with it in the next section.

### Reading the Tone Mapping persona

In *Figure 17.11*, I have added the **Background** image. Here, you will see the typical things we see in other personas:

Figure 17.11 – Parts of the Tone Mapping interface

The preceding figure shows the major areas of the Tone Mapping layout. As you can see, it is very similar to the interface of the Photo persona:

- **Tools**: These can be found along the left-hand side. We will cover them in depth in the next section.

- **Right studio**: Similar to the Photo persona, there is a right-hand studio.

- **Left studio**: When using Tone Mapping, my preference is to have the left studio of **Presets** out. If this is not your preference, you can hide the left studio by going to **Window | Studio | Hide Left**. If you do not see the left studio, go to the same place and click on **Left Studio** to bring it up (this window option is shown in a box in *Figure 17.11*).

Next, let's look at the different tools that can be found under this persona.

## The tools

The following figure shows the tools that are available for the Tone Mapping persona:

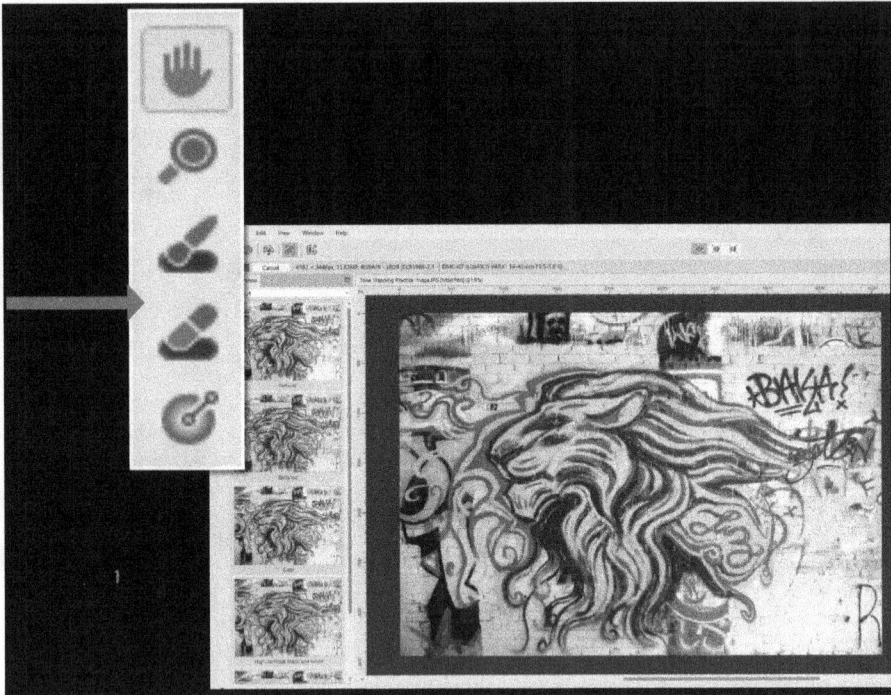

Figure 17.12 – Location and icon for the overlay tools

Some tools are different in the Tone Mapping persona than in the Photo persona:

- **Overlay paint tool**: You can use this tool to paint on the overlay for the area you want to affect with the adjustments you make.

- **Overlay erase tool**: Removes areas of the overlay that you want to change

- **Overlay gradient tool**: Applies a gradient overlay, exactly like the gradient tool we used in the Photo persona. Just click and drag to apply gradients and follow the same process to add and subtract nodes to/from the gradient line.

Now that we know the tools and what they do, let's explore the studio panels present in the Tone Mapping persona.

## The studio panels

In my professional opinion, the three most important panels are as follows:

- The **Overlays** panel
- The **Tone Map** panel

- The **Presets** panel (notice that mine is in the left studio in my layout):

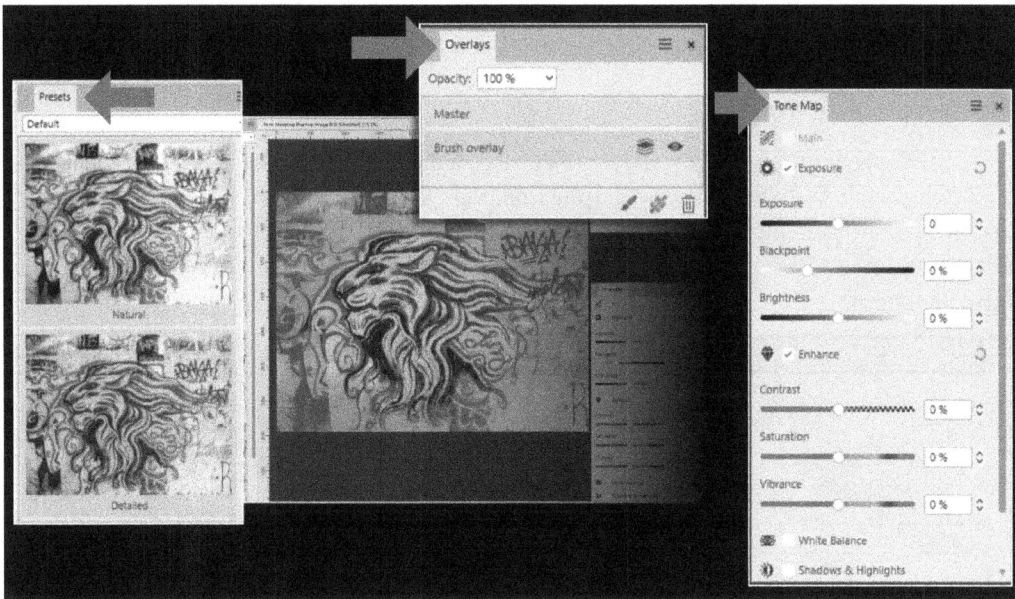

Figure 17.13 – Panel location

So, these three panels are the ones we will cover in the next few subsections. I have never used the **Scopes** panel, the **32-Bit Preview** panel, or the **History** panel, so if these interest you, you can dig into those personally.

## The Overlays panel

An overlay tells Affinity "*I want to apply all of the adjustments I am about to make to "this area".*" To make this work, there is a new panel in the studio for the Tone Mapping persona called the **Overlays** panel (shown in *Figure 17.13*). By default, the **Overlays** panel has one overlay in it called **Master**, meaning it applies adjustments to the entire image (see the **Master** overlay in *Figure 17.13*).

So, before we apply our first overlay, I want you to reaffirm your understanding that when it comes to overlays, we are making saved selections of the Tone Mapping persona.

### Applying a brush to an overlay

We will be working with the lion image I have been using to explain working with overlays for this example. The image can be found in the downloads for this section and is called `Tone Mapping Practice image`. Using the Overlay brush tool, I have painted the background of the image in *Figure 17.13* (see the red area in the **View** window). This has created a new overlay in the **Overlays** tab called **Brush overlay**. Notice in *Figure 17.13* that, below the **Overlays** panel, there are two overlays.

It is worth noting that the main adjustments can only be applied in the **Master** overlay; in any other overlay, this option is grayed out and can't be selected.

To fix an over selection, simply use the Overlay eraser brush to remove areas of red.

**Applying a gradient overlay**

The process of applying a gradient overlay is the same as in the Photo persona tool. In *Figure 17.14*, I added a gradient overlay to the image by clicking from right to left. I have changed the white balance to blue on the right-hand picture of *Figure 17.14* to show the application of the tool in an adjustment (this is not designed to be an edit, but simply a way to show you how the adjustment maps to the overlay you created):

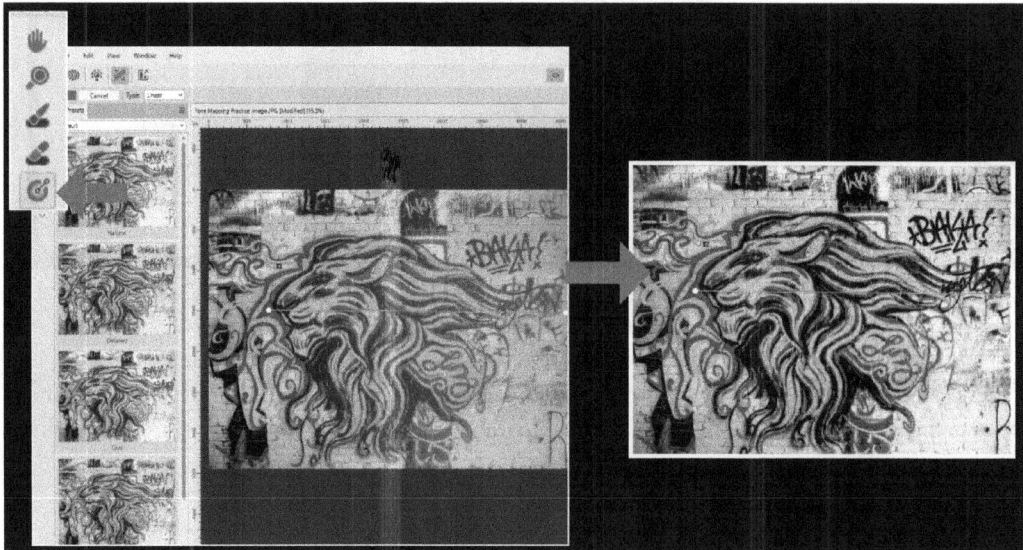

Figure 17.14 – Application of a Gradient overlay

I do not use overlays in edits. For me and the way I work with Tone Mapping, I apply the tone map to the entire image. I have never once applied only part of an adjustment to an image. However, I figured that as this would be a part of some of your workflows, I would include it. Now, let's look at some more interesting stuff, including how to utilize the Tone Mapping persona during an edit.

## The Tone Map panel

The **Tone Map** panel is one panel in the Tone Mapping persona and contains many of the same adjustments available in the Photo persona. You can access these adjustments by clicking on the larger category, which opens all the adjustments present in that category (see *Figure 17.15* for the structure of the panel):

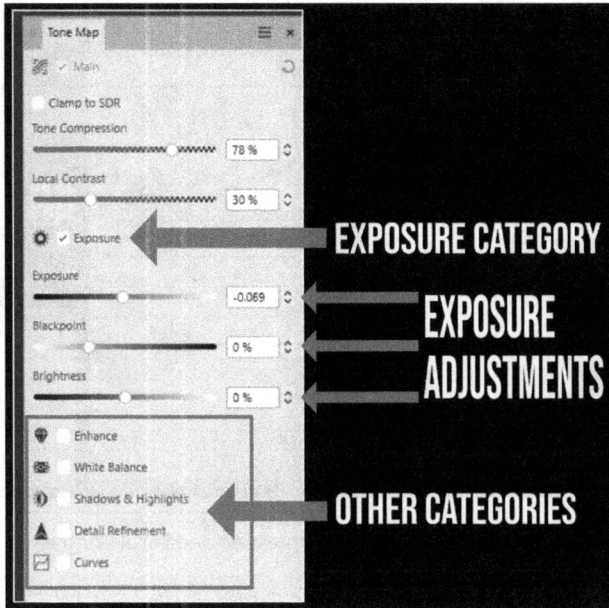

Figure 17.15 – Structure of the Tone Map panel

We won't be covering the area of main adjustment (see *Figure 17.15* for its location). Instead, we will be looking at the **Tone Compression** and **Local Contrast** options because these do not exist in other areas of Affinity Photo.

## Tone Compression

**Tone Compression** tries to set the correct tone across the image. In short, it tries to form a base image for all the other edits you plan on making – think of it as level setting the image so that you can build on the foundation. The effect of higher tone compression is a brighter image, with more saturation. We can observe this in the slider; as we move toward **100%** compression, we get a lighter, more saturated image, whereas as we approach **0%**, we get a darker, less saturated image.

I set **Tone Compression** as one of my first adjustments to get my base right.

## Local Contrast

**Local Contrast** increases or decreases contrast. I chose this image for the Tone Mapping section for the brick. If we increase **Local Contrast**, the washed-out pieces of the brick in the lower left become very visible and give the piece an entirely new look:

Figure 17.16 – Adjusting Tonal Compression and Local Contrast

In *Figure 17.6*, I have adjusted **Tonal Compression**, as well as **Local Contrast**. Notice the difference in the amount of detail present in the brick using just these simple adjustments. This is what makes Tone Mapping so powerful.

Next, let's look at how you can save the edits you make through the power of presets.

### The Presets panel

Presets are preloaded adjustments that you can use, create, or import to add pre-selected looks to your artwork. Out of the box, most of them will look goofy and extreme. You will have to adjust each preset once you apply it using the **Tone Map** panel as there is no way they will ever know what type of image you plan to apply them to. Lastly, repeating a term I brought up earlier – I like to play *Preset roulette* because I may miss one, so I like to see what they all look like before committing.

### Applying a preset

To apply a preset, simply choose a category and then click on the preset of your choice. The preset will be temporarily applied to the image, giving you a preview of it. You can adjust it from there using the **Tone Map** panel. If you change your mind and want to try another one, simply click on another preset and it will update. If you decide you do not want to tone map anything, simply hit **Cancel** – you will be brought back to the Photo persona.

To commit the preset once you are happy with it, simply click on **Apply** in the top-left corner.

In *Figure 17.17*, I have chosen a high-contrast black-and-white preset. This is a standard part of my black-and-white workflow and is key to the contrast technique, which I am going to show you later in the application portion:

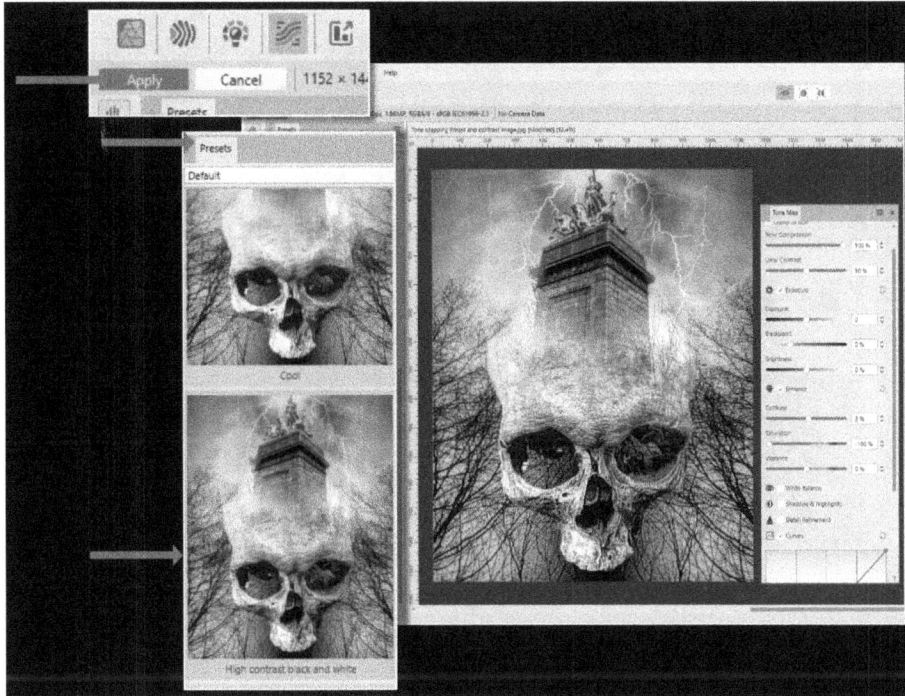

Figure 17.17 – Preset location and the Apply button

From here, you can adjust the baseline before returning to the Photo persona.

Next, let's see how we can develop our own presets.

### Developing your own presets

As you develop your adjustments, it may be beneficial to save presets, perhaps because you either shoot in a particular way where you apply the adjustment frequently, or you have that signature style you like. Either way, presets can be saved and shared using the menu in the top-right corner of the studio tab (see *Figure 17.18*).

If you would like to save your preset in the existing category (**Default**, **Crazy**, and so on), simply click on the menu in the top right and choose **Add Preset**. You can also rename it by choosing **Rename Category…**. To delete an individual preset you saved, right-click on the individual preset and choose **Delete Preset**:

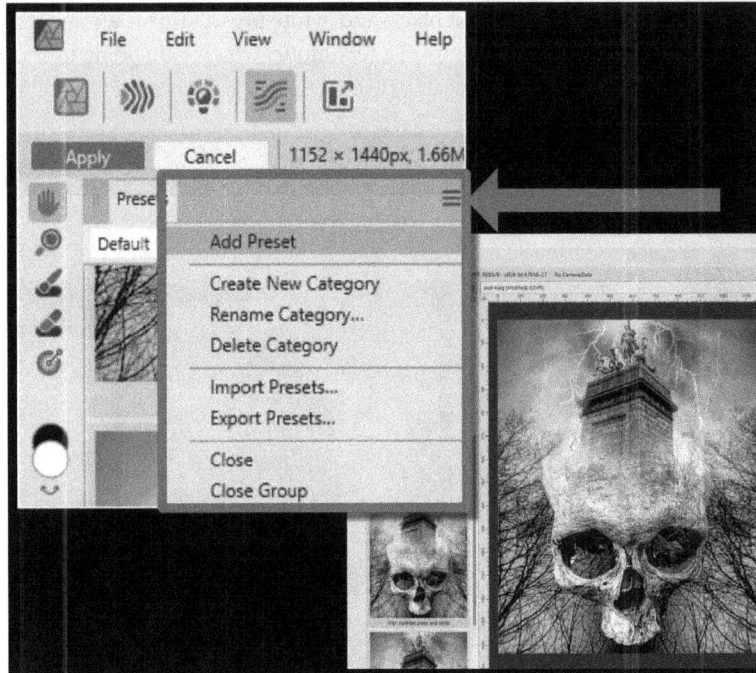

Figure 17.18 – Saving presets

To make a category for a group of your own presets, simply click on the menu bar and **Create New Category**.

To import or export preset categories to share with others, simply select the option to **Import Presets…** or **Export Presets…** – Affinity will generate a file that you can share with other Affinity enthusiasts.

To remove the category, simply click on **Delete Category**.

Now that you know about Tone Mapping, the question becomes, where do I use it? This is a valid question. Each artist may have different uses for it, but I want to show you my process for utilization that you can use as a basis for your methods as you grow as an editor.

## Applying Tone Mapping

In this section, we will cover the two methods of applying Tone Mapping and utilize them to finish my work using the Tone Mapping persona. I know that some of you out there will be using it as an incremental step in your work, but for me, these are the two most frequent uses in my workflow – that is, finishing various versions of an image, flattening it, and changing blend modes to add contrast and interest.

## Method 1 – finishing multiple versions of an image

One way I use Tone Mapping is as a finishing tool to explore different color schemes and feels for my work. So, after I am happy with the image, I export a flat version of it, such as in `.jpg` format, and then try various looks. In *Figure 17.19*, I have taken the same base image and generated a few variations:

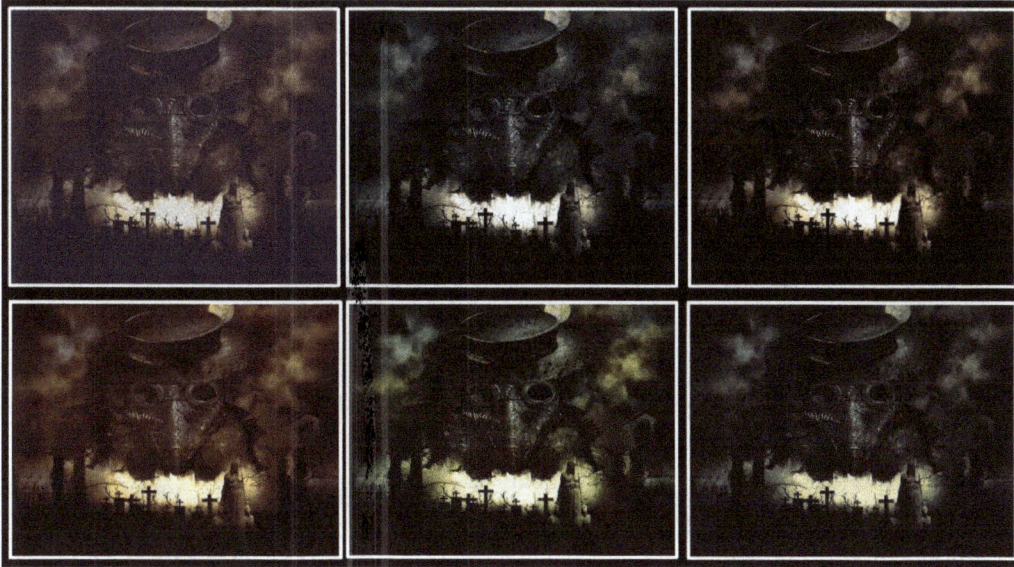

Figure 17.19 – Different Tone Mapping effects on the same .jpg

This gives me a very creative look at what I may not have thought about. Presets should be used to kick off your creative process relating to finishing as well. If I have a good image that I want to add to my portfolio, I always run a few tone-mapped adjustments to see if I left something creative on the table.

## Method 2 – kicking up the contrast

The second way I use Tone Mapping is to kick up the contrast and rock the color scheme for my edits. I create a high contrast, tone-mapped copy of the image and then apply it with various blend modes to up the contrast and color in my image in a very unusual way; I would like to share this with you in this chapter. I have included a CiA video in the *Technical requirements* section of this chapter so that you can see the process.

I did not include a working file for this section as it would not open in V1, but there is an application project that we'll look at that uses this technique. Let's get started with the steps:

1.  Work with your image until you feel you are comfortable with it. In *Figure 17.20*, I have taken this image to a point where I am ready to apply the aforementioned technique.

2. To get a flat copy of your image during a project, right-click on the **Layers** panel and select **Merge Visible** (see the left-hand side of *Figure 17.20*). This creates a new layer called **Pixel**, which is shown on the right-hand side of *Figure 17.20*:

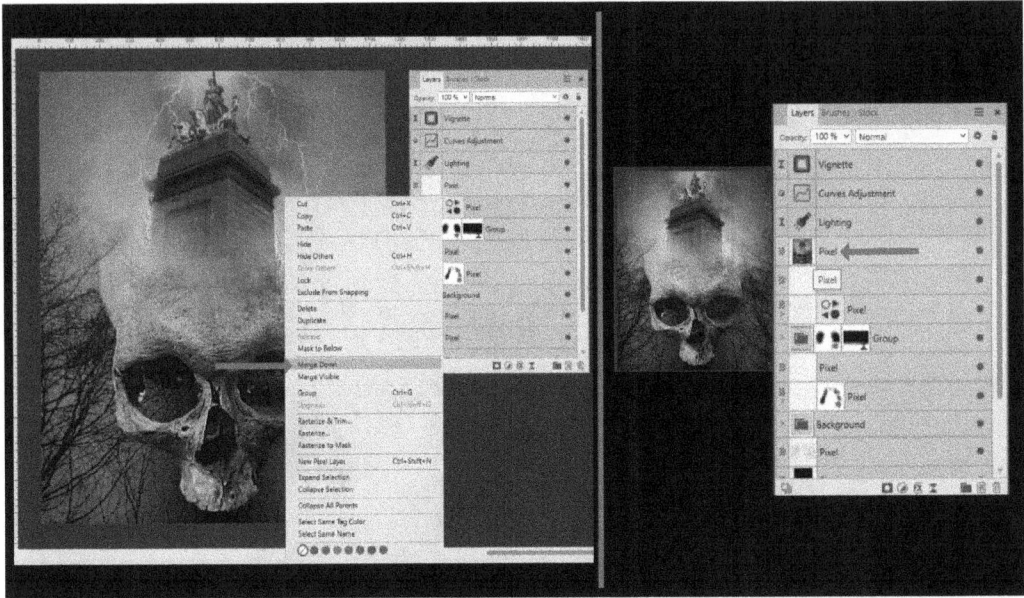

Figure 17.20 – Merging down for contrast

3. Now, move this flat layer into the tone map and adjust it to your taste (notice that I am using a local contrast preset).

4. Hit **Apply**. This will take you to the Photo persona. I have labeled this Tone Mapping layer to help you see where it exists.

5. Adjust the blend mode. This can be seen on the right-hand side of *Figure 17.21*.

6. Turn down the **Opacity** setting:

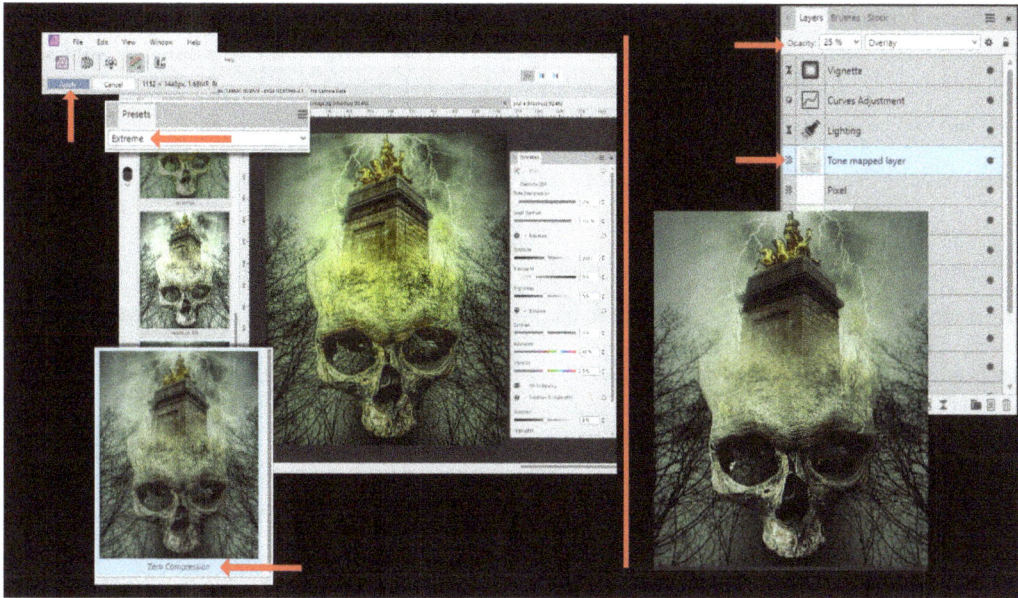

Figure 17.21 – Tone mapping and applying it

Look at the crazy amount of contrast we were able to generate and some of the bolder color values! This is known as a Monte Carlo simulation and is the unknown and exciting part of this process. It gives you a very unique finishing signature look that not many people think to apply. Take a look at the before and after pictures:

Figure 17.22 – Before and after

To replicate this (as I know we have V1 and V2 users, I have included a flattened image of the project (see the `Tone mapping preset and contrast image` file in the downloads). This flat layer represents the finished state of the project, so you can pick it up from here. Follow these steps to practice this technique:

1.   Open the file mentioned in the previous paragraph in the Photo persona and duplicate it.

2.   Select the duplicate layer and open the Tone Mapping persona.

3.   Choose the **Extreme – Zero compression** preset, turn up the saturation, and apply it.

4.   Back in the Photo persona, change the blend mode of the tone-mapped layer to **Overlay** and set **Opacity** to **25%**.

This will get you the same effect and allow you to see the benefit of using Tone Mapping as a way to increase contrast and extreme values in your edits for maximum creativity.

## Professional tips, tricks, and important points

When working with RAW, I use this early in my workflow and rarely, if ever, use overlays. Once I switch to the Photo persona, I assume I am never coming back to RAW. Whether you like it or not, I shoot RAW whenever I get the opportunity as there is so much more data to work with for a dynamic piece. Even your cell phone these days has a RAW capability.

Tone mapping is my secret to higher contrast extreme images – just make sure you merge down and then adjust the copied image. The preset will give you an idea of what is possible, but do not be afraid to try adjusting the blend mode, blend ranges, and opacities.

By combining your tone-mapped layers with masks to apply part of an image to focus areas, your viewer can have an entirely new experience with your piece and add visual interest.

## Summary

In this chapter, we focused on two distinct areas that we looked at in detail. You now know how to edit in RAW with Affinity Photo, along with why and how RAW works and what it is trying to do. By far, though, the work we did with the Tone Mapping persona as a creative tool was my favorite. The wild colors, the pushing of boundaries and limits, and the subtle dance of blend mode, opacity, and masking can make your image impossible to replicate and give you that signature style (if you haven't figured it out by now, mine is more extreme contrast and vivid colors).

With that, we have come to the last chapter in our beginner's journey, which involved exporting and then rounding home with some artistic techniques to get you out of the beginner stage and get you working like a professional.

# 18

# Exporting and Artist Efficiency Tips

As we close this book and complete the journey, there are a few more things that you may want to know about as a photo editor, and staying true to the idea of workflow, we have saved these for last (as we first had to show you how to create an image). In this chapter, we will cover the last step of the workflow, which is the idea of exporting. While we have been creating art all along the way, we need to be able to get that art out of Affinity and into a format that can be shared with the rest of the world. **Exporting** is the technical term for packaging this all up and presenting it in a way that the websites and printers understand.

Also in this chapter, we will be covering the panels and tools that I commonly refer to as **artist efficiency tips**. These are the panels that assist the editor in speeding up the workflow and meeting deadlines but, by themselves, do not necessarily have the utility of other panels we have looked at along the way. These panels are a great way to save the tools and images you have created during your learning and editing journey. As you mature as an editor, they will serve as a source of artistic flourishes that make your work more interesting and allow you to create faster.

So, let's begin the last step in our journey and open the door to the next chapter of your career as a proficient, confident photo editor with Affinity Photo. In this chapter, we will be covering the following topics:

- Fundamentals of exporting
- The Export persona
- Artist efficiency techniques
- Professional tips, tricks, and important points

## Technical requirements

The CiA video of this chapter can be found at `https://packt.link/6JQzu`.

You will also need the `Chapter 18` project to follow along with me.

# Fundamentals of exporting

As previously mentioned, exporting is the act of taking an `.AFPHOTO` file format you have created and converting it into a format that the rest of the world can read outside of Affinity Photo. Popular formats for exporting include `.jpg` and `.png` (more on the individual formats later). Choosing the right format is important, as each method takes the data that is present in the `.AFPHOTO` file and makes it readable in a different way. Some will compress the data and you could lose picture quality, and some maintain the image quality, but the file size is too large. So, in this section, we are going to look at the quickest way to export using the **File** menu and we will explore the different formats.

## Exporting in Affinity Photo

In the following figure, you will find the simple export function; this simple **Export…** option allows you to change the `.AFPHOTO` project you are working on into a format such as PNG or JPEG:

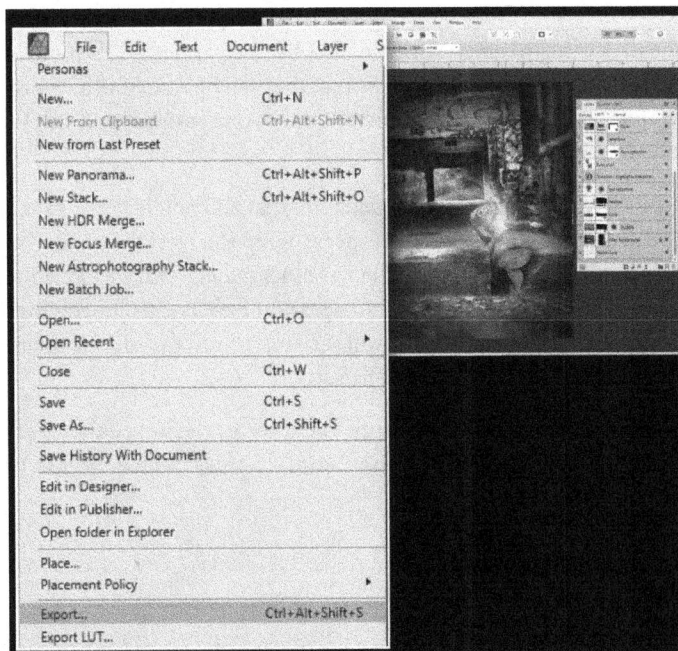

Figure 18.1 – Export menu location and format screen

To assist you in working with the export function, I have included a working file in the downloads for this section, (there is a V1 as well as a V2 file present, so even if you are following this in an older format, I've got you covered). We will, however, be working in V2 for the examples.

Exporting in Affinity Photo is very simple and is accomplished through the **File** menu. Simply select the **File** menu and then go down to **Export…**. The **Export** dialog page will come up and you will need to choose a format (see *Figure 18.1*). The various formats for exporting all treat the data differently, so it is important that we choose the *right* format for our intended application, but it really comes down to one very important concept and a question you have to answer:

*What is your end use for the image?*

This one question will largely define the format you are looking at using as a practical example, and the following scenarios will help illustrate this concept:

- If you are going to use the image on your website and you use a .png image, which is typically larger in file size, you will have longer loading times, as your web page needs more time to open these larger files. This is not an ideal situation for a website to have. Therefore .jpg may be a better choice.

- If you are going to share the file with another artist who does not have Affinity Photo and uses Photoshop, then the PSD format would be your choice.

These are some typical considerations and scenarios when you think about exporting the image. Now that we know what exporting is and where to find the different export options, let's look at the *how* portion of the equation.

## Reading the export window

This is what the export window looks like:

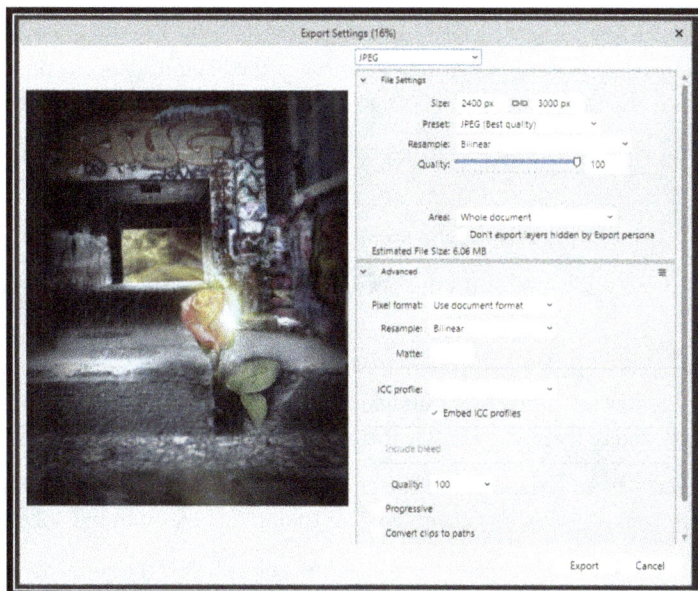

Figure 18.2 – The Export Settings window

In *Figure 18.2*, we see the **Export Settings** window, and I must admit that in my work, I use very few portions of this window. As you mature as an editor and find your formats, your style, and your products, you may change these settings more and more often, but for me, there are very few portions I tend to use (I stick to `.png` or `.jpg` for my work). Let's take a look at the different options found in the **Export Settings** window:

- **File Settings**:

  - **Size**: Displays the size of the document that was chosen when you created the image, so if you want to change it, now is the time. In *Figure 18.2*, the size is **2400 x 3000**. So, if we want to change the size with the aspect ratio maintained, make sure that the chain link (between the height and width in the **Size** option) is not broken.

  - **Preset** – If there are various presets for that format (i.e., **JPEG (Best quality)**), you can set that here (I personally do not mess with this ever because the file sizes are not that big with JPEG).

  - **Resample** – If you changed the size, this dictates the resampling method that will be used. Personally, I have never changed this.

  - **Quality** – This is a big one. You can change the quality to reduce the file size, but be aware there is a quality degradation to do this.

  - **Area** – This is one I use a lot. You can choose the entire document, a selected area, or the selection. The graphics in this book were exported largely by choosing **Selection**.

  - **Estimated File Size** – As you adjust the variables and format, the file size flexes as well. This gives you an estimate of the file size for the final export.

- **Advanced**:

  This section pulls again from the settings we chose at the beginning of the project, so for the most part, you won't have to adjust it. Let's go over the options under **Advanced**:

  - **Pixel format** – If you would like to export as 16-bit versus 8-bit, then this is where you change it.

  - **Resample** – Dictates the method it uses for the preceding.

  - **Matte** – If there is room around the border of the image, this is where you change the matte color.

  - **ICC profile** – Many printers come with downloadable profiles. If you have loaded them, this is where they show up. A practical example of this is I used to do sublimation printing on an Epson printer, and so there was a profile that would convert my values to the Epson equivalent.

  - **Include bleed** – Bleed is the area outside of the printable area and it is typically used to post trim if there is an edge-to-edge coverage. This option includes it in the document.

  - **Palletized** – Encodes the image to the pallets and colors below the adjustment. I personally have never used this.

In the next section, we will look at various exporting formats.

## Exporting formats

The following figure shows all the export formats available in Affinity:

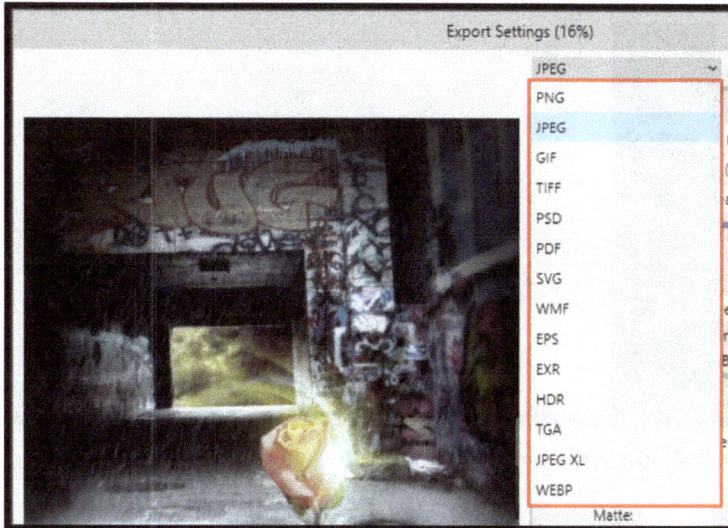

Figure 18.3 – Available formats

Each format has its own specific uses and, as previously mentioned, it is important that you understand the use of the finished image when you go to export. To go through every format here would make this chapter 100 pages long, and *most* of them you will never actually use. Once you dial in the type of work you typically do, that will largely dictate the type of exporting you need. Plus, hidden in the **Help** menu inside the Affinity software, there is a wonderful matrix written by the authors of the software. While we cannot reprint it here, I have included a screenshot of this table in *Figure 18.4*. It is not the intent of this figure to replace searching inside the **Help** tab, but rather to point out it exists *within* the software. To locate this screen, simply follow these steps:

1.  Move to the **Help** tab at the top of the screen.
2.  Type Export in the search bar.

    The export options will come up as a topic, and the following matrix will come up:

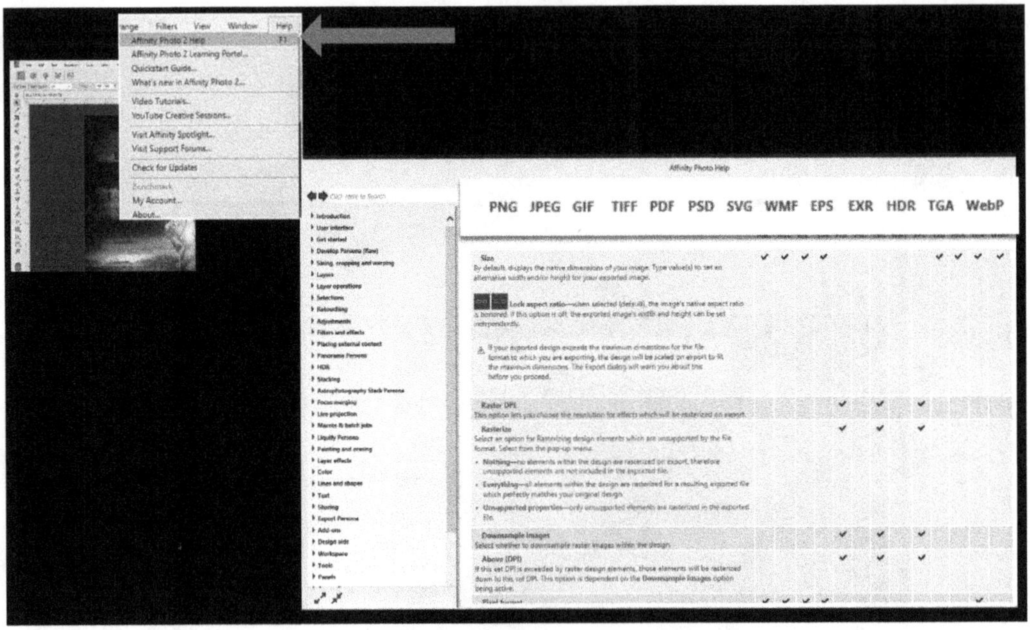

Figure 18.4 – Help function output in Affinity Photo for term "Export"

While we will not be covering every format in this list, I do want to share with you a general principle relating to exporting, and that is the concept of LOSS. For this discussion, we will use the .jpg format as an example. A .jpg format is an example of a LOSS style format because the way .jpg exports is through compression. It takes your image, the layers, and the colors and smashes them into one layer, and compresses the colors and adjustments to achieve the export.

The result is a small file size, desirable for websites that may have thousands of images, but a small file size comes from compressing the image down significantly. On the other hand, PNG is a LOSS format but achieves its compression differently and overall gives the same flat image, but with a higher quality level. For this printed book, we use PNG exported graphics to make sure the text and images are as sharp as they can be.

On a side note, the PNG file format also allows for transparency and is great for web work (file size is a bit larger) but JPG does not allow for transparency, so those logos and graphics will not look right in JPG if they have transparency. For me, PNG is my go-to unless file size is a critical part for this reason.

I have included a simple flowchart to help start and guide you on which format to use when; this will not be detailed (as you do need to know the application for the export) but it gives you an idea of where to start:

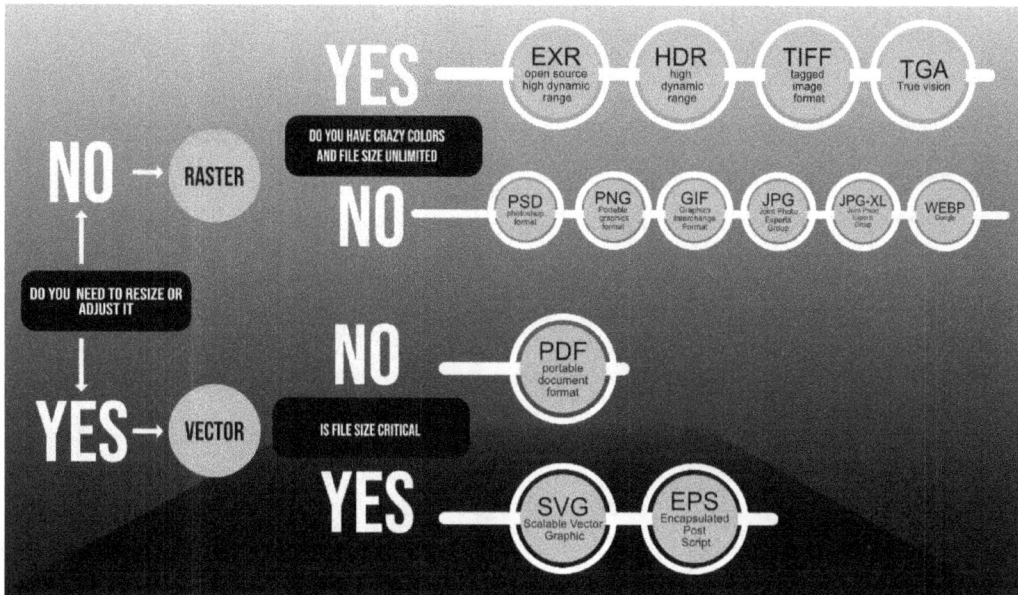

Figure 18.5 – Flowchart for formats

Now we know what exporting is and what formats exist. Next, we will turn our attention to the persona specifically created to assist in the exporting of documents.

## The Export persona

The Export persona is the final persona available in the Affinity Photo program, and it is in the upper-left corner of the interface. In order to understand this persona, we have to get a few terms down and an intent of what it is trying to do. This will make the mechanical portions of the program make sense:

- Intent (what is this persona trying to do?): This persona allows you to export multiple formats, sizes, and portions of your image all at once. For example, let's say you want to export the image in .png format at the original size, but you know that you will be placing an image preview on your website at 1500 x 1500 px. You can set up the Export persona to create both formats when you hit export.

- Terminology (slices): In the Export persona, we refer to the areas you want to export as **slices**, and a slice can be part of an image, such as a layer, or it can be a selected part of the entire image, such as the upper-left corner.

Now that we know some of the terms relating to the Export persona, let's look at the interface and see where these things occur.

## Exploring the interface

In the following figure, we can see the major portions of the Export persona:

Figure 18.6 – Interface of the Export person

Let's go over the options seen in the preceding figure:

- The Export persona option is in the upper-left corner
- The slice tool is shown in the left side tool area
- The three studio panels (pulled out for illustration):
  - **Layers** – Shows the available layers in the project (exactly transferred from the Photo persona)
  - **Slices** – Shows the slices you have created for export
  - **Export Options** – Allows you to set options and presets for the slices

To begin the exploration of the persona, we will begin with the simplest activity – that is, creating a slice.

## Creating a slice from the flat image

We will be using the working file placed inside the `Export Practice files` folder included in this chapter for the example, as it is a complex composite with many layers and gives us the opportunity to practice multiple techniques.

To begin, we see that we already have one slice in the **Slices** panel by default as soon as we open up the persona (see the **Slices** panel in *Figure 18.7*), and that is the complete image of `Export Practice - V2`.

Figure 18.7 – Exporting from a unified image

By default, the persona will include the entire image as the first slice.

### Selecting a slice

Now, if we wanted to create a slice from the part of our image (let's say we want to include only the rose, as we are making a square image preview for a website). We will take the **Slice** tool and click and drag over the area. Notice the things present in *Figure 18.7*:

- The size of the drag is present in the blue text on the image; in the example, it is **1273x1273** and is titled **slice 1**.

- In the **Slices** panel, there is a new slice created titled **slice 1**.

- If we twirl down the carat on the left-hand side of the slice, we can see the options for export. This is shown in the far right-hand corner of *Figure 18.7*.

Note, if you want to change the name of **slice 1**, we can just click on the slice name and rename it just like any other layer. I will be changing the name of this slice in the further examples to `preview`.

## Choosing the export file type

Notice, in the following figure, we see the dialog box for the slice called **Preview**:

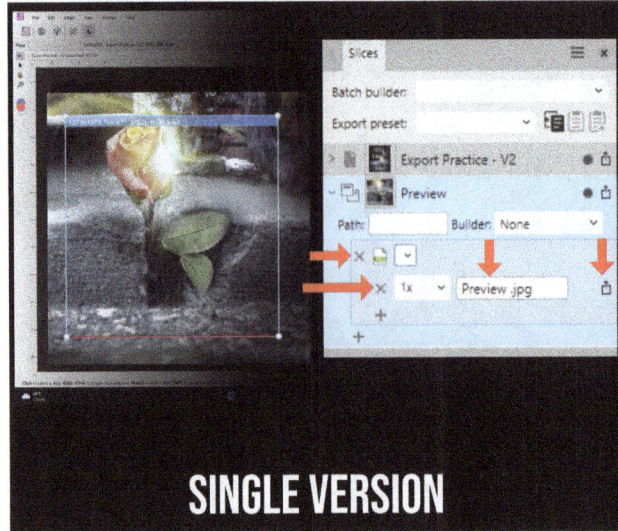

Figure 18.8 – Single version export options for a slice

Let's see what this dialog box contains. I have included red arrows highlighting important areas. We will match this left to right with the following bullet points:

- Preset selection: If you have set presets (covered in artist efficiency techniques), you can simply select a quick preset

- Multiplier: This magnifies the image, currently set to a **1x** size

- Name: Currently set to `Preview .jpg`

- Export single: This exports a single image, not all of them, so if you have multiple images, this will export the single version only.

Now, if this is what you want, you are all set, but what if you want to export one version at the regular size, and one at **3x** the size? For this, we have to add an export option.

## Adding the export file types

If you would like to export more than one version (size/file) of the slice, this is achievable in the Export persona as well. Simply click the + button in the lower left and another dialog will open for the same slice (see *Figure 18.9* for this second dialog box displayed). It will have the same options as the primary, but this is how you set up to export multiple versions of the same slice.

Notice in *Figure 18.9* that the name has changed to reflect the **3x** size:

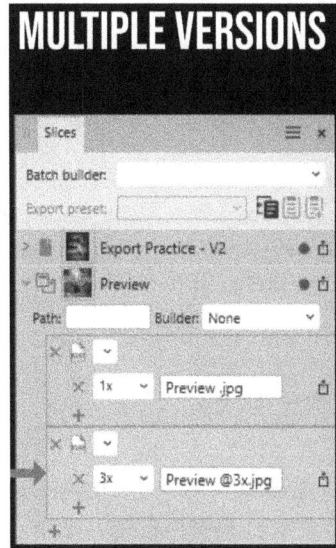

Figure 18.9 – Multiple version export options for a slice

## Exporting the slices

To export the slices, you have two options:

- You can export using just one option (say the **1x** size of the slice) by clicking on the export icon
- You can export using all the options for the image by clicking on the export icon on the slice layer (see *Figure 18.10*)

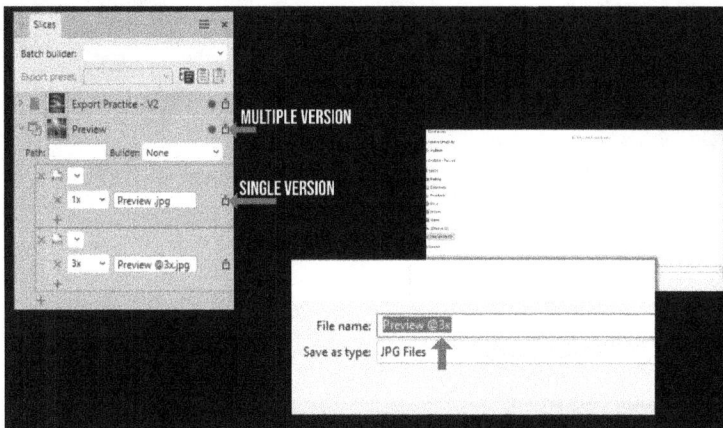

Figure 18.10 – Exporting location and options for the slice

Once you hit **Export**, Affinity will bring up a dialog box, and you will tell Affinity where you want to save the exported slices. It is at this step that you can rename the file as well and then simply hit **Save**.

Now that we know how to export a slice, let's look at how to export an entire layer.

## Creating a slice from a single layer

In the preceding examples, we dealt with the entire image, but sometimes you have a composition and you just want to save a single image from a single layer. The process for this is largely the same, with a few tweaks. We are going to continue with the `Export practice file` because it contains multiple layers and assets, making it the perfect candidate for this example.

For this example, the desire is to export only the rose layer, so to do this, we simply do the following:

1. Select the rose layer from the **Layers** panel.
2. Go to the bottom and select **Create Slice**.

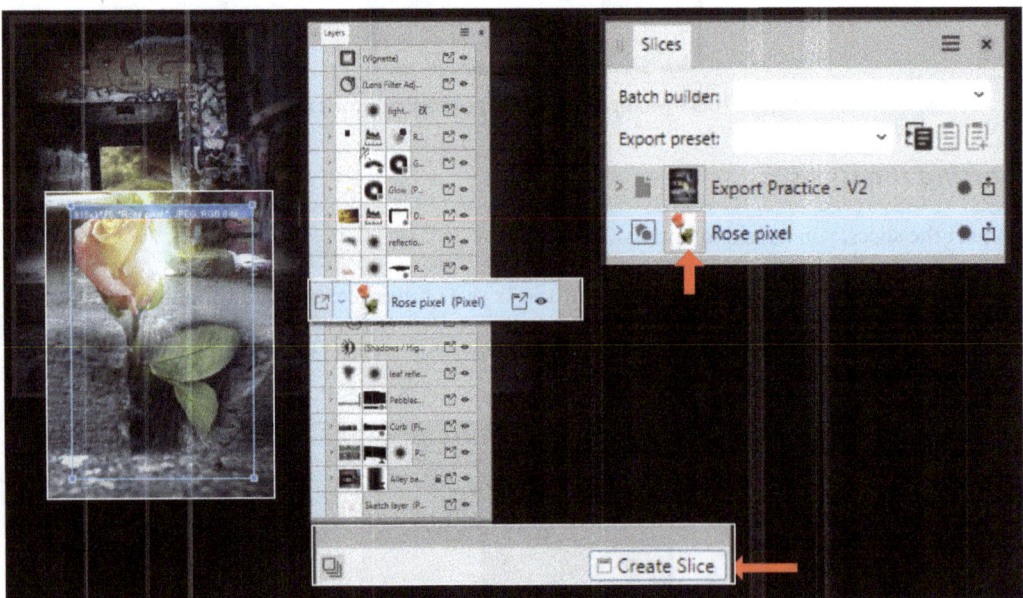

Figure 18.11 – Exporting a single layer as a slice

And then you will see the new rose layer in the **Slices** menu ready to export or add other versions.

Now, sometimes you want to save presets that Affinity may not have natively loaded; to do this, we can save presets, so let's look at how we do that.

## Saving a custom preset

Presets are saved and made in the **Export Options** Studio tab, and for this example, we will just be working on exporting the single layer. To follow along, simply click on the slice titled **Export practice - V2** (still using the same file we have been working with). As a reminder, this means we are just exporting the entire image, as shown.

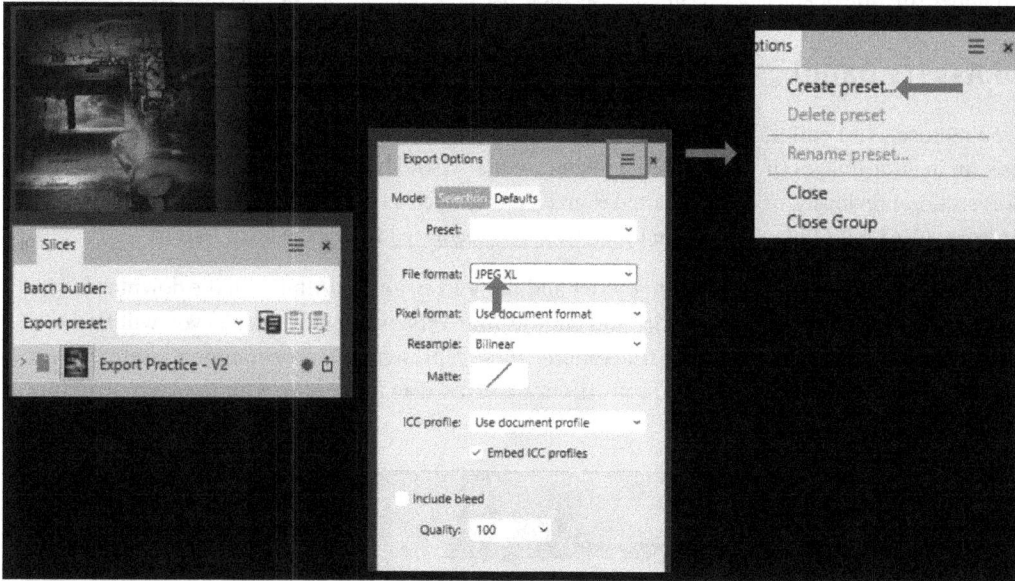

Figure 18.12 – Export options

When you click on the slice, the **Export Options** tab displays the options for this export (see *Figure 18.12* for the default). Now, if these work for you, great, but here is where you can change the settings based on your needs. In this case, I am going to make an adjustment to illustrate a change that can drive a preset.

All of the options for the export have been covered previously in this chapter, so we will not relist them here. For this example, however, I will be converting this export to the **JPEG XL** format (this format is only available in **V2**, so if you are following along in **V1**, simply change to another format to continue). See *Figure 18.12* for reference.

To create a preset, simply click on the menu in the upper-right corner of the panel and then select **Create preset…** (see *Figure 18.12*), and that is it.

We have officially covered the topic of exporting at a very basic and functional level, and as such, completed the workflow of a competent editor using Affinity Photo. In the next sections, we will deal with the efficiency upgrades of a professional editor.

# Artist efficiency techniques

As we wrap up this book, the panels discussed in this section are usually reserved for the artists that work with Affinity Photo on a regular basis, and we have chosen in this manual to save them for last, as it is essential to know the fundamentals prior to attempting to improve your workflow efficiency. In this section, we will be covering the finishing touches…the things that make life easier, and we have grouped them under the title of *efficiency techniques*, as their use is beneficial, but not necessarily required.

## The Assets panel

As you mature as an editor, there will be things you create that will be beneficial to have on hand. These can include images (if you are a compositor), certain overlays such as the rain overlay from the export section previously, or even certain design elements such as borders and shapes that you use in your work. In Affinity Photo, these can be saved and categorized under the term **Assets**.

We have included a practice file for both the V1 and the V2 versions of Affinity in the downloads (see the `Asset practice files` folder and choose your version). Inside the file, you will find two layers, one titled **Rain** and one titled **Atmosphere**. These represent two very common asset types for me, as these can be reused in various projects once created (see *Figure 18.13* for the layer structure of the file).

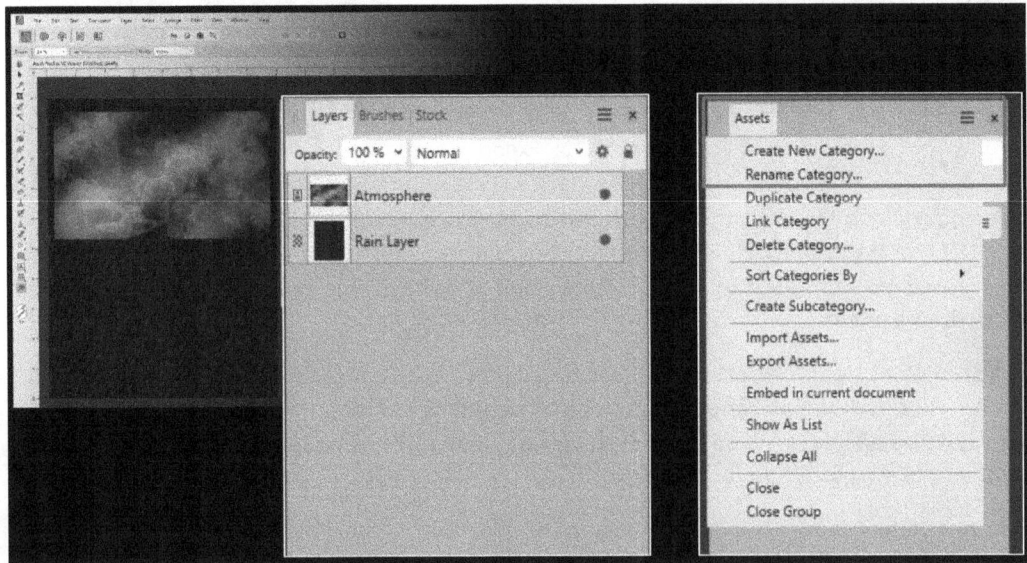

Figure 18.13 – Assets tab and category detail

Assets have their own studio panel (as shown in *Figure 18.13*), and this panel operates on a series of categories; so to add assets, it is simply creating a category and adding them to the category.

## Working with categories

Adding a category is simple and begins with clicking on the upper-right corner of the panel and selecting **Create New Category....** I have included a CiA video (see the *Technical requirements* section of this chapter for the link) to show the process, but have defined it here as well:

1.  Click on the four-line menu icon in the upper-right corner.

2.  Click on **Create New Category....**

3.  Rename the category to something you like.

Now that we have a category, we can create subcategories, so let's do that. I am going to create two, titled **Rain Layers** and **Atmosphere**, so I need a total of two subcategories (see *Figure 18.14*). To add a subcategory, follow these steps:

1.  Go to the **Assets** menu and click on **Create Subcategory....**

2.  Within the new category, click on the menu icon for that category and simply rename it.

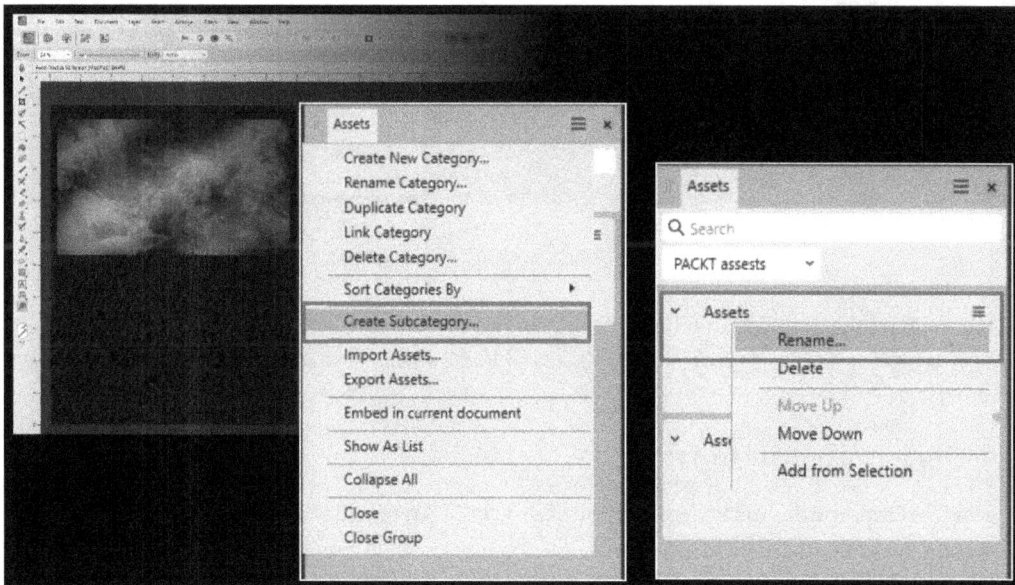

Figure 18.14 – Creating and renaming subcategories

Lastly, to add to the category, we need to work on the selection. Whatever you select will be added to the category as the asset. I typically work on single layers, or in the event of multiple layers, I will group them into one folder prior to adding them. For this example, these are single layers, which, in my opinion, is the best way to go to keep file sizes down.

To add a selection in a category, click on the menu for the category (the three little lines next to the category name), and then with the selection made, click on **Add from Selection**. You will see the thumbnail in the asset subcategory (see *Figure 18.15*).

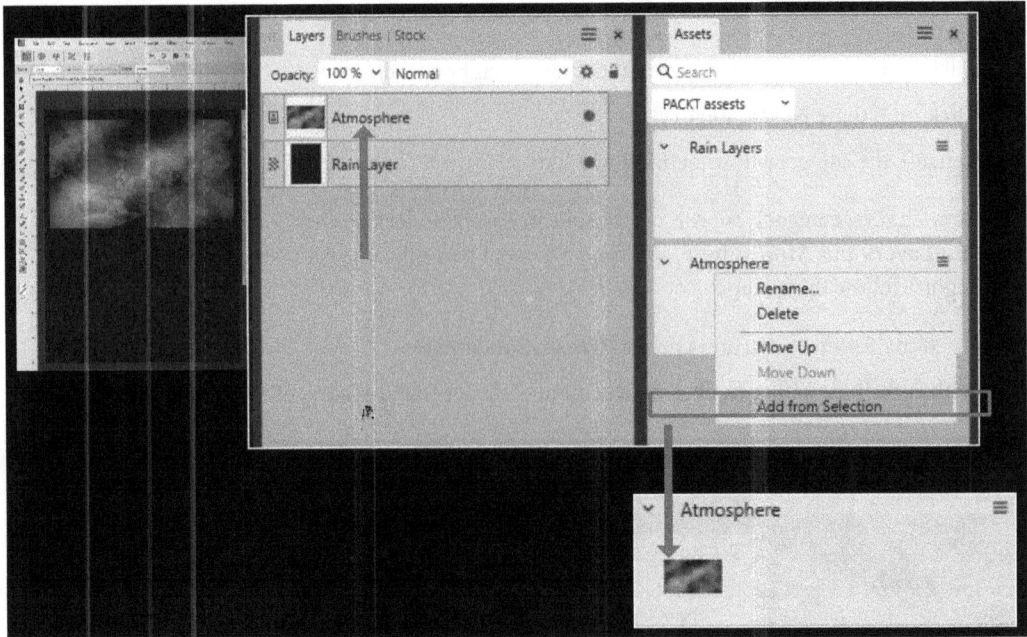

Figure 18.15 – Assets added into categories

Now that we have the assets created in the categories, we can export them, or if you have assets given to you, you can import assets from other places, so let's do that next so you can begin collecting from others.

### Importing and exporting assets

Importing and exporting assets is simple, and to assist you, I have included an importable set of assets you can practice with. I have included some assets that I have made along the way for you to practice with (see the `Christmas package assets` file in the `Asset Practice files` folder in the downloads).

To import a file, simply go to the studio menu in the **Assets** tab (in the upper-right corner), select **Import Assets…**, and then select the file.

To export assets, go to the **File** menu and, with the category selected, select **Export Assets…**, save the file, and then send it just like you would any other file.

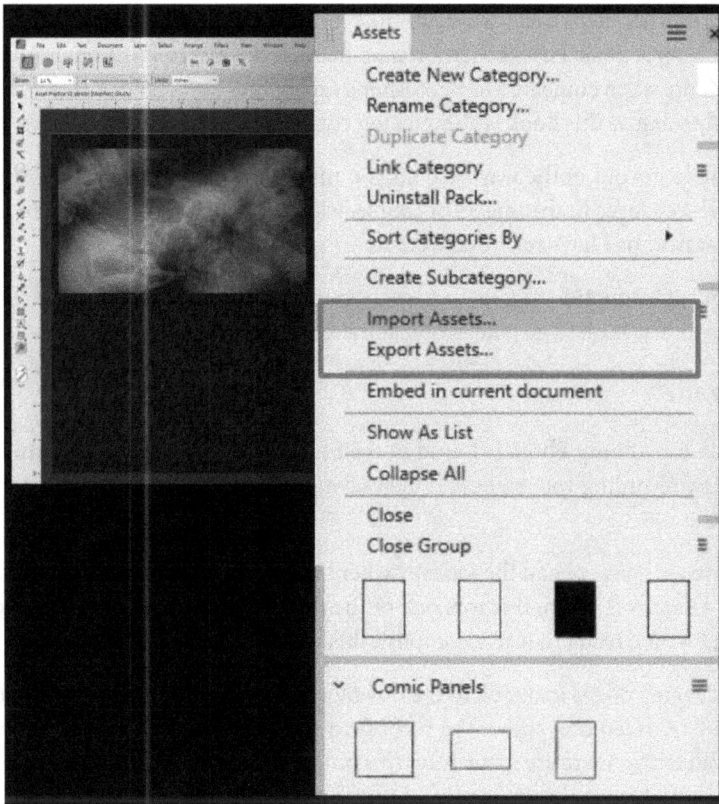

Figure 18.16 – Importing and exporting assets

Now that you have assets in your library, it is time to see how to use them… so let's dig in.

### Using assets in your work

To bring the assets into your projects, simply click and drag them into the project and then adjust them from there.

In the case of the Christmas package asset pack, you can arrange them as you see fit; however, with the rain and atmosphere overlays, we will have to change the blend modes and mask them to make them fit the individual scene on a case-by-case basis.

Now that we know about assets, let's explore another useful studio panel, **Library**, and the use of macros.

## Working with macros and the Library panel

The **Library** panel and the **Macro** panel work together in Affinity Photo. The way you want to think about this is that you use the **Macro** panel to record a macro and then you store the macro in the library for later use. At this point, you may be asking yourself…what is a macro?

A macro is a set of recorded steps that saves your time for common tasks. For Photoshop users, these are similar to Photoshop actions. I use the term *similar* because the macro function in Affinity Photo is a very immature feature when compared to Photoshop's capabilities for actions. I personally do not use this feature very often, but as this book is about being complete, I did not want to leave the topic out.

A professional note to get out of the way early: if you make a macro, you *must export it before you close the window*. The changes to the image, of course, will be saved, but the redoing of the steps will be lost. Many clever macros I have recorded were never saved learning this lesson.

So, wrapping up this introduction where we began, you record the macro using the studio tab and then store it in the library. So, let's go take a look at these two tabs.

### The Macro panel

The macro function for Affinity Photo is found in the **Studio** tab under **Window | Studio**. The way this panel works is by recording your steps and then saving them so you can apply the same sequence in each photo.

For this example, we are going to end the journey where we began, creating that old-time sepia-tone image (remember *Chapter 3* where this was one of the first techniques we learned). Now, we have come full circle and we are ready to just accomplish this with a recorded macro.

The image of our sleeping dog is included in the downloads for this section so you can follow along, and there is also a CiA video that shows the recording of the macro, and for this activity, I would recommend you watch it, as there are many ways this process can go wrong. The link of the video can be found in the *Technical requirements* section of the chapter.

So, let's see how to record your first macro action.

### Recording a macro

Begin by opening the image in Affinity Photo and then follow along:

1. Using the **Move** tool, make sure you do not have any layers selected.

2. Hit the red record button in the **Macro** panel. This will begin tracking your adjustments.

3. In the **Layers** panel, add a **Black & White Adjustment** layer and adjust to taste. Notice that this action has been logged in the **Macro** panel.

4. In the **Layers** panel, add a **Lens Filter Adjustment** layer. Notice in the **Macro** panel that this will be added as a step.

5. Lastly, add in a Live Filter layer for noise and add a bit of noise; this will also be added to the macro.

6. Hit the square stop button in the **Macro** panel. This will complete the macro and will end the recording.

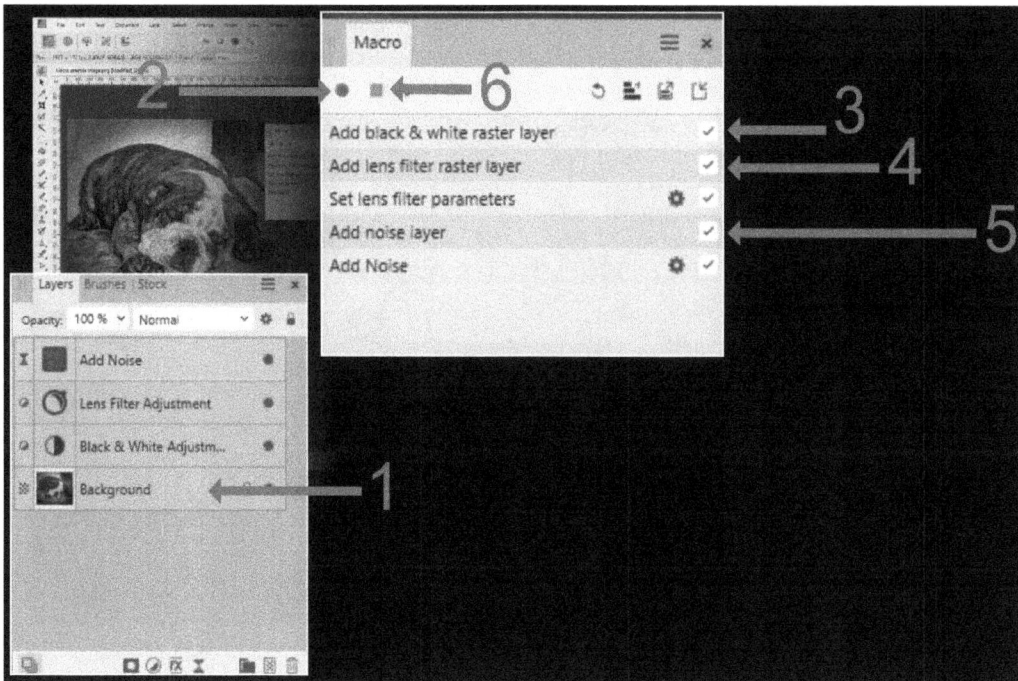

Figure 18.17 – Recording a macro

Now, it's time to test your macro and learn how to adjust the settings for the macro.

### Playing the macro back and adjusting the settings in the Macro panel

Testing your macro is simple; simply remove the adjustments from your image, and hit the triangle play button in the **Macro** panel. This will apply the steps to the image again and you *should* end up where you want to be with a few exceptions.

In *Figure 18.17*, you will see cog icons next to the **Add Noise** and **Lens Filter** steps. These are steps where we made adjustments to the default options when we recorded the macro. If you applied the filter and did not make any adjustments, Affinity assumes you just want to apply the default (in this case, the black and white adjustment has no gear icon because, when we recorded, we made no adjustments to the default).

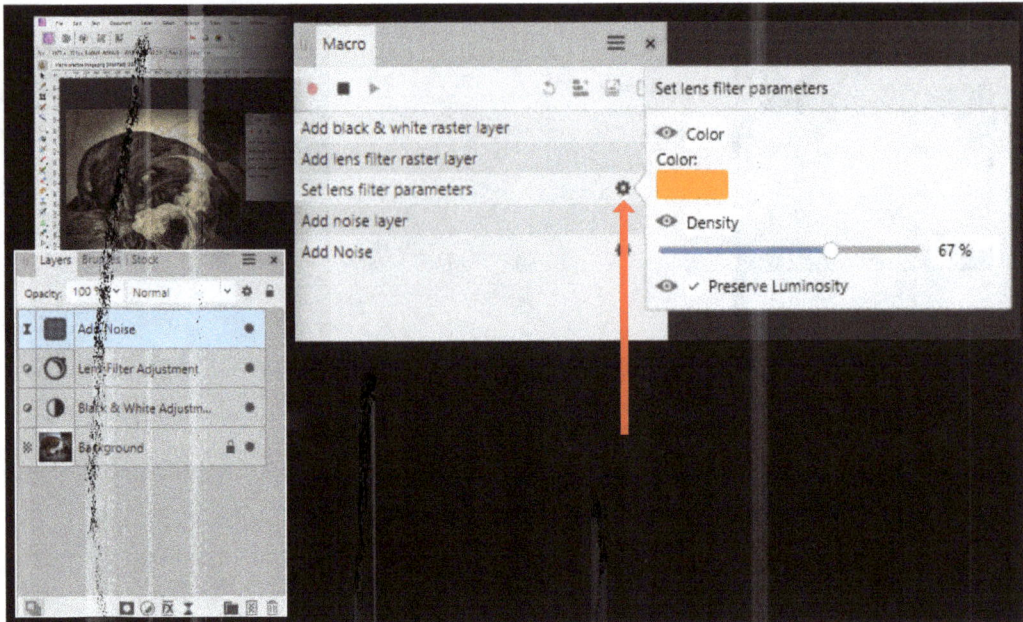

Figure 18.18 – Adjustments in the Macro panel

In the event you want a step to be modifiable, make an adjustment during the recording to the default settings.

In *Figure 18.18*, we see the lens filter adjustments that we can adjust post-macro application by clicking on the cog icon.

Now that we have a macro, we can export it to share with others (see the import export options next to the play option), but for this book, we are going to export it to the library, so let's learn about the library in the next section.

## The Library panel

As the name implies, the **Library** panel is where you can save and organize all of the macros you create. It works similarly to other panels with the same terms, such as *categories*, and then you nest the individual macros into categories. So, this is something we have seen with similar areas, such as when we created styles or saved presets in the Tone Mapping persona. Let's create a category and fill it with the macro we created in the previous section.

## Creating and adding to categories

To add a new category in Affinity Photo, simply click on the menu in the upper-right corner of the panel and choose **Create New Category...** (see *Figure 18.19*).

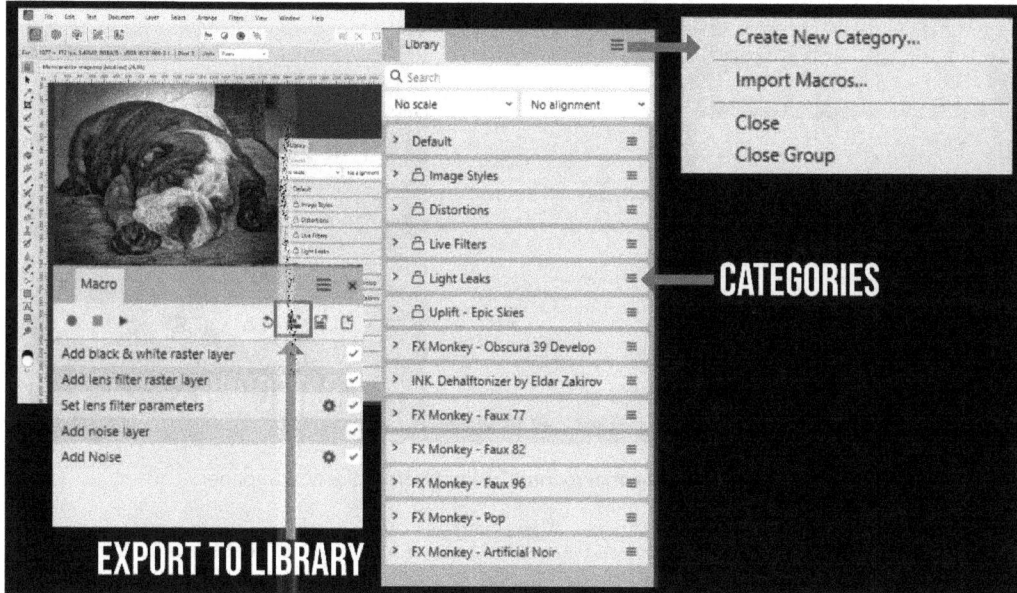

Figure 18.19 – Adding a category in Affinity Photo

This will create a new category in the **Library** tab. Rename the category to something you like (I am going to use `Packt` for the example).

This category is now created, and we can add to it. There are two ways to do so:

- With a macro created, simply hit **Export to Library** (this method implies you are creating the macros and then moving them before you close the program).

- We can import macros. This assumes you exported them when you created them at an earlier time (I have imported the **Sepia Old Timey Photo** macro into the **Packt** category).

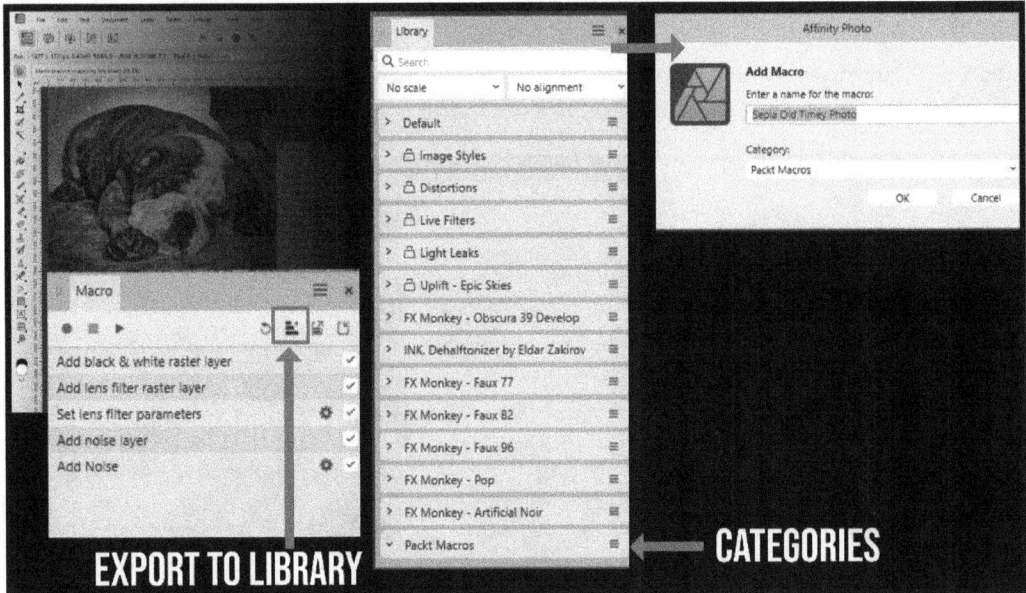

Figure 18.20 – Adding the macro into the library from the Macro panel

Now that you have created a new category of time-saving macros, you can export them so that if you ever had to reboot the version of Affinity Photo, you could reinstall them. Let's learn how to export in the next section.

## Importing and exporting categories into your library

To export your library categories with the category menu selected, not the panel menu, simply click on **Export Macros…**. This will create an exportable file that you can later import into the library through the panel menu.

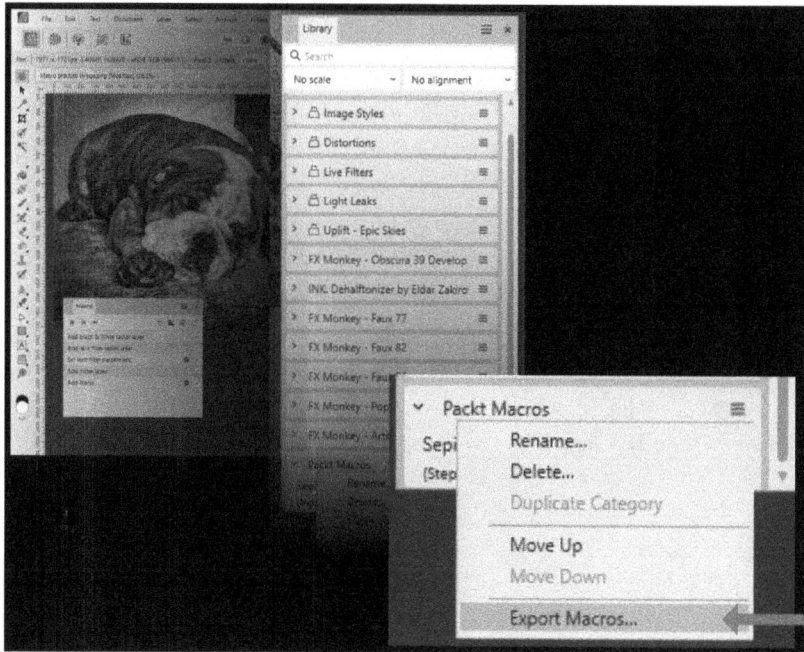

Figure 18.21 – Exporting categories

We have exported **Packt Macros** as a file for the downloads in this section, so simply import the macro into the **Library** panel and you are in business.

## Professional tips, tricks, and important points

I use the **Assets** panel daily and fill it with the things that I use most often for the type of work I do compositing. And as part of the standard flow, I always run through the assets and ask, *Can any of these make the piece better?*

Whether it is an atmosphere, a lightning strike, or an article texture layer, the goal is not to use them every time, but to remember they are there. If you bury these things in the folders on your computer, you may forget they exist and miss out.

For websites, I make macros for common adjustments that I use; that way, they all look uniform, and it saves me a ton of time. As an example, if you wanted the images to all have a blue tone to them with desaturation, instead of adjusting all of them by hand, simply create a macro to achieve the result. Make sure you make the macro modifiable for adjustments when the macro runs, though.

Many macros are destructive, so make sure you make a copy of the base image before applying them.

# Summary

In this chapter, we learned how to export, record macros, and manage our library of assets, and all of them will come in handy as your body of work grows and matures. Many beginners focus on the advanced topics and are in a hurry to gain mastery, but imagine trying to understand the **Assets** panel if you didn't know how to mask or select the right portions of the image…it would be impossible. The same thing could be said for the idea of macros; if you bought a macro pack and just hit **Apply**…all of the steps and layers on your image would make no sense and you would have to learn what Live Adjustment layers are and be right back at the fundamental stage.

As this is the last chapter of the book, there is no next move for me, no next chapter, so let me express my appreciation for spending the time with me and allowing me to pass on what I know and love to another person. I firmly believe life is measured by what you give back to others and so for me, education is the way in which I do that. Thank you for the journey, and I wish you luck on the next leg of yours.

# Index

# ‹packt›

Subscribe to our online digital library for full access to over 7,000 books and videos, as well as industry leading tools to help you plan your personal development and advance your career. For more information, please visit our website.

## Why subscribe?

- Spend less time learning and more time coding with practical eBooks and Videos from over 4,000 industry professionals

- Improve your learning with Skill Plans built especially for you

- Get a free eBook or video every month

- Fully searchable for easy access to vital information

- Copy and paste, print, and bookmark content

Did you know that Packt offers eBook versions of every book published, with PDF and ePub files available? You can upgrade to the eBook version at packtpub.com and as a print book customer, you are entitled to a discount on the eBook copy. Get in touch with us at customercare@packtpub.com for more details.

At www.packtpub.com, you can also read a collection of free technical articles, sign up for a range of free newsletters, and receive exclusive discounts and offers on Packt books and eBooks.

# Other Books You May Enjoy

If you enjoyed this book, you may be interested in these other books by Packt:

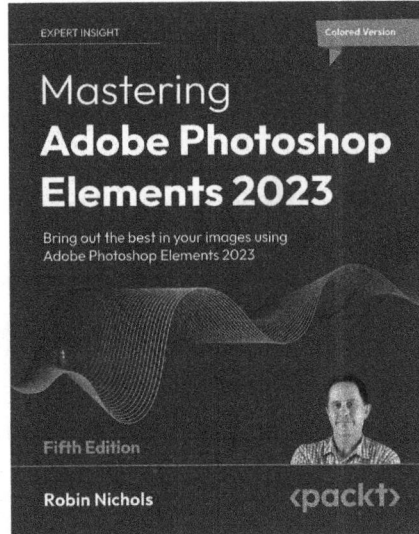

**Mastering Adobe Photoshop Elements 2023 - Fifth Edition**

Robin Nichols

ISBN: 978-1-80324-845-5

- How to retouch images professionally, replace backgrounds, remove people, and resize your images
- Animate parts of your photos to create memes to wow your social media fans
- Showcase your photos and videos with all-new collage and slideshow templates
- Use image overlays to create unique depth of field effects
- Discover advanced layer techniques designed to create immersive and powerful illustrations
- Take your selection skills to the next level for the ultimate in image control
- Develop your illustration skills using the power of Elements' huge range of graphics tools and features
- Easily create wonderful effects using Adobe's awesome AI technology

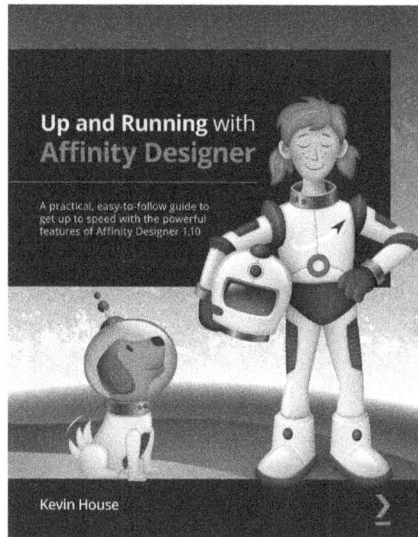

**Up and Running with Affinity Designer**

Kevin House

ISBN: 978-1-80107-906-8

- Explore the interface and unique UX characteristics of Affinity Designer
- Discover features that allow you to manipulate and transform objects
- Apply color, shading, and effects to create unique compositions
- Employ layers to organize and simplify complex projects
- Use grids, guides, and snapping features as design aids
- Adapt to Affinity Designer's custom workspaces and keyboard shortcuts
- Explore the workflow and design best practices for more predictable and successful outcomes
- Identify potential stumbling blocks in your design process and learn how to avoid them

## Packt is searching for authors like you

If you're interested in becoming an author for Packt, please visit authors.packtpub.com and apply today. We have worked with thousands of developers and tech professionals, just like you, to help them share their insight with the global tech community. You can make a general application, apply for a specific hot topic that we are recruiting an author for, or submit your own idea.

# Share Your Thoughts

Hi!

I am Jeremy Hazel, author of *Professional Image Editing Made Easy with Affinity Photo*. I really hope you enjoyed reading this book and found it useful for increasing your productivity and efficiency in Affinity Photo.

It would really help me (and other potential readers!) if you could leave a review on Amazon sharing your thoughts on *Professional Image Editing Made Easy with Affinity Photo*.

Go to the link below or scan the QR code to leave your review:

https://packt.link/r/1800560788

Your review will help me to understand what's worked well in this book, and what could be improved upon for future editions, so it really is appreciated.

Best Wishes,

# Download a free PDF copy of this book

Thanks for purchasing this book!

Do you like to read on the go but are unable to carry your print books everywhere? Is your eBook purchase not compatible with the device of your choice?

Don't worry, now with every Packt book you get a DRM-free PDF version of that book at no cost.

Read anywhere, any place, on any device. Search, copy, and paste code from your favorite technical books directly into your application.

The perks don't stop there, you can get exclusive access to discounts, newsletters, and great free content in your inbox daily

Follow these simple steps to get the benefits:

1.  Scan the QR code or visit the link below

https://packt.link/free-ebook/9781800560789

2.  Submit your proof of purchase
3.  That's it! We'll send your free PDF and other benefits to your email directly

www.ingramcontent.com/pod-product-compliance
Lightning Source LLC
Chambersburg PA
CBHW080131220326
41598CB00032B/5026